Developmental Toxicology: Mechanisms and Risk

Row 1: J. Hanson; J. Scandalios, P. Iannaccone, E.M. Johnson
Row 2: R. Pratt, J. Manson; F. Welsch, R. Morrissey, M. Solursh
Row 3: A. Haney; R. Brent, J. McLachlan; N. Kaplan

Developmental Toxicology: Mechanisms and Risk

Edited by

JOHN A. McLACHLAN
National Institute of Environmental Health Sciences

ROBERT M. PRATT
National Institute of Environmental Health Sciences

CLEMENT L. MARKERT
North Carolina State University

COLD SPRING HARBOR LABORATORY
1987

Banbury Report 26: Developmental Toxicology: Mechanisms and Risk

Printed in the United States of America
Cover and book design by Emily Harste

Library of Congress Cataloging-in-Publication Data

Developmental Toxicology: Mechanisms and Risk

(Banbury report, ISSN 0198-0068 ; 26)
Based on a meeting held at the Banbury Center of Cold Spring Harbor Laboratory, Oct. 1986, sponsored by Abbott Laboratories and others.
Includes index.
1. Developmental toxicology—Congresses.
I. McLachlan, John A. II. Pratt, Robert M.
III. Markert, Clement L. (Clement Lawrence),
1917- . IV. Cold Spring Harbor Laboratory.
V. Abbott Laboratories. VI. Series. [DNLM:
1. Abnormalities—embryology—congresses.
2. Toxicology—congresses. W3 BA19 v.26 / QS 675 D4894
1986]
RA1224.45.D48 1987 615.9 87-18244
ISBN 0-87969-226-X

All Cold Spring Harbor Laboratory publications may be ordered directly from Cold Spring Harbor Laboratory, Box 100, Cold Spring Harbor, New York 11724. (Phone: 1-800-843-4388) in New York State (516) 367-8425.

Banbury Report Series

Banbury Report 1: Assessing Chemical Mutagens
Banbury Report 2: Mammalian Cell Mutagenesis
Banbury Report 3: A Safe Cigarette?
Banbury Report 4: Cancer Incidence in Defined Populations
Banbury Report 5: Ethylene Dichloride: A Potential Health Risk?
Banbury Report 6: Product Labeling and Health Risks
Banbury Report 7: Gastrointestinal Cancer: Endogenous Factors
Banbury Report 8: Hormones and Breast Cancer
Banbury Report 9: Quantification of Occupational Cancer
Banbury Report 10: Patenting of Life Forms
Banbury Report 11: Environmental Factors in Human Growth and Development
Banbury Report 12: Nitrosamines and Human Cancer
Banbury Report 13: Indicators of Genotoxic Exposure
Banbury Report 14: Recombinant DNA Applications to Human Disease
Banbury Report 15: Biological Aspects of Alzheimer's Disease
Banbury Report 16: Genetic Variability in Responses to Chemical Exposure
Banbury Report 17: Coffee and Health
Banbury Report 18: Biological Mechanisms of Dioxin Action
Banbury Report 19: Risk Quantitation and Regulatory Policy
Banbury Report 20: Genetic Manipulation of the Early Mammalian Embryo
Banbury Report 21: Viral Etiology of Cervical Cancer
Banbury Report 22: Genetically Altered Viruses and the Environment
Banbury Report 23: Mechanisms in Tobacco Carcinogenesis
Banbury Report 24: Antibiotic Resistance Genes: Ecology, Transfer, and Expression
Banbury Report 25: Nongenotoxic Mechanisms in Carcinogenesis
Banbury Report 26: Developmental Toxicology: Mechanisms and Risk

Corporate Sponsors

Abbott Laboratories
American Cyanamid Company
Amersham International plc
Becton Dickinson and Company
Cetus Corporation
Ciba-Geigy Corporation
CPC International Inc.
E.I. du Pont de Nemours & Company
Eli Lilly and Company
Genentech, Inc.
Genetics Institute
Hoffmann-La Roche Inc.
Monsanto Company
Pall Corporation
Pfizer Inc.
Schering-Plough Corporation
Smith Kline & French Laboratories
Tambrands Inc.
The Upjohn Company
Wyeth Laboratories

Core Supporters

The Bristol-Myers Fund, Inc.
The Dow Chemical Company
Exxon Corporation
International Business Machines Corporation
The Procter & Gamble Company
Rockwell International Corporation Trust
Texas Philanthropic Foundation Inc.

Special Program Support

James S. McDonnell Foundation
Alfred P. Sloan Foundation

Participants

Nicole Bournias-Vardiabasis, Division of Neurosciences, City of Hope Beckman Research Institute, Duarte, California

Andrew G. Braun, Department of Applied Biological Sciences, Massachusetts Institute of Technology, Cambridge

Robert L. Brent, Jefferson Medical College, Thomas Jefferson University, Philadelphia, Pennsylvania

William J. Breslin, Mammalian and Environmental Toxicology Research Laboratory, Dow Chemical USA, Midland, Michigan

Lennart Dencker, Department of Toxicology, University of Uppsala, Sweden

Gerald M. Edleman, Department of Developmental and Molecular Biology, Rockefeller University, New York, New York

Arthur F. Haney, Division of Reproductive Endocrinology and Infertility, Department of Obstetrics and Gynecology, Duke University Center, Durham, North Carolina

James William Hanson, Division of Medical Genetics, Department of Pediatrics, University of Iowa, Iowa City

Philip M. Iannaccone, Department of Pathology, Northwestern University Medical School and Northwestern University Cancer Center, Chicago, Illinois

E. Marshall Johnson, Department of Anatomy, Jefferson Medical College, Philadelphia, Pennsylvania

Norman Kaplan, Biometry and Risk Assessment Program, National Institute of Environmental Health Sciences, Research Triangle Park, North Carolina

Carole A. Kimmel, Reproductive Effects Assessment Group, U.S. Environmental Protection Agency, Washington, D.C.

David Kochhar, Department of Anatomy, Jefferson Medical College, Philadelphia, Pennsylvania

Edward James Lammer, Embryology-Teratology Unit, Massachusetts General Hospital, Boston

Michael Levine, Department of Biological Sciences, Columbia University, New York, New York

Jeanne M. Manson, Reproductive and Development Toxicology, Smith Kline & French Laboratories, Philadelphia, Pennsylvania

Clement L. Markert, Department of Animal Science, North Carolina State University, Raleigh

John A. McLachlan, Laboratory of Reproductive and Developmental Toxicology, National Institute of Environmental Health Sciences, Research Triangle Park, North Carolina

Richard Miller, Department of Obstetrics and Gynecology, University of Rochester School of Medicine, New York

Richard E. Morrissey, National Institute of Environmental Health Sciences, Research Triangle Park, North Carolina

Godfrey P. Oakley, Center for Environmental Health, Centers for Disease Control, Atlanta, Georgia

Roger A. Pedersen, Laboratory of Radiobiology and Environmental Health and Department of Anatomy, University of California, San Francisco

Robert M. Pratt, Laboratory of Reproductive and Developmental Toxicology, National Institute of Environmental Health Sciences, Research Triangle Park, North Carolina

Jerry M. Rice, National Cancer Institute, Frederick Cancer Research Facility, Maryland

Thomas D. Sabourin, Environmental Sciences Department, Battelle Columbus Division, Ohio

Lauri Saxén, Department of Pathology, University of Helsinki, Finland

John George Scandalios, Department of Genetics, North Carolina State University, Raleigh

William J. Scott, Jr., Children's Hospital Medical Center, Cincinnati, Ohio

Thomas Hill Shepard, Central Laboratory for Human Embryology, Department of Pediatrics, School of Medicine, University of Washington, Seattle

Michael Solursh, Department of Biology, University of Iowa, Iowa City

Janet A. Springer, Division of Mathematics, Food and Drug Administration, Washington, D.C.

Ralf Stahlmann, Institute for Toxicology and Embryopharmacology, Free University of Berlin, Federal Republic of Germany

Frank Welsch, Cell Biology Division, Chemical Industry Institute of Toxicology, Research Triangle Park, North Carolina

This volume is dedicated to

Sergio Egidio Fabro (1932–1986)

Bridge devotee, photographer, soccer star,
sports car driver, physician, scientist,
teacher, and friend.
Teratology has not been the same before or since.

Preface

The purpose of the Banbury Conference on Developmental Toxicology: Mechanisms and Risk, held on October 19–22, 1986 at the Banbury Center, Cold Spring Harbor Laboratory, was twofold: (1) to explore mechanisms and systems pertinent to normal and abnormal development and (2) to bring together workers in diverse areas to share their insights. This was the first meeting to address the mechanisms of induction of abnormal embryonic and fetal development from a molecular, as well as risk assessment, viewpoint. Moreover, other meetings on related topics have not provided common ground for developmental biologists and teratologists. It was thought that such a meeting would stimulate workers in different disciplines to recognize common scientific problems, as well as to explore the scientific basis of risk assessment.

To accomplish these objectives, the meeting was organized into five primary topics: In Vitro Approaches; Molecular and Experimental Embryology; Nonmammalian Models; Experimental Animal—Human Comparisons; and Risk Assessment. Each topic was addressed by leading experts in their fields. The underlying theme was development of models to understand the mechanisms of abnormal embryonic and fetal development in humans. Thus, the topic, "In Vitro Approaches," addressed the current attempts to evaluate abnormal cell differentiation and perturbation of normal cell processes by chemicals in vitro. The value of these studies to in vivo teratology was explored. The most recent advances in "Molecular and Experimental Embryology" were highlighted; these new concepts in cell differentiation and expression should stimulate new approaches to the study of mechanisms of abnormal development. The topic of "Nonmammalian Models" dealt with the question of complexity and pertinence of species. Although the ultimate question was the mechanisms underlying abnormal human development, the question was asked, Must all experimental model studies be conducted in mammals or would, perhaps, more simple systems suffice for models? Are there common mechanisms that shed insight on crucial developmental steps? The closeness of fit for models, from a mechanistic standpoint, was explored in the topic, "Experimental Animals–Human Comparisons." Similarities and differences in laboratory animals and humans were explored. The utility of such models for cross-species prediction and mechanistic propositions was critically evaluated. Finally, diverse viewpoints on the mechanisms and utility of assessment of teratogenic risk to humans were presented and discussed in the session on "Risk Assessment." The basic biology presented earlier in the meeting was contrasted to the real needs to make quantitative decisions about hazards to human developmental health associated with environmental chemicals.

This volume should provide a unique resource in the area of abnormal development and the assessment of its risk in human populations. This area has lagged behind that of carcinogenic risk assessment; we hope that this volume will provide a much needed stimulus for this important area of public health concern.

The organizers gratefully acknowledge the funding provided by the James S. McDonnell foundation. We are especially appreciative of the Banbury staff for their help in the organization of this meeting.

J.A. McLachlan
R.M. Pratt
C.L. Markert

* * *

During the final preparation of this book, our friend and coeditor, Dr. Robert Pratt, died. This is a tragedy almost beyond comprehension to us. Bob was one of the driving forces in organizing this Banbury Conference and these proceedings which demonstrates his interest in developmental mechanisms and human health.

All of us who participated in this Banbury Conference remember how enthusiastic and buoyant Bob was during the meeting. His ability to illuminate scientific points and provide humor and friendship to everyone at the meeting reflected Bob's broader role as a scientist and friend to so many.

We are deeply saddened by the loss of Bob Pratt, but we hope that this book has, in some small part, captured and retained some of the energy that he so freely shared.

Contents

In Vitro Approaches

Altered Differentiation and Induction of Heat Shock Proteins in *Drosophila* Embryonic Cells Associated with Teratogen Treatment

NICOLE BOURNIAS-VARDIABASIS* AND CAROLYN H. BUZIN[†]
*Division of Neurosciences
City of Hope Beckman Research Institute
[†]Division of Cytogenetics
City of Hope Medical Center
Duarte, California 91010

OVERVIEW

Utilization of *Drosophila* embryonic cultures has allowed us to develop an in vitro assay for assessing the effects of putative teratogenic agents on embryonic cell differentiation. Because in vitro differentiation of *Drosophila* embryonic cells, such as myotubes and ganglia, has been investigated extensively in our laboratory, we have used these endpoints, among others, to determine the potential teratogenic hazard of over 150 agents. In this group, a number of drugs and physical agents that have already been shown to act as teratogens in mammals have also been found to inhibit muscle and neuron differentiation in *Drosophila* embryonic cultures. Along with determining cellular-level-type changes, we have also examined the effects of such teratogens on protein synthesis in *Drosophila* embryonic cells. We found a strong correlation between inhibition of muscle and/or neuron differentiation after teratogen exposure and the induction of a number of proteins. These teratogen-induced proteins are identical to the heat shock proteins. Exposure of *Drosophila* embryos, larvae, or pupae to hyperthermia or a variety of other environmental stresses results in the enhanced synthesis of a number of so-called heat shock proteins. In fact, heat shock protein induction is one of the most basic regulatory mechanisms in living organisms. Virtually all organisms from *Escherichia coli* to man have heat shock proteins. Considerable molecular information has already accumulated on the identity of these proteins, but their function has yet to be determined. The consensus so far points to an important role in structural integrity, as well as in cell survival and protection after exposure to some types of environmental stress. This consensus has directed our current studies, which are aimed at identifying the role of heat shock proteins in teratogenesis and investigating the differential regulation of heat shock genes in response to different stimuli.

Banbury Report 26: Developmental Toxicology: Mechanisms and Risk

INTRODUCTION

Teratology has certainly made rather substantial strides in recent years, and one of the areas that has experienced tremendous growth is the development and use of a number of in vitro systems to identify possible teratogens and to determine the types of mechanisms involved in the teratogenic response. It is well known that teratogens can have a variety of effects on the developing embryo and at various organizational levels (i.e., cell morphogenesis, nucleic acid synthesis, protein synthesis). Various models have already been used, including cell, organ, and whole-embryo cultures derived from invertebrates and vertebrates (Kimmel et al. 1982).

Drosophila developmental biologists have had an early start examining various developmental questions such as morphogenesis, differentiation, determination, pattern formation, and how gene mutations or environmental changes can interfere with normal developmental pathways. As early as 1935, Goldschmidt described the production of a number of developmental abnormalities found in *Drosophila* adults after embryonic exposure to hyperthermic conditions (37°C for 20–30 min). Because the abnormalities observed were similar to those induced by genetic mutations, he termed these abnormalities "phenocopies" (Goldschmidt 1935). Besides *Drosophila*, developmental defects after heat shock treatments have also been reported in sea urchins (Roccheri et al. 1981), in *Xenopus* (Elsdale et al. 1976), and in various mammals (Edwards and Wanner 1977; Shepard 1985). Phenocopies have also been produced by a variety of chemical agents (borate salts and purine and pyrimidine analogs [Sang and McDonald 1954; Rizki and Rizki 1965]) and, more recently, in our laboratory (Bournias-Vardiabasis 1983) and in others (Schuler et al. 1982) by more "classical" teratogens. It is hypothesized that the defects produced are the result of disrupting stage-specific processes (including reduction of normal mRNA and protein production). Thirty years after Goldschmidt's observations came the discovery by Ritossa on the induction of puffing at specific sites on salivary glands after exposure of third instar larvae to hyperthermia and other agents (Ritossa 1962). Much later came the finding that this increased rate in specific message transcription translated into the induction of a set of proteins otherwise not present in the cell (Tissieres et al. 1974). This heat shock response has been observed in bacteria, yeast, plants, and a wide variety of higher eukaryotic organisms. In addition to heat shock, the proteins can be induced by anoxia, inhibitors of oxidative phosphorylation, transition series metals, teratogens, ionophores, ether, and steroid hormones (for extensive reviews, see Nover 1984; Craig 1985; Petersen and Mitchell 1985). In *Drosophila*, which is the most thoroughly investigated organism, seven major heat shock or stress proteins have been identified. They range in size from 80,000 daltons to 20,000 daltons and are designated hsp 83, hsp 70, hsp 68, hsp 27, hsp 26, hsp 23, and hsp 22.

Extensive information is now available on the genes encoding these heat shock proteins; monoclonal and polyclonal antibodies have been used to identify their cellular distribution, but very little is known about their functions. It is this question about the functional significance of heat shock proteins that has propelled our current investigation in examining teratogen-induced heat shock proteins and how they might relate to teratogenesis.

RESULTS

Drosophila Assay

The *Drosophila* in vitro assay is based on the principle that teratogenic effects can be caused by abnormal cell death, failure of cell interaction, reduced biosynthesis, or impeded morphogenetic movement (Wilson 1977). The endpoints used in assessing the teratogenic potential of a given substance involve detection of interference with normal muscle and/or neuron differentiation. Briefly, the assay is as follows (for more detailed methodology, see Seecof 1980): *Drosophila* embryos are collected routinely on plates of standard corn meal medium supplemented with live yeast from population cages containing SOR wild-strain flies. Embryos are synchronized to ±1 hour by collecting for 2 hours. Synchronized embryos are allowed to develop at room temperature to the gastrulation stage; they are then dechorionated and gently homogenized, and the embryonic cells are pelleted. The cells are resuspended, counted, and plated out in 35-mm tissue culture dishes at 1.6×10^6 cells per dish in modified Schneider's medium supplemented with 18% fetal calf serum. Muscle and neuron differentiation in vitro has already been characterized in some detail (Seecof et al. 1973a,b; Donady et al. 1975; Gerson et al. 1976a,b). Both neuron and muscle cells are terminally differentiated by 24 hours after oviposition. They are functionally and structurally identical to in-vivo-generated ones. They present temporal, sequential, and morphological characteristics, which are a suitable measure of differentiation applicable to teratogenicity evaluation. It is assumed that the mechanisms of organ formation are similar in all multicellular animals and, therefore, the results from testing with *Drosophila* cells would be applicable to higher organisms, including man.

After the cells have differentiated, they are fixed, stained, and prepared for automated scoring by an image analyzer. A chemical is classified as eliciting a teratogenic response if it results in a statistically significant reduction in the number of myotubes and ganglia when compared with controls (Fig. 1). A 50% reduction in the number of myotubes and/or ganglia is taken as a positive response. The same chemical is tested in three or more separate trials before being classified as acting as a teratogen or nonteratogen in the

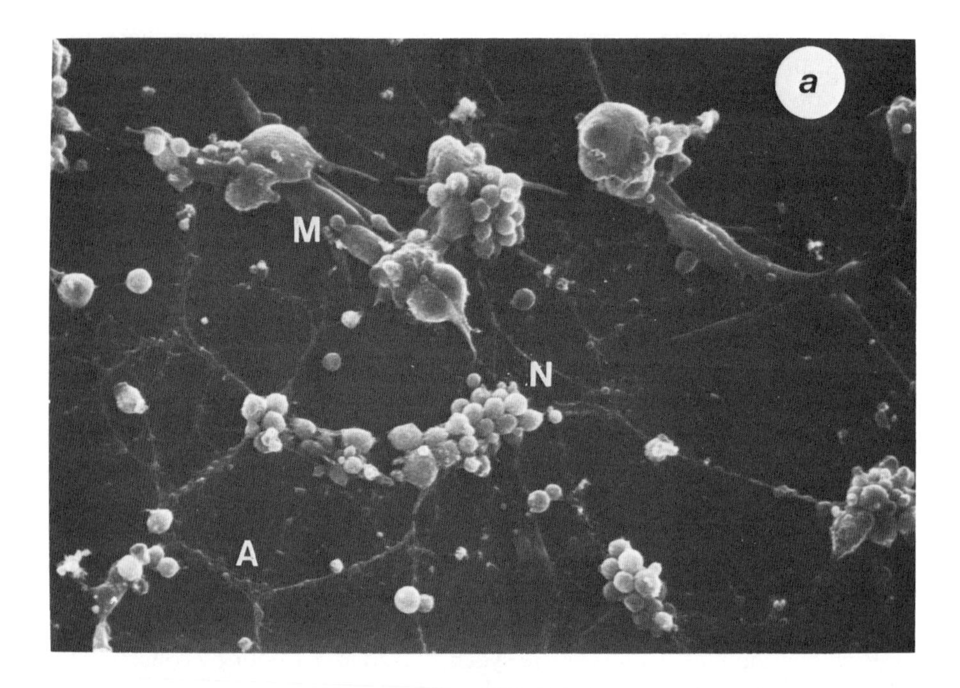

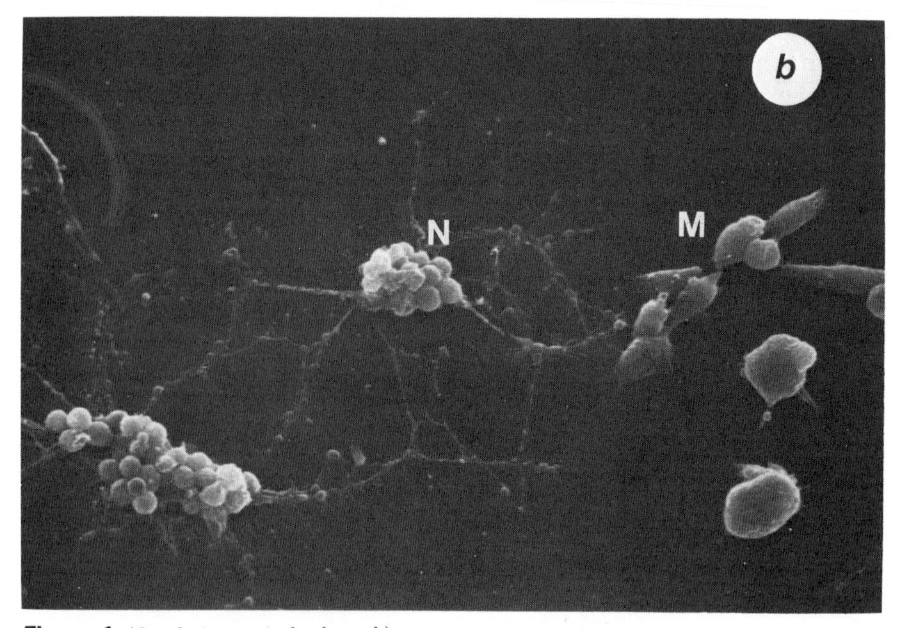

Figure 1 (*See facing page for legend.*)

Drosophila cultures. A rather large number (> 150) of chemicals has been tested in this assay (Bournias-Vardiabasis and Flores 1983; Bournias-Vardiabasis et al. 1983a), with a relatively small number of false positives (nonteratogens that test positive) and false negatives (drugs that are teratogens in a variety of animal assays that test negative in this assay). Table 1 summarizes results on a small but representative sample of chemicals tested

Table 1
Comparison of In Vitro Assay Results and In Vivo Teratology Studies

Compound	Teratogenicity[a]	*Drosophila* assay[b]
1 Acetazolamide	+/− animal; ND/human	−
2 Acetylsalicylic acid	+/animal; ND/hum	
3 Chlorambucil	+/animal; suspect/human	−
4 Coumarin	−/animal; +/human	+
5 Cyclophosphamide	+/animal; suspect/human	+
6 Dexamethasone	+/animal; ND/human	+
7 Diethylstilbestrol	weak +/animal; + human	+
8 Diphenylhydantoin	+/animal; +/human	+
9 Diphenhydramine	−/animal; ND/human	−
10 Ethyl alcohol	weak +/animal; weak +/human	−
11 Hydroxyurea	+/animal; ND/human	+
12 Hyperthermia	+/animal; suspect/human	+
13 Isoniazid	−/animal; ND/human	−
14 Methyl mercury	+/animal; +/human	+
15 Mirex	weak +/animal; ND/human	−
16 Penicillin G	−/animal; ND/human	−
17 Phthalimide	−/animal; ND/human	−
18 Retinoic acid, all *trans*	+/animal; ND/human	+
19 Sodium arsenate	+/animal; ND/human	+
20 Thalidomide	+/− animal; + human	+

In vitro assay results from Smith (1983).
[a]ND, not demonstrated.
[b]The designation plus (+) or minus (−) refers to comparison of the neuron and muscle count to controls; + indicates a significant (>50%) decrease in the neuron and/or muscle count. Effects of some of these compounds on *Drosophila* cultures have been reported previously (Bournias-Vardiabasis et al. 1983).

Figure 1
Scanning electron micrograph of *Drosophila* embryo cells. (*a*) Normal 24-hr culture. Myocytes have fused to form multinucleated myotubes (M), whereas neuronal cells (N) are aggregated in clusters (ganglia). Note also the extensive axonal (A) connections. Magnification, 787.5×. (*b*) Teratogen-treated 24-hr culture (1 mM acetylsalicylic acid). Note incomplete fusion of myocytes (M) and decreased number of neuron clusters (N) and axonal connections. Magnification, 787.5×. (Reprinted, with permission, from Bournias-Vardiabasis et al. 1983b.)

to date and how our data compare with the classification assigned by Smith et al. (1983) for in vitro validation of test methods. We have also been able to incorporate a *Drosophila*-derived microsomal fraction (S-27) in the assay, which is capable of activating proteratogens such as cyclophosphamide (Bournias-Vardiabasis 1983). A further refinement of the assay involved testing a number of wild-type *Drosophila* strains; dose/response differences were observed when diethylstilbestrol, diphenylhydantoin, imipramine, testosterone, and tolbutamide were added to the cultures (Bournias-Vardiabasis and Flores 1983).

Teratogen-induced Heat Shock Proteins

In our efforts to investigate possible teratogenic mechanisms at the molecular level, we decided to determine whether any differences existed in the number or level of cellular proteins between teratogen-treated and untreated *Drosophila* cells. Using standard two-dimensional gel electrophoresis procedures, we examined the effects of a number of teratogens and nonteratogens on protein synthesis. The conditions for labeling were as follows: After cells are prepared for tissue culture, they are incubated at 25°C for 20 hours and then labeled in the presence of the drug for 1 hour with 200 μCi of [^{35}S]methionine (New England Nuclear; sp. act., 1000 Ci/mmole) per milliliter. Heat treatments are given to primary embryonic cells at 20 hours after plating when muscle and neuron differentiation is essentially completed. Cells are incubated for 10 minutes at 43°C and labeled with 200 μCi of [^{35}S]methionine per milliliter during the entire heat treatment. Ether treatment involves placing cultures in a saturated ether chamber for 60 minutes. After labeling, the cells are washed, solubilized, and counted by methods described previously (Buzin and Seecof 1981). Samples containing between 500,000 and 1,000,000 cpm are loaded for gel electrophoresis.

Samples are separated by two-dimensional gel electrophoresis (Buzin and Petersen 1982), pH 5–7 or pH 6–8 gradient for isoelectric focusing, and a 9–15% gradient acrylamide slab for SDS–polyacrylamide gel electrophoresis. Gels are stained with Coomassie blue, fluorographed, dried, and exposed to Kodak XAR film. Integrated optical densities of all major heat shock proteins are measured from fluorographs with an Omnicon image analysis system (Bausch and Lomb). At least three independent labelings are performed.

Teratogen-treated cells exhibited a dramatic increase in the synthesis of two small heat shock proteins, hsp 22 (a and b isoforms) and hsp 23. We found a strong correlation between drugs that inhibit embryonic differentiation in vitro and those that induce the synthesis of the two small heat shock proteins. Reports on the testing of a large number of these drugs have already been published (Buzin and Bournias-Vardiabasis 1982, 1984), but a table with some of the representative published data plus more recent results is included in this paper (Table 2). As Table 2 indicates, drugs that inhibit normal embryonic cell differentiation in *Drosophila* cultures also induce a subset of the heat shock proteins (hsp 22 [a + b] and hsp 23). What is quite interesting is that treatments such as ether, heat shock, or exposure to some of the metal ions induce the full complement of the heat shock proteins (Table 2; Fig. 2; N. Bournias-Vardiabasis, unpubl.). These initial results indicated that the induction of the heat shock proteins is under variable regulatory control.

DISCUSSION

Extensive knowledge has been accumulated thus far dealing with the morphological effects and, more recently, the molecular effects of environmental stresses on the developing fetus. Using the in vitro *Drosophila* assay, a high degree of concordance has been found between agents deemed teratogenic in the *Drosophila* system and other systems such as in vivo animal studies and human epidemiology studies. Because the assay monitors cell death and cell proliferation events, the results so far indicate that measurement of effects on a final differentiated cellular state is an efficient criterion for teratogenicity evaluation. Interference with any preceding step in the differentiation process, including cell viability and rate of proliferation, will affect the outcome.

Table 2
Effect of *Drosophila* Teratogens on the Synthesis of hsp 22, hsp 23, and hsp 70

Treatment	Concentration (mM)	Induction of hsp 22 and hsp 23	hsp 70[a]
Arsenate	0.05	+	+
5-azacytidine	0.03	+	–
Cadmium	0.1	+	+
Coumarin	1.0	+	–
Dexamethasone	0.1	+	–
Diethylstilbestrol	0.01	+	–
Diphenylhydantoin	0.1	+	–
β-ecdysterone	0.01	+	–
Ether (60 min)		+	+
Heat shock (43°C, 10 min)		+	+
Methyltestosterone	0.01	+	–
Pentobarbital	1.0	+	–
Thalidomide	1.0	+	–
Tolbutamide	1.0	+	–

Partially adapted from Buzin and Bournias-Vardiabasis (1984).
[a]A plus (+) also indicates induction of all seven major heat shock proteins.

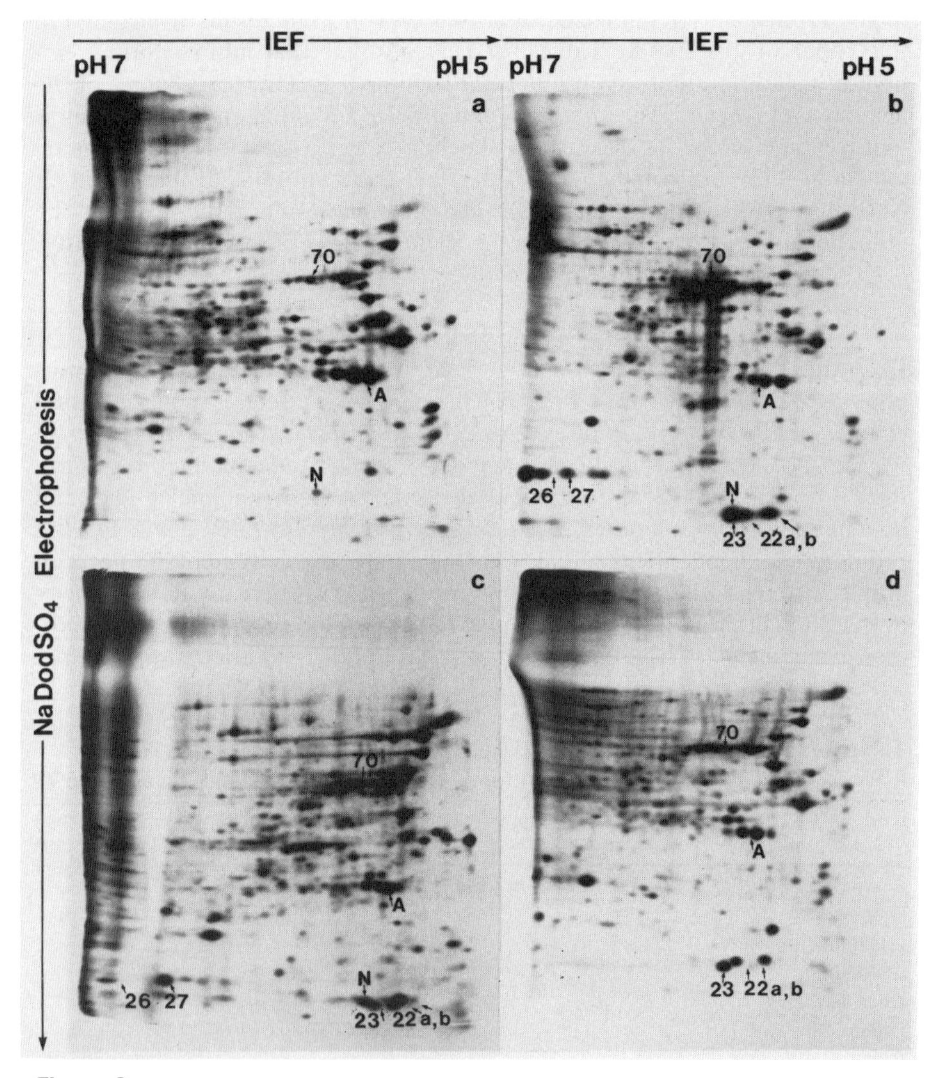

Figure 2

Protein synthesis in teratogen-treated *Drosophila* primary embryonic cells. Drugs used and exposure times (cpm × days) were control (980,000) (*a*); 43°C for 10 min (860,000) (*b*); cadmium (0.1 mM) (820,000) (*c*); thalidomide (1 mM) (780,000) (*d*); (A) actins; (IEF) isoelectric focusing. (Reprinted, with permission, from Bournais-Vardiabasis et al. 1983b.)

We are currently investigating the effects of false negatives at the biochemical level by looking at differences in the expression of cholinergic neurotransmitters. There have been various isolated reports on the effect of neuron-specific teratogens and how neurotransmitters may play an important role in regulating cell movement during tissue morphogenesis (McBride and Vardy 1983; Zimmerman 1985). Thus far, our preliminary experiments suggest that we can identify such biochemical alterations (Bournias-Vardiabasis et al. 1986; N. Bournias-Vardiabasis, unpubl.).

The data obtained thus far indicate that the *Drosophila* assay can, in addition to being a potential tool for tier I screening, be used for understanding some of the mechanisms of teratogenesis. The vast biochemical, molecular, and developmental knowledge on *Drosophila* should provide us with some important clues as to the roles of genetic and biochemical variables in the process of teratogenesis.

Information from the heat shock protein induction response of *Drosophila* has also provided us with at least some possibilities on how heat shock protein induction and developmental abnormalities can be interrelated. Though details are still sketchy, a number of recent results suggest that heat shock proteins function to protect the cell and, consequently, the whole organism against environmental stresses. Three lines of evidence point to the contention that heat shock proteins are at least involved in thermotolerance:

1. Heat shock protein induction is a prerequisite for development of thermotolerance (Mitchell et al. 1979; McAlister and Finkelstein 1980; Buzin and Bournias-Vardiabasis 1982; Li and Werb 1982).
2. Environmental stresses other than heat shock that induce heat shock proteins are also active in the induction of thermotolerance (Li and Werb 1982; Li 1983).
3. Early *Drosophila* embryos (which cannot induce heat shock proteins in response to heat) (Graziosi et al. 1980; Roccheri et al. 1981), *Dictyostelium* mutants defective in the induction of heat shock proteins (Loomis and Wheeler 1982), and addition of cycloheximide (Loomis and Wheeler 1980; Henle and Leeper 1982) preclude induction of thermotolerance.

The last and, for us, the most interesting, line of thought to be dealt with is the relationship between induction of heat shock proteins and their protective function, if any, in developmental abnormalities. Phenocopy formation has been postulated to be due to the absence of a specific translation product needed at a particular time in a developmental program. A sudden high heat shock inhibits transcription quickly, resulting in a lack of mRNA for products usually produced at that time in development. However, a preheat treatment does not inhibit transcription but allows production of both heat shock and normal mRNAs, which are exported as ribonucleoproteins to the cytoplasm. It is possible that an inactive ribonucleoprotein pool containing mRNAs is sequestered in the cytoplasm. This gradual accumulation of mRNAs in an inactive pool provides the basis for late survival of a potentially lethal heat shock and allows resumption of the normal developmental pathway when the heat shock is released (Mitchell et al. 1979).

Our interest and subsequent experiments will focus on determining whether the small heat shock proteins provide a protective function. Conditions will be sought that allow *Drosophila* cells to produce only the small heat shock proteins without damage to the host cells. Experiments will be carried out to determine whether these cells can be protected from the effects of various teratogens. Another way to study the function of the small heat shock proteins is to specifically prevent their synthesis. We will inject embryos with antisense RNA specific for the small heat shock proteins and study the effects of heat shock plus teratogen exposure on subsequent embryonic development. This technique has already been used by McGarry and Lindquist (1986) to transform normal *Drosophila* cells into ones in which the synthesis of hsp 26 was dramatically reduced. No protection or developmental studies of these cells was carried out, however.

ACKNOWLEDGMENTS

We thank Josephine Flores and Cheryl Clark for excellent technical support, John Hardy for the scanning electron micrographs, and Vicki Cogley for secretarial assistance. Part of the work presented here was supported by Johns Hopkins Alternatives to Animals grants and March of Dimes—Hazards in the Workplace grants to N.B.V. Scanning electron micrographs were prepared in the Shared Instrumentation Laboratory using the Philips SEM 505 purchased with funds from the National Institutes of Health (S10 RRO 1467-01).

REFERENCES

Bournias-Vardiabasis, N. 1983. On the teratogenic effect of coumarin and hydroxycoumarin in *D. melanogaster*. *Drosophila Inform. Serv.* **59:** 24.

Bournias-Vardiabasis, N. and J. Flores. 1983. Drug metabolizing enzymes in *Drosophila melanogaster:* Teratogenicity of cyclophosphamide *in vitro*. *Teratog. Carcinog. Mutagen.* **3:** 255.

Bournias-Vardiabasis, N., C.H. Buzin, and J.G. Reilly. 1983a. The effect of 5-azacytidine and cytidine analogs on *Drosophila melanogaster* cells in culture. *Wilhelm Roux's Arch. Dev. Biol.* **192:** 299.

Bournias-Vardiabasis, N., P. Salvaterra, and T. Nair. 1986. Neuronal development in *Drosophila* embryonic cultures. In *Progress in development biology,* part A (ed. H.C. Slavkin), p. 123. A.R. Liss, New York.

Bournias-Vardiabasis, N., R.L. Teplitz, G.P. Chernoff, and R.L. Seecof. 1983b. Detection of teratogens in the *Drosophila in vitro* test: Assay of 100 chemicals. *Teratology* **28:** 109.

Buzin, C.H. and N. Bournias-Vardiabasis. 1982. Teratogens induce small heat-shock proteins. In *Heat shock: From bacteria to man* (ed. M.J. Schlesinger et al.), p. 387. Cold Spring Harbor Laboratory, Cold Spring Harbor, New York.

———. 1984. Teratogens induce a subset of small heat shock proteins in *Drosophila* primary embryonic cell cultures. *Proc. Natl. Acad. Sci.* **81:** 4075.

Buzin, C.H. and N.S. Petersen. 1982. A comparison of the multiple *Drosophila* heat shock proteins in cell line and larval salivary glands by two-dimensional gel analysis. *J. Mol. Biol.* **158:** 181.

Buzin, C.H. and R.L. Seecof. 1981. Developmental modulation of protein synthesis in *Drosophila* primary embryonic cell cultures. *Dev. Genet.* **2:** 237.

Craig, E. 1985. The heat shock response. *CRC Crit. Rev. Biochem.* **18:** 239.

Donady, J.J., R.L. Seecof, and S.A. Dewhurst. 1975. Ultrastructural differentiation during *Drosophila* neurogenesis *in vitro. Differentiation* **4:** 9.

Edwards, M.J. and R.A. Wanner. 1977. Extremes of temperature. In *Handbook of teratology* (ed. J.G. Wilson and F.C. Fraser), p. 421. Plenum Press, New York.

Elsdale, T., M. Pearson, and M. Whitehead. 1976. Abnormalities in somite segmentation following heat shock in *Xenopus* embryos. *J. Embryol. Exp. Morphol.* **35:** 626.

Gerson, I., R.L. Seecof, and R.L. Teplitz. 1976a. Ultrastructural differentiation during embryonic *Drosophila* myogenesis *in vitro. In Vitro* **12:** 615.

———. 1976b. Ultrastructural differentiation during *Drosophila* neurogenesis *in vitro. J. Neurobiol.* **7:** 447.

Goldschmidt, R. 1935. Gen und Auszeneigenschaft (Untersuchung an *Drosophila*) und II. *Z. Vererbungsl.* **69:** 38.

Graziosi, G., F. Micall, R. Marzari, F. de Cristini, and A. Savoini. 1980. Variability of response of early *Drosophila* embryos to heat shock. *J. Exp. Zool.* **214:** 141.

Henle, K.J. and D.B. Leeper. 1982. Modification of the heat shock response and thermotolerance by cycloheximide, hydroxyurea, lucanthone in CHO cells. *Radiat. Res.* **90:** 339.

Kimmel, G.L., K. Smith, D.M. Kocchar, and R.M. Pratt. 1982. Overview of *in vitro* teratogenicity testing: Aspects of validation and application to screening. *Teratog. Carcinog. Mutagen.* **2**(3/4)**:** 221.

Li, G.C. 1983. Induction of thermotolerance and enhanced heat shock protein synthesis in Chinese hamster fibroblasts by sodium arsenite and ethanol. *J. Cell. Physiol.* **115:** 116.

Li, G.C. and Z. Werb. 1982. Correlation between synthesis of heat shock proteins and development of thermotolerance in Chinese hamster fibroblasts. *Proc. Natl. Acad. Sci.* **79:** 3218.

Loomis, W.F. and S.A. Wheeler. 1980. Heat shock response of *Dictyostelium. Dev. Biol.* **79:** 399.

———. 1982. Chromatin-associated heat shock proteins in *Dictyostelium. Dev. Biol.* **90:** 412.

McAlister, L. and D.B. Finkelstein. 1980. Heat shock protein and thermal resistance in yeast. *Biochem. Biophys. Res. Commun.* **93:** 819.

McBride, W.G. and P.H. Vardy. 1983. Pathogenesis of thalidomide teratogenesis in the marmoset: Evidence suggesting a possible trophic influence of cholinergic nerves in limb morphogenesis. *Dev. Growth Differ.* **25(4):** 361.

McGarry, T.J. and S. Lindquist. 1986. Inhibition of heat shock protein synthesis by heat inducible antisense RNA. *Proc. Natl. Acad. Sci.* **83:** 399.

Mitchell, H.K., G. Moller, N.S. Petersen, and L. Lipps-Sarmiento. 1979. Specific protection from phenocopy induction by heat shock. *Dev. Genet.* **1:** 181.

Nover, L. 1984. *Heat shock response of eukaryotic cells.* Springer-Verlag, New York.

Petersen, N. and H. Mitchell. 1985. Heat shock proteins. In *Comprehensive insect physiology, biochemistry, and pharmacology* (ed. G.A. Kerkut and L.I. Gilbert), p. 347. Pergamon Press, New York.

Ritossa, F. 1962. A new puffing pattern induced by heat shock and DNP in *Drosophila. Experientia* **18:** 571.

Rizki, R.M. and T.M. Rizki. 1965. Morphogenetic effects of 6-azauracil and 6-azauridine. *Science* **150:** 223.

Roccheri, M.C., M.G. DiBernardo, and G. Guidize. 1981. Synthesis of heat shock proteins in developing sea urchins. *Dev. Biol.* **83:** 173.

Sang, J.H. and J.M. McDonald. 1954. Production of phenocopies in *Drosophila* using salts, particularly metaborate. *J. Genet.* **52:** 392.

Schuler, R.L., B.D. Hardin, and R.W. Niemeier. 1982. *Drosophila* as a tool for the rapid assessment of chemicals for teratogenicity. *Teratog. Carcinog. Mutagen.* **2:** 293.

Seecof, R.L. 1980. Preparation of cell cultures from *Drosophila melanogaster* embryos. *TCA Man.* **5:** 1019.

Seecof, R.L., J.J. Donady, and R.L. Teplitz. 1973a. Differentiation of *Drosophila* neuroblast to form ganglion-like clusters of neurons *in vitro. Cell Differ.* **2:** 143.

Seecof, R.L., I. Gerson, J.S. Donady, and R.L. Teplitz. 1973b. *Drosophila* myogenesis *in vitro.* The genesis of small myocytes and myotubes. *Dev. Biol.* **35:** 250.

Shepard, T.H. 1985. *Catalog of teratogenic agents.* Johns Hopkins University Press, Baltimore, Maryland.

Smith, M.K., G.L. Kimmel, D.M. Kochhar, T.H. Shepard, S.P. Spielberg, and J.G. Wilson. 1983. A selection of candidate compounds for *in vitro* teratogenesis test validation. *Teratog. Carcinog. Mutagen.* **3:** 461.

Tissieres, A., H.K. Mitchell, and U.M. Tracy. 1974. Protein synthesis in salivary glands of *Drosophila* melanogaster: Relation to chromosome puffs. *J. Mol. Biol.* **84:** 389.

Wilson, J.G. 1977. Current status of teratology. In *Handbook of teratology* (ed. J.G. Wilson and F.C. Fraser), p. 309. Plenum Press, New York.

Zimmerman, E.F. 1985. Role of neurotransmitters in palate development and teratologic implications. In *Developmental mechanisms: Normal and abnormal,* p. 283. A.R. Liss, New York.

COMMENTS

McLachlan: All of the treatments and chemicals shown in your first slide induce heat shock proteins. Are there any studies that show suppression or depression of heat shock proteins that are already expressed of any molecular weight with any kind of treatment? There was a paper in

Nature some years ago showing that some heat shock protein expression came down over time of development during embryogenesis in the rat.

Bournias-Vardiabasis: The small heat shock proteins, which we see as being developmentally regulated, come and go. If you look at 22K, 23K, 26K, and 28K proteins and embryonic development (starting from oogenesis) and larval and pupal stages, you see modulation. There are treatments, but not in *Drosophila,* that do suppress heat shock proteins. In *Drosophila,* we have not seen any treatments that have suppressed heat shock protein induction.

Morrissey: You mentioned that a number of teratogens produce the heat shock proteins. What about high levels of nonteratogens? Will toxic responses also induce heat shock proteins?

Bournias-Vardiabasis: I don't know that. We did dose/response studies with arsenic, and if the cells were killed, there were problems with incorporation; there was not enough label to see anything. When doing the kinetics, for example, with diphenylhydantoin, you can do a concentration that is tenfold higher and leave it there for a short time. You can then see the expression of the heat shock proteins much earlier than you would using the normal teratogen dose. I would imagine it is toxic at those doses, and you would see them turned on. The big question is, although it is obviously good for the organism to have heat shock proteins, is it an indication of a protective function or injury to the cell or both, and which happens first?

Welsch: My question relates to using this assay as a prescreening method to detect the embryotoxic hazard potential of chemicals with unkown biological activity.

Bournias-Vardiabasis: The LD_{50}s were done at the time for us to have a way of knowing that we were not dealing with toxicity. It was not thought of as a way of determining embryo risk assessment.

Johnson: The system with which she is working has a high probability of being capable of evaluating the developmental hazard potential of agents. It is a very clean system, and it has a lot of endpoints.

Saxen: Do the cells within the aggregate behave differently from single dispersed cells?

Bournias-Vardiabasis: There is a concentration requirement. It is necessary to have a certain number of cells in proximity to each other in order for the cells to undergo differentiation. If the cells are plated below 10^5 per dish, they will not differentiate properly.

Scandalios: Do your data clearly show that you induce heat-shock-type proteins?

Bournias-Vardiabasis: Yes. They have been identified as heat shock proteins.

Scandalios: Do you have any evidence that you are affecting transcription or translation and not activation of sequestered or membrane-bound proteins?

Bournias-Vardiabasis: We know that these are heat shock proteins because we have done Cleveland digests, and the peptide fragments are identical. We have also done Western blots that identify them as being identical to the heat shock proteins.

Scandalios: Your induction could, in fact, be activation of proteins. What I am asking is whether or not a lot of the heat shock proteins that are induced by many different agents would appear under normal developmental conditions.

Bournias-Vardiabasis: I think that is a distinct possibility.

Kimmel: Can you differentiate between the protein patterns in a situation where you have pretreatment with a nontoxic level of a chemical versus treatment with a level that will affect the neuron development?

Bournias-Vardiabasis: No. It's an all-or-none thing, although the level of induction varies. The teratogen 5-azacytidine is the most potent inducer of heat shock.

Kimmel: In the case of hyperthermia, if a 35°C pretreatment exposure is used, versus 43°C, what is the difference?

Bournias-Vardiabasis: The difference is in the level of induction. The proteins are all turned on at the same time. What you see at 43°C is the depression of the rest of the proteins. If you look at the 43°C treatment at different times, only the heat shock proteins are there.

Rice: Is there anything distinctive about the amino acid content of any of these proteins?

Bournais-Vardiabasis: No. They have conserved sequences. The only homology that has given us some insight into the possible structural role is their having a 40% homology with α-crystalline proteins found in the lens. But, other than that, there is nothing that is very striking.

Teratogen Metabolism

ANDREW G. BRAUN
Department of Applied Biological Sciences
Massachusetts Institute of Technology
Cambridge, Massachusetts 02139

OVERVIEW

Many teratogens inhibit the attachment of ascites tumor cells to lectin-coated plastic surfaces. Using attachment inhibition as an assay, it is possible to detect biologically active metabolites of thalidomide and diphenylhydantoin (DPH). The metabolites are generated by the standard hepatic microsome–NADPH system. Inhibitors of cytochrome P-450 monooxygenases also inhibit the production of these metabolites. The metabolites can be partially purified by organic solvent extraction. Biologically active thalidomide metabolites may be of great value in embryology as temporary inhibitors of cell–cell or cell–extracellular–matrix (ECM) interactions. They may also be of value in clinical settings due to their ability to modify the immune response.

INTRODUCTION

Substantial epidemiological evidence indicates that most birth defects are caused by nongenetic environmental factors. For example, identical twin studies have found a lack of concordance in over 60% of twins in which one member was born with a congenital malformation (Myrianthoplolous 1976). Rapid temporal changes in specific defect incidence (Layde et al. 1980; Owens et al. 1981) and changes in specific malformation rates in migrant populations (Naggan and MacMahon 1967) also suggest that environmental factors play a major role in birth defect etiology.

Although a variety of environmental factors may be implicated in birth defects, the increased level of maternal exposure to man-made chemicals demands that toxicologists evaluate potentially teratogenic chemicals before they are released.

Conventional animal testing is very expensive, in terms of both financial and technical resources. With the large number of chemicals developed annually, alternative means of teratologic evaluation are necessary. During the last decade, several groups have tried to develop simple, inexpensive, in vitro systems appropriate for the detection of chemical teratogens. Several of these systems are discussed elsewhere in this volume.

Banbury Report 26: Developmental Toxicology: Mechanisms and Risk

Beyond their potential utility as monitors of teratogenic risk, these assay systems may be of great value as tools to investigate basic mechanisms of teratogenesis and development.

RESULTS AND DISCUSSION

My co-workers and I have developed and tested a teratogen assay system based on the observation that many nongenotoxic teratogens inhibit the attachment of tumor cells to lectin-coated plastic disks. The rationale behind the system has been discussed previously (Braun et al. 1982b). Figure 1 is a schematic depiction of the assay.

In a survey of 135 chemicals tested by others for their teratogenic activity in animals, we found the assay to have a 79% accuracy (Braun 1985). Among inhibitory teratogens, a crude quantitative correspondence between in vivo and in vitro activity was found (Braun et al. 1982a). Recently, we have turned our attention to human teratogens that appear to require metabolic activation for inhibitory activity.

Figure 2 is a simple representation of the toxicologist's view of the embryonic–maternal system. Ingested chemicals enter the circulation from the gut and are subject to enzymatic modification on passage through the maternal liver. Further transformations are possible at the embryo. Potential teratogens may be inactivated or proteratogens activated near target tissue. Strickler et al. (1985) have developed evidence that major malformations in response to DPH are related to the biochemical activity of the child rather than the mother. It also seems possible that teratogen site specificity may be due to differences in embryonic metabolic activity from tissue to tissue.

Our studies of teratogen metabolism have concentrated on two human teratogens: thalidomide and DPH. When thalidomide was incubated with hepatic microsomes from a number of animals (+NADPH), an inhibitory metabolite was formed (Braun and Dailey 1981). Metabolite generation was time- and temperature-dependent (Fig. 3). Hepatic microsomes from Aroclor 1254- or methylcholanthrene-induced rats yielded lower levels of inhibitory metabolites than microsomes from uninduced animals. Metabolite formation was reduced by several classic inhibitors of cytochrome P-450 monooxygenase activity (Braun et al. 1986). Yield was not affected by epoxide hydrolase or its inhibitor tetrachloropropane oxide (TCPO). Beyond a partial purification (Braun et al. 1986), efforts to fully purify and identify the inhibitory metabolite(s) have been unsuccessful. Thus, it appears that thalidomide is a prototype proteratogen requiring cytochrome P-450 monooxygenase transformation for biological activity.

A strength of in vitro assay systems is their ability to dissect complex phenomena into simple elements. For example, two analogs of thalidomide, EM12 and EM87, differing only at a single site in the phthalimide moiety of

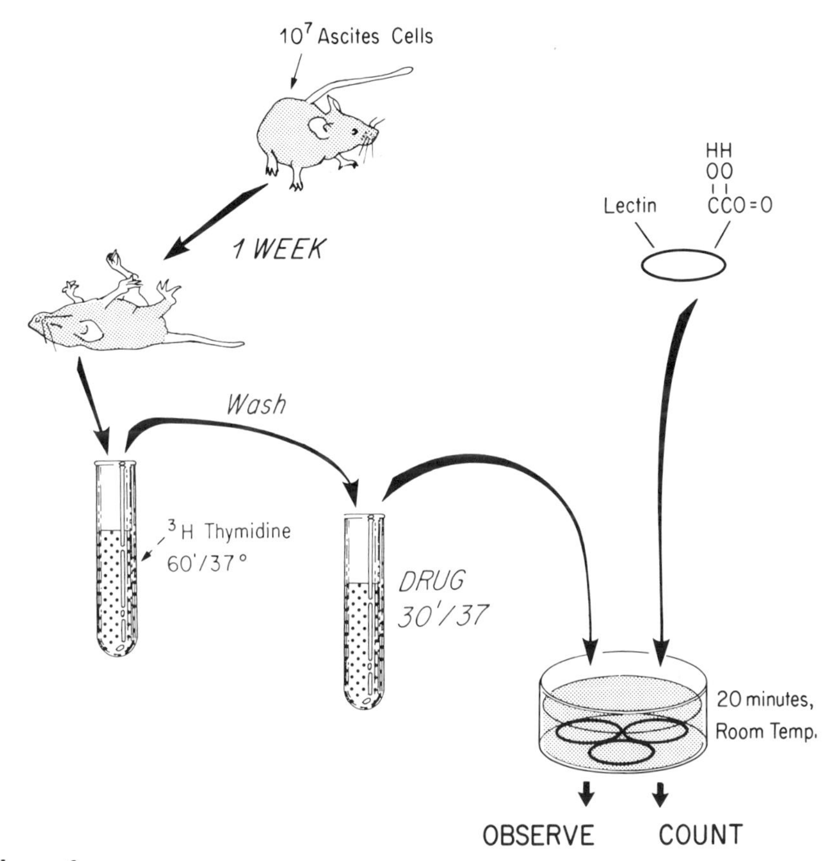

Figure 1

Schematic of assay system. MOT ascites tumor cells grown in C3H/HeJ mice are harvested, incubated with tritiated thymidine for 1 hr, washed free of label, and resuspended in phosphate-buffered saline. An aliquot of cells is then incubated with test agent for 30 min and poured over 1.25-cm polyethylene disks previously coated with concanavalin A or another plant lectin. After 20 min, the coated disks are removed, washed in saline, and the number of attached cells is quantitated by scintillation counting.

the molecule, have very different teratogenic activity in vivo (Helm et al. 1981; Helm and Frankus 1982). EM87 is far less teratogenic, in vivo, than the parent compound, whereas EM12 is considerably more active. Both analogs inhibited attachment if incubated with microsomes and NADPH (Braun and Weinreb 1984). However, if the analogs, or thalidomide itself, were preincubated at pH 7.4, a time-dependent decrease in activity was found (Fig. 4). Thalidomide is known to undergo spontaneous hydrolysis at alkaline pH.

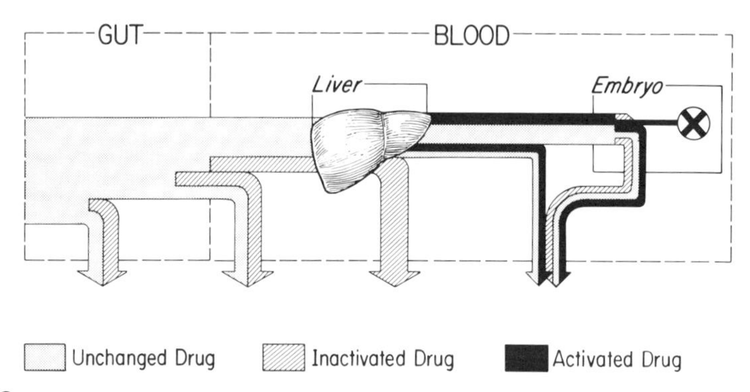

Figure 2
A toxicologist's view of the embryo.

From the data of Figure 4, it appears that the hydrolysis products cannot be metabolically activated to inhibitory products. The rapid decline of EM87 activity on alkaline incubation suggests that the instability of this compound is responsible for its low teratogenic activity. On the other hand, the relative stability of EM12 suggests that higher levels of metabolite can accumulate in the embryo, thereby increasing its teratogenic activity. In vitro testing may also unravel the question of why optically inactive analogs of thalidomide are nonteratogenic in vivo (Helm and Frankus 1982).

These questions are of more than academic interest. Thalidomide appears to have powerful effects on the immune system. Several reports indicate that the drug prolongs skin engraftment (Hellmann et al. 1965) and prevents graft

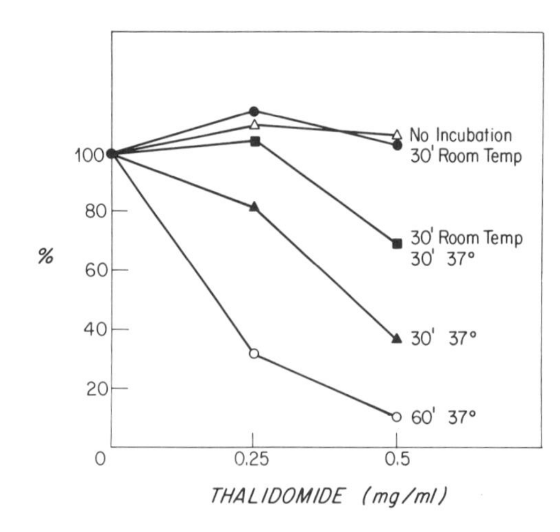

Figure 3
Thalidomide activation is time- and temperature-dependent. Thalidomide at the indicated concentration was incubated for 0, 30, or 60 min, as indicated, with NADPH and canine hepatic microsomes at room temperature or at 37°C. (●) NADPH was omitted from the incubation.

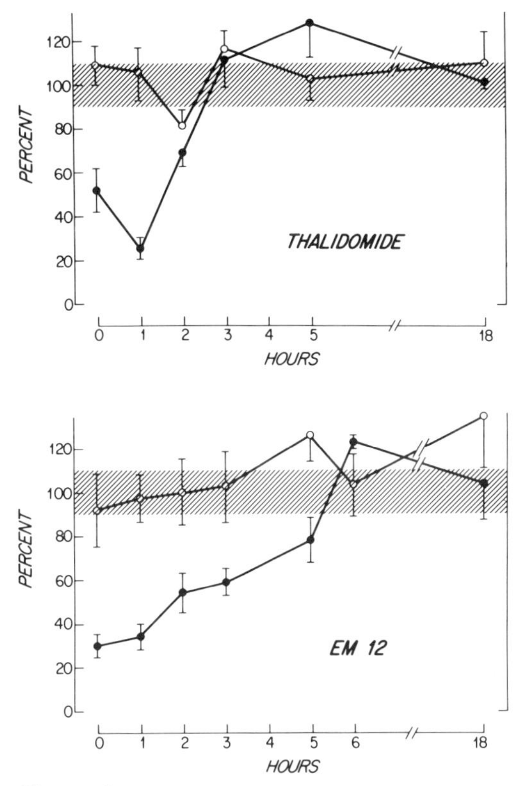

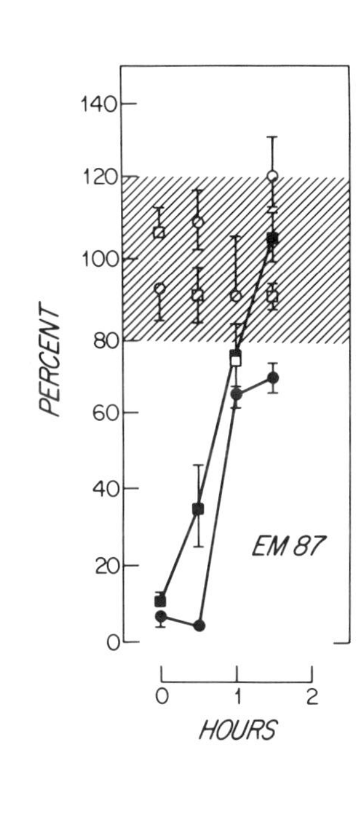

Figure 4

Decay of thalidomide, EM12, and EM87 on alkaline incubation. The three drugs were incubated in phosphate-buffered saline for the indicated time at 37°C; MOT ascites cells, canine hepatic microsomes, and NADPH were added and incubated for an additional 30 min at 37°C; and the attachment of MOT cells to concanavalin-A-coated disks was measured. Hatched area indicates the range (±1 S. E. M.) of negative controls; (○) dimethylsulfoxide (DMSO) vehicle; (●) drug at 500 μg/ml. S.E.M. of individual points indicated. (Reprinted, with permission, from Braun and Weinreb 1985.)

versus host disease (Vogelsang et al. 1986). Thalidomide has powerful anti-inflammatory activity in humans (Sheskin et al. 1983; Gutierrez-Rodriguez 1984). An understanding of thalidomide's mechanism of action may well lead to the development of medically useful drugs having no teratogenic side effects.

DPH (also called phenytoin and Dilantin), a widely used antiepileptic medication, has been implicated in the etiology of a complex syndrome of congenital malformations. The drug is not mutagenic and does not appear to have any in vitro activity at concentrations (20 μg/ml) normally found in patients. There is evidence that children having major malformations born of mothers taking DPH have a defect in arene oxide metabolism (Strickler et al.

1985), an indication that metabolic processes have a role in the drug's teratogenicity.

DPH, by itself, did not inhibit attachment at concentrations of less than 100 μg/ml. There was some evidence of slight inhibitory activity at higher concentrations. When incubated with canine hepatic microsomes (+NADPH), strong inhibitory activity was detected at less than 25 μg/ml (Fig. 5). Hepatic microsomes from uninduced and induced rats have produced considerably less inhibitory activity (A.G. Braun and F.A. Harding, unpubl.). Microsomes from human placenta also activate DPH to an inhibitory metabolite (A.G. Braun and M.D. Collins, unpubl.). Other experiments have shown that oxygen is required for DPH activation and that inhibitors of cytochrome P-450 also prevent or reduce activation.

These drugs and their metabolites can be valuable tools for the developmental biologist. For example, the thalidomide analog EM87 can be quite useful in the examination of morphogenic interactions. EM87 and its hydrolysis products have no apparent biological activity. The drug is almost insoluble. Metabolically activated EM87 has a lifetime of less than 30 minutes. Thus, using activated EM87, it may become possible to interfere with morphogenic cell–cell or cell–ECM interactions for a defined interval at a defined location.

ACKNOWLEDGMENTS

The work reported here was carried out with the invaluable help of many technicians to whom I am greatly indebted: Mr. Peter Horowicz, Mr. Bradley

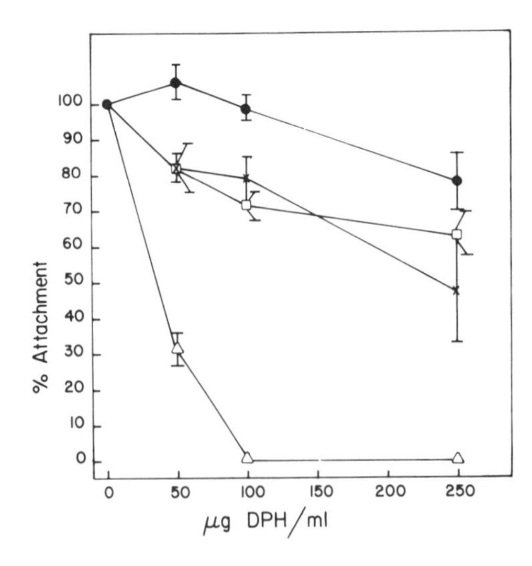

Figure 5

DPH activation. DPH at the indicated concentration was incubated with microsomes and NADPH for 60 min at 37°C, the microsomes were removed by centrifugation, and the MOT cells were added. After an additional 30 min at 37°C, the suspension was poured over concanavalin-A-coated disks, and attachment was measured. (□) No microsomes; (△) uninduced rat hepatic microsomes; (×) phenobarbital-induced rat hepatic microsomes; (●) Aroclor 1254-induced rat hepatic microsomes. S.E.M. indicated.

Nichinson, Mr. Steven Weinreb, Mr. David Emerson, Ms. Christine Buckner, and Ms. Fiona Harding. We were supported by Department of Energy contract DE-ACO2-83ER60174 and Environmental Protection Agency (EPA) grant CR-807809. The contents of this paper do not necessarily reflect the policies of the EPA.

REFERENCES

Braun, A.G. 1985. New perspectives in tests for teratogenicity. In *Safety evaluation and regulation of chemicals* (ed. F. Homberger), p. 230. Karger, Basel.

Braun, A.G. and J.P. Dailey. 1981. Thalidomide metabolite inhibits tumor cell attachment to concanavalin A coated surfaces. *Biochem. Biophys. Res. Commun.* **98:** 1029.

Braun, A.G. and S.L. Weinreb. 1984. Teratogen metabolism: Activation of thalidomide and thalidomide analogues to products that inhibit the attachment of cells to concanavalin A coated surfaces. *Biochem. Pharmacol.* **33:** 1471.

———. 1985. Teratogen metabolism: Spontaneous decay products of thalidomide and thalidomide analogues are not bio-activated by liver microsomes. *Teratog. Carcinog. Mutagen.* **5:** 149.

Braun, A.G., F.A. Harding, and S.L. Weinreb. 1986. Teratogen metabolism: Thalidomide activation is mediated by cytochrome P-450. *Toxicol. Appl. Pharmacol.* **82:** 175.

Braun, A.G., B.B. Nichinson, and P.B. Horowicz. 1982a. The inhibition of tumor cell attachment to concanavalin A coated surfaces as an assay for teratogenic agents: Approaches to validation. *Teratog. Carcinog. Mutagen.* **2:** 343.

Braun, A.G., D.J. Emerson, B.B. Nichinson, and C.A. Buckner. 1982b. Quantitative correspondence between the in vivo and in vitro activity of teratogenic agents. *Proc. Natl. Acad. Sci.* **79:** 2056.

Gutierrez-Rodriguez, O. 1984. Thalidomide: A promising new treatment for rheumatoid arthritis. *Arthritis Rheum.* **27:** 1118.

Hellmann, K., D.I. Duke, and D.F. Tucker. 1965. Prolongation of skin homograft survival by thalidomide. *Br. Med. J.* **2:** 687.

Helm, F.-C. and E. Frankus. 1982. Chemical, structural and teratogenic activity of thalidomide-related compounds. *Teratology* **25:** 47A.

Helm, F.-C., E. Frankus, E. Friedrichs, I. Graudums, and L. Flohe. 1981. Comparative teratological investigation of compounds structurally and pharmacologically related to thalidomide. *Arzneim. Forsch.* **31:** 941.

Layde, P.M., K. Dooley, J.D. Erickson, and L.D. Edmonds. 1980. Is there an epidemic of ventricular septal defects in the USA? *Lancet* **I:** 407.

Myrianthoplolous, N.C. 1976. Congenital defects in twins: Epidemiological study. *Birth Defects Orig. Artic. Ser.* **8**.

Naggan, L. and B. MacMahon. 1967. Ethnic differences in the prevalence of anencephaly and spina bifida in Boston, Massachusetts. *N. Engl. J. Med.* **277:** 1119.

Owens, J.R., F. Harris, E. McAllister, and L. West. 1981. 19-year incidence of neural tube defects in area under constant surveillance. *Lancet* **II:** 1032.

Sheskin, J., H. Muckter, and E. Frankus. 1983. Supidimid, ein nicht teratogenes Thalidomide zur Behandlung der Leprareaktion? *Hautarzt* **34:** 168.

Strickler, S.M., L.V. Dansky, M.A. Miller, M.H. Seni, E. Andermann, and S.P. Spielberg. 1985. Genetic predisposition to phenytoin-induced birth defects. *Lancet* **II:** 746.

Vogelsang, G.B., A.D. Hess, G. Gordon, and G.W. Santos. 1986. Treatment and prevention of acute graft–versus host disease with thalidomide in a rat model. *Transplantation* **41:** 644.

COMMENTS

Manson: You have a 5-log scale there. A classic inhibitor of cell-to-cell communication would probably be working in the 10^{-6} to 10^{-8} range. If an agent has an IC_{50} of 10^{-1}, do you think it's an important response? Is it indicative that the compound has developmental toxicity?

Braun: I don't know. It is possible for a compound that has very little maternal toxicity to work at extremely high concentrations. And so, one could argue, by the A/D ratio, that the absolute value is irrelevant; however, the relative value between toxicity and inhibitory activity is relevant.

Welsch: At a recent conference, Dr. Oliver Flint stated that he has indications from the rat embryo limb-bud mesenchyme chondrogenic differentiation assay and the rat midbrain neuron differentiation assay that metabolism of thalidomide is not required to detect activity that can be observed only in the limb-derived cells. Apparently, hydrolysis products form spontaneously in the culture medium and interfere with chondrogenic differentiation. It seems to me that we still don't know whether metabolism is required for biological activity. You can state, with some confidence, that the phenomenon of attachment of tumor cells is inhibited after treatment with thalidomide incubated with mammalian microsomes.

Braun: I don't think there is much question that a metabolite is involved in the attachment–inhibition phenomenon. There are a lot of factors, for example, inhibitors and the oxygen requirement, that are not involved in hydrolysis. That doesn't mean that thalidomide's function or activity is necessarily due to the metabolite. The interpretation of Dr. Flint's results and mine may be in conflict, but the two observations are reproducible. If I contend that the metabolite has something to do with actual teratogenesis, that is an open question. But certainly, the fact of interference with cell-cell interaction seems to be mediated by a metabolite.

Iannaccone: Have any of the metabolites been identified by chromatography?

Braun: I have tried but haven't been able to do it yet.

Miller: You indicated that EM87 would be very good for local interaction studies because it is so insoluble, but metabolism would still be important. Could you please elaborate?

Braun: Some metabolite interferes with an interaction, and it will presumably interfere with other interactions as well. If the metabolite itself, and not the parent compound, is isolated and is then applied to regions of interest, it will have the desired effect. The parent compound would not.

Miller: Is there much variation in the human placental preparations in terms of their inhibitory effects from one to another?

Braun: There is a substantial difference, an order of magnitude difference, in almost every assay of human placental activity. Some of our preparations are very, very close to being inactive; others are quite active. This was clearly the most active preparation. We have no idea why.

Miller: Are you measuring the P-450?

Braun: Yes, an aryl hydrocarbon hydroxylase (AHH) assay has been made, and it is unrelated to this activity. We find that hepatic microsomes from Aroclor-induced animals are inactive in this assay, or very nearly inactive. Apparently, the entire metabolic process is accelerated in Aroclor-induced liver, and the compounds of interest are relatively unstable under those conditions.

Miller: Would you screen phenytoin and thalidomide with the same preparation, and do they correspond?

Braun: We have. I should say that phenytoin is inactive in this assay, except at very high concentrations. In the presence of Aroclor-induced microsomes, it is inactive. Induced microsomes from Aroclor- or phenobarbital-induced animals are inactive in both cases. The pattern of cytochrome P-450 inhibitor response for phenytoin is different from that of thalidomide, and it appears that different pathways are involved.

Kochhar: You mentioned that many other cell types can be used in your assay. Have you used human or primate cells in the attachment assay to determine their response to thalidomide?

Braun: I haven't done it with thalidomide, but we have used human lymphoblast and CHO cells, and they function normally in the assay. The phenomenon seems to be relatively independent of the cell type, and the sensitivity of cells to a particular drug doesn't seem to be very different. I think what is involved in this assay is membrane response to drugs and how receptors are affected and their mobility in the membrane, and how the actin monofilaments are attached to the receptors.

Kochhar: If you were to interfere with the receptors for a particular drug, would that interfere with your test?

Braun: No. Any drug that requires a particular receptor, for example, dexamethasone (DEX), is much more active in the systems that have specific DEX receptors. Our system does not have DEX receptors and, therefore, only very high concentrations are successful in inhibiting attachment. For that reason, DEX is a false negative in this assay.

Morrissey: For the last 3 years, the National Toxicology Program has had the mouse ovarian tumor cell assay and Dr. Pratt's human embryonic palatal mesenchyme cell assay under evaluation as in vitro teratology screening systems. We recently held a workshop. We selected our chemicals from the Smith list, which is probably weighted toward chemicals that have a growth-inhibitory, rather than a cell-attachment, effect. What criteria did you use when you selected your chemicals?

Braun: Chemicals were tested that we believed had sufficient data from animal testing. Our selection was based on Dr. Shepard's listing, Shardein's compendium, and the general literature. That was the primary consideration in our selection. A secondary criterion was availability and cost. We knew that mutagens would have no effect, so we made an effort to include mutagens in the list. So, X-rays, for example, are in the list, as well as a variety of drugs that we predicted would fail.

Manson: In the thalidomide work, do you think it's possible that the mixed function oxidase system is working on a hydrolysis product to create the active metabolite?

Braun: I can't speak about the other two compounds, EM12 and EM87, but with thalidomide, the kinetics of hydrolysis and the kinetics of disappearance suggest that the parent drug is involved. It would require a remarkable hydrolysis product, one whose stability in this slightly alkaline medium is identical to thalidomide. It's conceivable, but it seems unlikely. The hydrolysis products tend to have very different half-lives. The thalidomide structure has four amino groups and they can hydrolyze at all four sites. It appears from structural studies that the glutarimide end is irrelevant to thalidomide's biological activity. So, the whole story is at the phthalimide end. The phthalimide amino sites are the two relevant hydrolysis sites. If both are hydrolyzed, the product is stable. If one site is hydrolyzed, the kinetics of the second hydrolysis would be determinative. We have tested all of the hydrolysis products available to us, and none are inhibitory. Our system is more sensitive than Dr. Flint's system to very different compounds. I think resolution of this quandary requires the extraction of thalidomide-related compounds from embryos.

Brent: Are you familiar with the work of Wuest?

Braun: No.

Brent: He was an organic chemist at Sloan–Kettering in the early 1960s. He synthesized about 17 of the analogs of thalidomide. I was surprised at your statement that some thalidomide derivatives were more potent teratogens than thalidomide, because his conclusion was that the most potent teratogen was thalidomide. These compounds are so insoluble, how would you use them in an embryo experiment? I think thalidomide is 1 part per 10,000 soluble.

Braun: Our idea is that thalidomide metabolites function on the exterior surface of the cell, so they don't have to migrate through the cell. I think their extreme insolubility is a strength for a developmental test because they are localized, although their local concentration may be difficult to measure. One can also solubilize them in DMSO and get water–DMSO emulsions that have very high concentrations. So, it is technically possible to increase the concentration.

Johnson: In validating in vitro systems, some of us tend to look at groups of agents, the great majority of which are "teratogenic." The way to really test these systems is to use a chemical list where there are fewer teratogens. This concept has become lost in our methods of trying to validate in vitro assays. The critical thing is the nature of the population of the chemicals tested.

Somite-stage Mammalian Embryo Culture—Use in Study of Normal Physiology and Mechanisms of Teratogenesis

THOMAS H. SHEPARD, ALAN G. FANTEL, AND PHILIP E. MIRKES
Central Laboratory for Human Embryology
Department of Pediatrics
School of Medicine
University of Washington
Seattle, Washington 98195

INTRODUCTION

In vitro culture of somite-stage embryos was reintroduced by New in 1967. Earlier work by Nicholas and Rudnick (1934, 1938) had shown that successful results from whole-embryo culture of rodents could be obtained. New's method utilized rat embryos attached to a raft and maintained in a circulating medium of rat serum. The circulation was driven by oxygen and carbon dioxide flowing through filters packed with rouge. This technique was extremely difficult because of the need to attach Reichert's membrane of the parietal yolk sac to the raft, the skill required to transfer the raft into the glass apparatus while under fluid, and the construction of the rouge filters to allow for the proper flow of gas. Three embryos could be grown in each circulator, and a maximum of about 9–12 embryos could be explanted in one session. Growth of the embryos was very good, but the technical difficulties noted above delayed the application of the method to teratologic problems.

As a crude measurement of interest by teratologists in the whole-embryo culture method, the number of abstracts describing its use, presented at the annual Teratology Society for each year, have been plotted in Figure 1. New presented his first method to the annual Teratology Society meeting in 1969. Relatively little scientific activity occurred until 1973 when the method was modified and simplified (New et al. 1973). The simplification consisted of a closed tube that was rotated, allowing for the overlying oxygen and carbon dioxide to be dissolved in the medium and delivered to the embryo. This modification allowed 10–15 embryos to be grown in each tube, and a statistical examination of survival rate, size, somite number, and protein content became possible. The increased number of embryos also allowed investigators to perform concentration/response studies. The old glass circulator method was abandoned, except for an apparatus developed by Robkin (Rob-

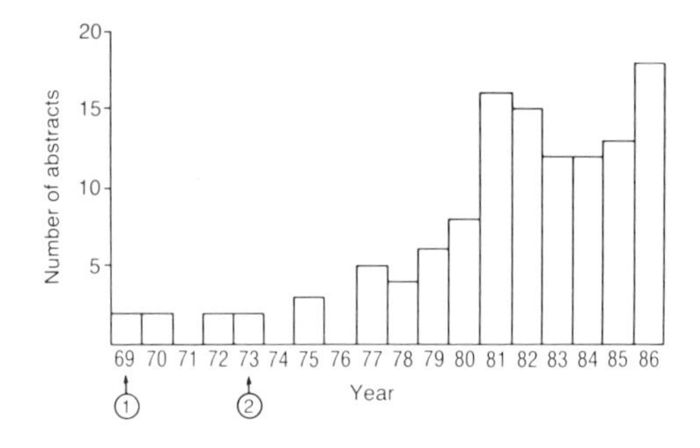

Figure 1
Number of abstracts dealing with whole-embryo culture presented at annual meetings of the Teratology Society. (1) Dennis New's lectures at the Teratology Society; (2) introduction of the rotary tube method by New and associates.

kin et al. 1972), which allowed for direct heart rate recording by impulses transmitted via a red (HE-NE) laser to a photo-multiplying tube. New's rotating-tube embryo culture method is now used widely, and the number of abstracts at the yearly meeting of the Teratology Society ranges from 15 to 20. The "New" technique is used by at least ten laboratories in the United States, as well as laboratories in Europe, Japan, and Australia, and over 100 agents have been studied.

Explant Technique

The technique of explantation and culture is very similar for the mouse and rat embryo. The animals are usually sacrificed by cervical dislocation. After removal of the sites, the decidua is teased off, leaving the yolk sac within which is an inverted embryo. A small amount of ectoplacental cone remains attached. The parietal yolk sac and Reichert's membrane are opened and retracted, and the conceptus is transferred to medium. The entire operation is done under a physiologic salt solution using sterile technique. In experienced hands, 15–25 sites can be removed per hour. Details along with illustrations are available in publications by New (1967, 1978), New et al. (1976), Cockroft (1976), and Kitchin et al. (1986).

Nutrients and Gassing

The percentage of oxygen to which the embryo is exposed is critical, depending on the stage of development. At the presomite or early head-fold stage, 5–10% oxygen is optimal, whereas after neural tube closure, 80% or more is necessary. The gas mixture always contains 5% CO_2. Data on oxygen requirement are available in publications by New (1967, 1978), New and Coppola (1970a), Shepard et al. (1969), and Tarlatzis et al. (1984). Concentrations of 20–40% oxygen interfere with neural tube closure (Morris and New 1979). After closure of the neural tube, hyperbaric oxygen has been used and produces some improvement in growth (New and Coppola 1970b; Cockroft 1973).

Culture medium is composed mainly of serum with varying amounts of physiologic saline or tissue culture medium added. Immediately centrifuged, heat-inactivated rat serum is a requirement, but a saving of time and money is possible by substituting some of the rat serum with human serum in up to 50% of the medium. At least 20% of the medium must be composed of rat serum. Data on nutrient requirements are published in papers by Steele and New (1974), Kochhar (1975), Lear et al. (1983), Priscott (1983), and Priscott et al. (1983). Sanyal (1980) has made a study of changes in the content of medium in which embryos are grown.

The embryos may be grown from the presomite stages up through the late somite stage, and the results by light microscopy, protein determination, and morphology are very close to the growth that is obtained in vivo (Cockroft 1976; New et al. 1976). Morphological scoring systems have been described previously (Brown and Fabro 1981; Sadler et al. 1982). Growth of mouse ova up through the early somite stages has been achieved by Hsu et al. (1974), but only 3–5% of the ova cultured survive to the early somite stage.

Alternate Methods for Exposure of Embryos to Teratogens

Test agents may be added directly to the culture medium, or serum from animals treated with the agent may be used (Kochhar 1975; Beaudoin and Fisher 1981). In the case of the human, serum drawn from patients treated with various drugs can be used as an additive (Chatot et al. 1980). More precise studies of the bioactivation of chemicals can be achieved by adding liver homogenate supernatant (S-9) or liver microsomes to the medium (Fantel et al. 1979). Microinjection of the agent directly into various embryonic compartments has also been used (Satish et al. 1985).

RESULTS

Agents Studied

At the outset we intended to list all the agents studied in embryo culture, but because 106 agents have been identified, their listing would be too lengthy for this publication. As an alternative, we are making this list available to the participants of the Banbury Conference or to any scientist requesting it.

Nutrition of the Embryo

Work at the University of Washington (Aksu et al. 1968; Shepard et al. 1970; Tanimura and Shepard 1970; Mackler et al. 1973) has demonstrated some of the pathways of energy metabolism in rat embryos of early-somite to late-limb-bud stages. Uniformly labeled [^{14}C]glucose was added to the nutrient serum, and after 3 hours of culture the metabolic fate of the glucose was determined by measuring the amount of liberated $^{14}CO_2$, [^{14}C]lactate released into the medium, and [^{14}C]glucose (or its products) incorporated into the tissues. The results showed a very high rate of glucose consumption in early-somite embryos and their membranes, and about 90% of the glucose consumed at this stage was converted to lactic acid (see Fig. 2). During the following 2 days of development, glucose consumption per gram of protein dropped precipitously, and this drop was associated with a decrease in percent conversion to lactic acid. This finding strongly suggests that the early somite-stage embryos have an energy-yielding metabolic pattern characterized by high rates of glycolysis, with relatively little activity of the Kreb's cycle and electron transport system. During neural tube closure and establishment of the circulatory system, glycolysis falls while the Kreb's cycle and electron transport activity increase. Studies of mitochondrial metabolism during these stages of development confirm these findings (Mackler et al. 1973).

New has summarized this work and the other contributions made by Neubert et al. (1971) and other laboratories. Ellington (1980) has studied the effect of addition of glucose to serum of fasted rats and determined that levels

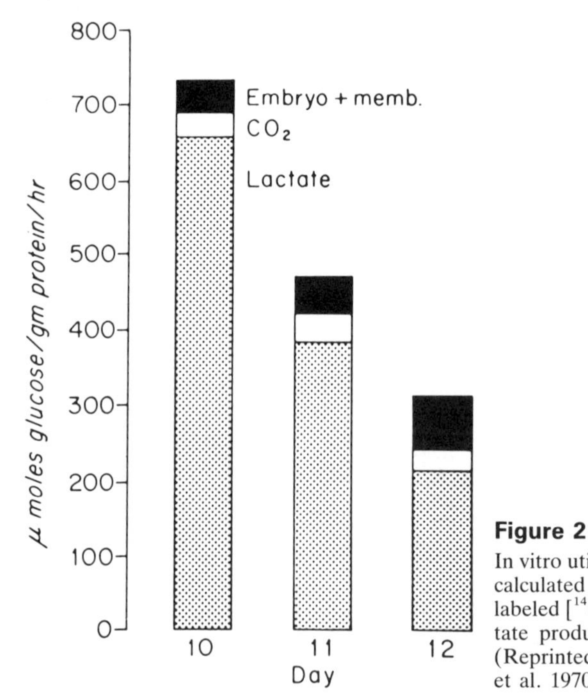

Figure 2

In vitro utiliziation of glucose by rat embryos calculated from studies using uniformly labeled [^{14}C]glucose. Note sharp drop in lactate production after day 10 (11 somites). (Reprinted, with permission, from Shepard et al. 1970.)

below 100 mg/dl are associated with poor growth. An ingenious study by Freinkel et al. (1984) has shown that the addition of D-mannose to the medium produces an inhibition of glycolysis that is associated with growth retardation and faulty neural tube closure. About 10 years ago, we were looking for a way of killing our embryos at the end of the culture period, and we attempted to use carbon monoxide. Much to our surprise, carbon monoxide at levels that were promptly lethal to the adult rat had no effect on the growth or development of the rat embryo (Robkin et al. 1976). This was subsequently explained by Robkin and Cockroft (1978) by the finding that glucose consumption increased and the embryos converted back to a glycolytic metabolism when exposed to carbon monoxide.

In an interesting series of papers, the role of the yolk sac placenta in the nutrition of the embryo has been elucidated. This work, concisely summarized by some of the scientists responsible (Lloyd et al. 1985), started with the following question: How does trypan blue cause malformations in an embryo when it is accumulated only in the visceral yolk sac? Through the use of a number of lysosomally active macromolecules, they were able to hypothesize that the visceral yolk sac captures the protein molecules by pinocytosis and accumulates them in the lysosomes where they are broken down into amino acids and subsequently transferred to the embryo for protein synthesis. The proof of this hypothesis came from the following experiments in which [^{3}H]leucine-labeled rat serum was used in an in vitro whole-embryo culture (Freeman et al. 1981). The labeled macromolecules appeared in the visceral yolk sac, and the radioactive leucine was transferred to the embryo for protein synthesis. It would be of interest to know whether the visceral yolk sac accumulates certain types of proteins selectively from the serum or perhaps acts as a scavenger for the older (ragged) maternal proteins. Another question might be answered by an in vitro culture: What is the contribution of fatty acids and fats to the nutrition of a growing embryo?

Temperature Effects

The embryonic heart rate, as measured directly, is closely correlated with the temperature of the surrounding medium (Robkin et al. 1972). The heart rate in an embryo of 25 somites is about 160 beats per minute at 38°C; lowering the temperature reduces heart rate about 7% for each degree drop in temperature (see Fig. 3).

Normal growth temperature for rodent embryos is 37°C–38°C. Embryos grown in a temperature of 40.5°C or above were found to be growth retarded, and this retardation was specifically localized to the brain area (Cockroft and New 1975). A good example of the benefit derived by direct access to the embryo is the data of Mirkes (1985), who studied the time necessary to induce

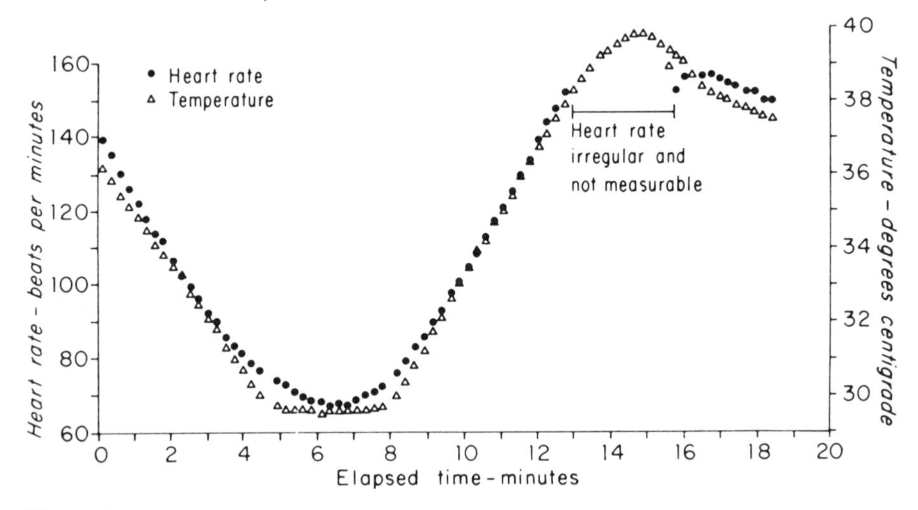

Figure 3

Correlated heart rate and temperature for a day-11 rat embryo. (Reprinted, with permission, from Robkin et al. 1972.)

abnormal morphogenesis when embryos are incubated in vitro at 41°C, 42°C, and 43°C. As can be seen in Figure 4, an embryonic effect was seen after exposures of 15 minutes at 43°C, 30 minutes at 42°C, and 2 hours at 41°C. Studies by Germain et al. (1985) on pregnant mice immersed in hot water of different temperatures for varying times gave similar findings. These data may be useful for establishing risks to the human fetus, which is probably also adversely affected by hyperthermia (summarized by Warkany 1986). The role of heat shock proteins produced by these embryos and other heated tissues is

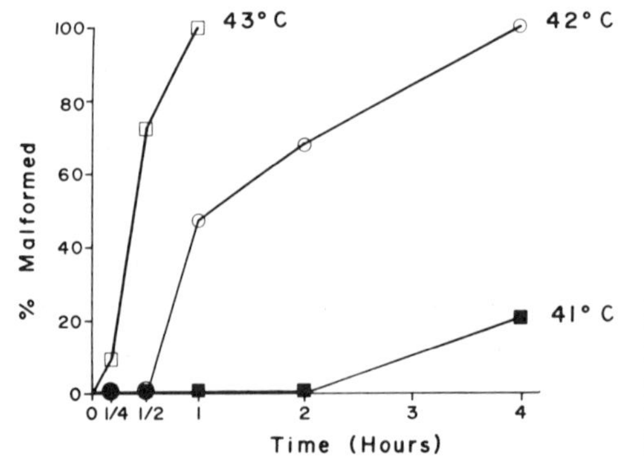

Figure 4

Effect of time of exposure and temperature increase above basal on malformations of day-10 rat embryos grown for 24 hr in vitro. (Data from Mirkes 1985.)

of increasing interest as a general mechanism of teratogenesis, and a number of laboratories are involved in work with such proteins.

Pharmacologic Studies of Teratogenesis

In vitro culture of embryos can give partial answers to two major questions: (1) Must a tertogen by bioactivated in order to affect the embryo, and (2) is the bioactivation a major contribution of the mother, the embryo, and/or the placenta? Enzymes catalyzing oxidation, reduction, and hydrolytic reactions are implicated most frequently in bioactivation. These reactions are referred to as phase-I reactions because they must often precede subsequent conjugating reactions categorized as phase II. Most studies to date have dealt mainly with phase-I monoxygenase systems (P-450) in hepatic tissue. It has been found that a 9000 supernatant fraction prepared from liver could bioactivate cyclophosphamide in embryo culture medium (Fantel et al. 1979). About one half of the chemicals studied to date have a requirement for bioactivation. Others, such as sodium salicylate and ethanol, appear to act directly on the isolated embryo (summarized by Shepard et al. 1983). One chemical, cytochalasin D, has been shown to be bioinactivated in the presence of S-9 (Fantel et al. 1981). By utilizing either phenobarbital (PB) or 3-methylcholanthrene (3-MC) to induce animals used as the source of S-9, Faustman-Watts et al. (1983) and Greenaway et al. (1982) showed inducer-specific bioactivation for 2-acetylaminofluorene (3-MC) and cyclophosphamide (PB). They were also able to show that embryos exposed to 3-MC in vivo were sufficiently induced to enable them to bioactivate 2-acetylaminofluorene in the absence of exogenous P-450 (Juchau et al. 1985). This finding added further support to the observation of Filler and Lew (1981) that preimplantation embryos had the capacity to metabolize some drugs. The relative role of bioactivation by the mother and the embryo is under study in many laboratories. It is likely that different compounds are metabolized at different sites and that species and time of embryonic development and gestational age are important determinants of these roles. Data have been uncovered, indicating that oxygen concentration may play an important role in teratogenicity of salicylates, niridazole, cyclophosphamide, and phosphoramide mustard, presumably by modulating cytochrome P-450 activity (Greenaway et al. 1985). The kinetics of damage to embryonic cells by drugs and other agents can be studied in more precise detail by using whole-embryo cultures.

DISCUSSION

Advantages and Disadvantages of In Vitro Embryo Culture

Because more detailed discussion of this subject has been given by New (1975, 1978), Fantel (1982), and Shepard et al. (1983), these features will only be summarized briefly and discussed. The results of treatment become known after 24 or 48 hours, as compared to 21 days with the usual in vivo test in rats. These techniques permit more precise control over embryonic exposures and eliminate the maternal animal, avoiding the confounding features of maternal physiology. These include maternal variables such as stress, nutritional effects, and uterine position. The culture system allows concentration-dependent associations to be determined; these values may prove useful in risk assessment for humans.

To determine optimal doses for whole animal tests, effective concentration ranges may be learned from preliminary in vitro embryo cultures. Kitchin et al. (1986) have compared the minimal teratogenic concentrations of 18 chemicals and drugs in vivo with the in vitro level that produces malformations or growth retardation. As can be seen from the plot in Figure 5, the correlation is quite close (r = 0.82).

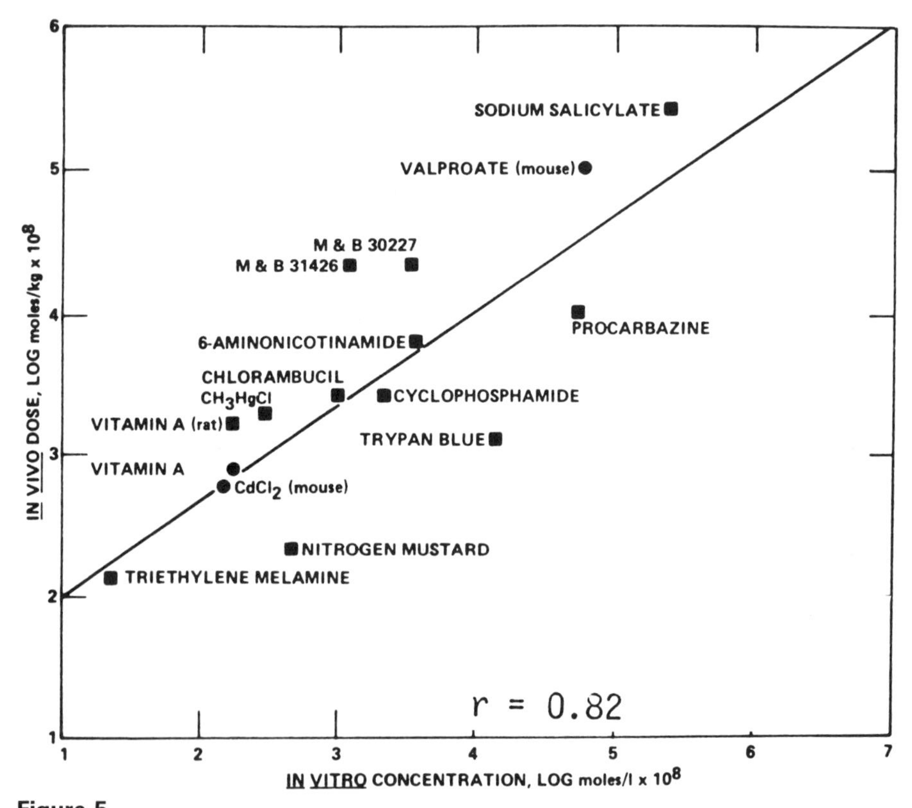

Figure 5

Correlation of effect levels of teratogen concentrations in vivo and in vitro. (Reprinted, with permission, from Kitchin et al. 1986.)

Other advantages are that small quantities of rare or expensive materials, such as radioactively labeled compounds, can be used. In addition, concentration responses can be calculated from fewer litters because the individual embryo, and not the litter, is the test unit. For our purposes, the most useful aspect has been the ability to add fairly well-defined metabolizing systems from different maternal species to the culture. Most of our studies have employed enzyme systems prepared from rat liver, but the ability to use those of human origin offers the possibility of extrapolation to pregnant women. Finally, and perhaps most important, direct observation of embryos during teratogenesis may lead to new and unexpected hypotheses as to mechanisms of teratogenic activity.

Disadvantages are that the cultures are technically complicated and moderately expensive; however, the expense is approximately the same as for in vivo testing (Fantel 1982). The period of organogenesis that can be studied is limited, excluding analysis of later critical morphogenetic endpoints such as those involving central nervous system, skeletal, or urogenital systems and palatal closure. The advantages of isolation from the maternal and uterine environment and the absence of the chorioallantoic placenta are also disadvantages because a number of potential teratogenic mechanisms may be omitted. The relationship between embryonic lesions seen following exposure in vitro and the subsequent anomalies of the fetus and newborn are often unclear. Elucidation of these relationships requires similar dosing of pregnant animals at comparable stages and examination of the embryo and fetuses throughout gestation. Many other factors may be involved in vivo, including repair processes, as well as resorption and embryonic loss.

Use of Embryo Culture for Mass Prescreening

It is probable that the whole-embryo culture system will not be adopted as a prescreen but, instead, will be used to study specific molecular and embryopathogenic mechanisms. The determination of minimally effective concentrations of parent chemicals and their metabolites and the role of interspecies variation in metabolizing enzymes appear to be the most useful at present.

SUMMARY

The development and methods for culture of postimplantation rodent embryos are reviewed. The use of this method has played an important part in elucidating the role of glucose, protein, and amino acids in the growth of embryos. The early-somite embryo is largely glycolytic, whereas the later-somite-stage embryos have converted to an aerobic system, using the terminal electron transport system and requiring less glucose. The role of the yolk sac placenta in the concentration and lysosomal digestion of protein macromolecules into amino acids is reviewed. Variables associated with hyperthermia, a possible human teratogenic agent, have been presented as an illustration of how direct studies might lead to more specific risk assessments. Illustrations are cited as to how bioactivation of teratogenic chemicals leads to a better understanding of molecular teratogenesis. The advantages and disadvantages of in vitro embryo culture are given.

ACKNOWLEDGMENTS

This work was supported by grants from the National Institutes of Health (HD00836, HD16287, and HD12717).

REFERENCES

Aksu, O., B. Mackler, T.H. Shepard, and R.J. Lemire. 1968. Studies of mechanisms underlying the development of congenital anomalies of embryos of riboflavin-deficient, galactoflavin-fed rats. II. Role of the terminal electron transport systems. *Teratology* **1:** 93.

Beaudoin, A.R. and D.L. Fisher. 1981. An in vivo/in vitro evaluation of teratogenic action. *Teratology* **23:** 57.

Brown, N.A. and S. Fabro. 1981. Quantitation of rat embryonic development in vitro: A morphological scoring system. *Teratology* **24:** 67.

Chatot, C.L., N.W. Klein, J. Piatek, and L.J. Pierro. 1980. Culture of rat embryos on human serum: Use in the detection of teratogens. *Science* **207:** 1471.

Cockroft, D.L. 1973. Development in culture of rat foetuses explanted at 12.5 and 13.5 days of gestation. *J. Embryol. Exp. Morphol.* **29:** 473.

———. 1976. Comparison of in vitro and in vivo development of rat foetuses. *Dev. Biol.* **48:** 163.

Cockroft, D.L. and D.A.T. New. 1975. Effects of hyperthermia on rat embryos in culture. *Nature* **258:** 604.

Ellington, S.K.L. 1980. In-vivo and in-vitro studies on the effects of maternal fasting during embryonic organogenesis in the rat. *J. Reprod. Fertil.* **60:** 383.

Fantel, A.G. 1982. Culture of whole rodent embryos in teratogen screening. *Teratog. Carcinog. Mutagen.* **2:** 231.

Fantel, A.G., J.C. Greenaway, M.R. Juchau, and T.H. Shepard. 1979. Teratogenic bioactivation of cyclophosphamide in vitro. *Life Sci.* **25:** 67.

Fantel, A.G., J.C. Greenaway, T.H. Shepard, M.R. Juchau, and S.B. Selleck. 1981. The teratogenicity of cytochalasin D and its induction by drug metabolsm. *Teratology* **23:** 223.

Faustman-Watts, E., J.C. Greenaway, M.J. Namkung, A.G. Fantel, and M.R. Juchau. 1983. Teratogenicity in vitro of 2-acetylaminofluorene: Role of biotransformation in the rat. *Teratology* **26:** 19.

Filler, R., and K.J. Lew. 1981. Developmental onset of mixed-function oxidase activity in preimplantation mouse embryos. *Proc. Natl. Acad. Sci.* **78:** 6991.

Freeman, S.J., F. Beck, and J.B. Lloyd. 1981. The role of the visceral yolk sac in mediating protein utilization by rat embryos cultured in vitro. *J. Embryol. Exp. Morphol.* **66:** 223.

Freinkel, N., N.J. Lewis, S. Akazawa, S.I. Roth, and L. Gorman. 1984. The honeybee syndrome—Implications of the teratogenicity of mannose in rat-embryo culture. *N. Engl. J. Med.* **310:** 223.

Germain M.-A., W.S. Webster, and M.J. Edwards. 1985. Hyperthermia as a teratogen: Parameters determining hypothermia-induced head defects in the rat *Teratology* **31:** 265.

Greenaway, J.C., A.G. Fantel, T.H. Shepard, and M.R. Juchau. 1982. The in vitro teratogenicity of cyclophosphamide in rat embryos. *Teratology* **25:** 335.

Greenaway, J.C., P.E. Mirkes, E.A. Walker, M.R. Juchau, T.H. Shepard, and A.G. Fantel. 1985. The effect of oxygen concentration on the teratogenicity of salicylate, niridazole, cyclophosphamide, and phosphoramide mustard in rat embryos in vitro. *Teratology* **32:** 287.

Hsu, Y.-C., J. Baskar, L.C. Stevens, and J.E. Rash. 1974. Development in vitro of mouse embryos from the two-cell egg stage to the early somite stage. *J. Embryol. Exp. Morphol.* **31:** 235.

Juchau, M.R., C.M. Giachelli, A.G. Fantel, J.C. Greenaway, T.H. Shepard, and E.M. Faustman-Watts. 1985. Effects of 3-methylcholanthrene and phenobarbital on the capacity of an embryo to bioactive teratogens during organogenesis. *Toxicol. Appl. Pharmacol.* **80:** 137.

Kitchin, K.T., B.P. Schmid, and M.K. Sanyal. 1986. Rodent whole-embryo culture as a teratogen screening method. *Methods Find. Exp. Clin. Pharmacol.* **8:** 291.

Kochhar, D.M. 1975. The use of in vitro methods in teratology. *Teratology* **11:** 273.

Lear, D., A. Clarke, A.P. Gulamhusein, M. Huxam, and F. Beck. 1983. Morphological, total nucleic acid, and total protein analyses of rat embryos cultured in supplemented and unsupplemented human serum. *J. Anat.* **137:** 279.

Lloyd, J.B., S.J. Freeman, and F. Beck. 1985. Embryonic protein nutrition and teratogenesis. *Biochem. Soc. Trans.* **13:** 82.

Mackler, B., R. Grace, B. Haynes, G.J. Bargman, and T.H. Shepard. 1973. Studies of mitochondrial energy systems during embryogenesis in the rat. *Arch. Biochem. Biophys.* **158:** 885.

Mirkes, P.E. 1985. Effects of acute exposures to elevated temperatures on rat embryo growth and development in vitro. *Teratology* **32:** 259.

Morris, G.M. and D.A.T. New. 1979. Effect of oxygen concentration on morphogenesis of cranial neural folds and neural crest in cultured rat embryos. *J. Embryol. Exp. Morphol.* **54:** 17.

Neubert, D., H. Peters, S. Teske, E. Kohler, and H.J. Barrach. 1971. Studies on the problem of "aerobic glycolysis" occurring in mammalian embryos. *Naunyn-Schmiedebergs Arch. Pharmacol.* **268:** 235.

New, D.A.T. 1967. Development of explanted rat embros in circulating medium. *J. Embryol. Exp. Morphol.* **17:** 513.

———. 1975. Studies on mammalian embryos in vitro during the period of organogenesis. In *Methods for detection of environmental agents that produce congenital defects* (ed. T.H. Shepard et al.), p. 159. Elsevier/New-Holland, New York.

———. 1978. Whole-embryo culture and the study of mammalian embryos during organogenesis. *Biol. Rev. Camb. Philos. Soc.* **53:** 81.

New, D.A.T. and P.T. Coppola. 1970a. Development of explanted rat fetuses in hyperbaric oxygen. *Teratology* **153:** 162.

———.1970b. Effects of different oxygen concentrations on the development of rat embryos in culture. *J. Reprod. Fertil.* **21:** 109.

New, D.A.T., P.T. Coppola, and D.L. Cockroft. 1976. Improved development of head-fold rat embryos in culture resulting from low oxygen and modifications of the culture serum. *J. Reprod. Fertil.* **48:** 219.

New, D.A.T., P.T. Coppola, and S. Terry. 1973. Culture of explanted rat embryos in rotating tubes. *J. Reprod. Fertil.* **35:** 135.

Nicholas, J.S. and D. Rudnick. 1934. The development of rat embryos in tissue culture. *Proc. Natl. Acad. Sci.* **20:** 656.

———. 1938. Development of rat embryos of egg cylinder to head-fold stages in plasma cultures. *J. Exp. Zool.* **78:** 205.

Priscott, P.K. 1983. Rat post-implantation embryo culture using heterologous serum. *Aust. J. Exp. Biol. Med. Soc.* **61:** 47.

Priscott, P.K., P.G. Gough, and R.D. Barnes. 1983. Serum protein depletion by cultured post-implantation rat embryos. *Experimentia* **39:** 1042.

Robkin, M.A. and D.L. Cockroft. 1978. The effect of carbon monoxide on glucose metabolism and growth of rat embryos. *Teratology* **18:** 337.

Robkin, M.A., D.W. Beachler, and T.H. Shepard. 1976. The effect of carbon monoxide on early rat embryo heart rates. *Environ. Res.* **12:** 32.

Robkin, M.A., T.H. Shepard, and T. Tanimura. 1972. New in vitro culture technique for rat embryos. *Teratology* **5:** 367.

Sadler, T.W., W.E. Horton, and C.W. Warner. 1982. Whole embryo culture: A screening technique for teratogens? *Teratog. Carcinog. Mutagen.* **2:** 243.

Sanyal, M.K. 1980. Development of the rat conceptus in vitro and associated changes in components of culture medium. *J. Embryol. Exp. Morphol.* **58:** 1.

Satish, J., R.M. Pratt, and M.K. Sanyal. 1985. Differential dysmorphogenesis induced by microinjection of an alkylating agent into rat conceptuses cultured in vitro. *Teratology* **31:** 61.

Shepard, T.H., T. Tanimura, and M. Robkin. 1969. In vitro study of rat embryos. I. Effects of decreased oxygen on embryonic heart rate. *Teratology* **2:** 107.

———. 1970. Energy metabolism in early mammalian embryos. *Dev. Biol.* (suppl.) **4:** 42.

Shepard, T.H., A.G. Fantel, P.E. Mirkes, J.C. Greenaway, E. Faustman-Watts, M. Campbell, and M.R. Juchau. 1983. Teratology testing: I. Development and status of short-term prescreens. II. Biotransformation of teratogens as studied in whole embryo culture. In *Developmental pharmacology* (ed. S. MacLeod et al.), p. 147. Alan R. Liss, New York.

Steele, C.E. and D.A.T. New. 1974. Serum variants causing the formation of double hearts and other abnormalities in explanted rat embryos. *J. Embryol. Exp. Morphol.* **31:** 707.

Tanimura, T. and T.H. Shepard. 1970. Glucose metabolism by rat embryos in vitro. *Proc. Soc. Exp. Biol. Med.* **135:** 51.
Tarlatzis, B.C., M.K. Sanyal, W.J. Biggers, and F. Naftolin. 1984. Continuous culture of postimplantation rat conceptus. *Biol. Reprod.* **31:** 45.
Warkany, J. 1986. Teratogen update: Hyperthermia. *Teratology* **33:** 365.

COMMENTS

Morrissey: Was the last graph you showed a comparison between the in vivo dose in a pregnant animal and the in vitro dose?

Shepard: Yes. But, as I recall, they assumed complete distribution in the whole animal, which is risky.

Morrissey: I think that the LD_{50} correlations are difficult if not done in pregnant animals. We have found that the in vitro IC_{50} corelates quite well with the in vivo embryo toxic dose for the chemical where that is known. In the studies that Dr. Johnson referred to, there were two-thirds teratogens and one-third nonteratogens. I would have to take issue with what Dr. Johnson said about calculating the percentages of reactive verus nonreactive teratogens. What we do for the National Toxicology Program in vitro teratology studies is to calculate the percentage of positives and negatives that were correctly identified on the basis of current animal and human data.

Welsch: Some 100 compounds have been tested in whole-embryo culture, and there seems to be a predominance of certain lesions. Among them are an abnormally closed neural tube, and quite commonly, a particular prosencephalic lesion, which is not seen in vivo. There seems to be a very narrow spectrum of abnormal differentiation that one can induce. How does one interpret these results obtained in whole-embryo culture?

Brent: It's probably a lethal effect. If you look at abnormal chromosome configurations in the early embryo and analyze the morphogenesis of those embryos, you see that 99% of them die with some type of serious central nervous system effect. Studies of unbalanced translocation and radiation-induced translocations indicate that if the insult occurs early in organogenesis, a severe central nervous system effect would result. That is one of the difficulties with embryo culture. Because you don't know what is going to happen at term, what you are calling a genetic defect may actually be a lethal effect. Dr. Shepard discussed the fact that this technique is very valuable in the study of pharmacokinetics and mechanisms. Do you want to comment on the fact that some people are using it for screening and are making interpretations with regard to teratogenic risk in humans. One organ that you can't study is the differentiating brain, and that is crucial from the standpoint of environmental risk, because brain development is a mid-gestational phenomenon.

Shepard: You wanted me to comment on the use of the whole-embryo culture as a screen. I think you can get some useful information from it, but, it is too expensive to screen 10,000 agents.

Pratt: In Europe, some drug companies (such as Sandoz in Basel) are using the whole-embryo culture as a primary screen for many of the chemicals and drugs that they are developing.

Oakley: Dr. Asher Ornoy presented some very interesting embryo culture data at the last meeting of the European Teratology Society, which could be relevant to human teratogenesis because he showed that the common ketoacids that increase in human diabetes are each teratogenic in rat embryo culture. More importantly, he showed that when the agents are combined, they are teratogenic at lower concentrations—concentrations similar to those seen in the human situation.

Brent: There is a difference between testing a compound without knowing what effect it has in the human and studying a disease like diabetes when you want to dissect out the metabolism and the pharmacokinetics. Dr. Shepard said that the embryo culture is very valuable when looking at the metabolic agents. But given an unknown situation, for example, where someone has lost a baby and you are trying to evaluate what caused it, I think the embryo culture is fraught with difficulty.

Hanson: Dr. Shepard, has anybody made any progress toward developing a primate system like this?

Shepard: The primate doesn't have an inverted yolk sac.

Hanson: I mean some kind of full embryo culture method.

Shepard: It would be more expensive.

Hanson: I understand that. But for studying mechanisms of teratogenesis in primates, it might be somewhat better than using a nonprimate.

Shepard: I don't know of anyone who has done that.

Iannaccone: Didn't Dr. Freinkel's data show that glucose loading was teratogenic?

Shepard: Yes.

Iannaccone: We know that mannose in this system is highly teratogenic, causing rotational malformations and other problems. You implied that it was because the embryo couldn't utilize glucose? Is mannose directly teratogenic?

Shepard: As I recall, mannose interfered with glucose metabolism.

Scott: Dr. Shepard, I wanted to expand on the point with which you started the discussion and that is the tremendous importance of aerobic glycolysis to the early embryo. There is a situation that is clearly the same in tumor cells. Rapidly proliferating cells, for some reason, choose aerobic glycolysis as the means to develop ATP. I think that if we understood why they were so dependent on that phenomenon, it would be of theoretical and practical significance.

Shepard: One thing I didn't mention is that the oxygen requirements for embryos in culture are just as specific. If, at the beginning of the culture period, you give the embryo more than about 10% oxygen, you get neural tube defects. Later on, you have to give at least 80% oxygen or they are malformed.

Stahlmann: I would like to make a short comment on a drug that we studied extensively with in vitro and in vivo systems. It is the virustatic agent, acyclovir. In routinely performed segment-II studies with rats (2 × 25 mg/kg [subcutaneously] day 6–15 of gestation), no prenatal toxicology was noticed by the company. Without further information, it is not reasonable to study higher doses, because it then crystallizes in renal tubules and causes maternal nephrotoxic effects due to the poor solubility of the drug. The first hint that it might be teratogenic, like some other virustatics, came from in vitro studies with a rat whole-embryo culture system that my colleagues Drs. Stephan Klug and Constanze Lewandowski performed. As a follow-up, we tested the substance in vivo at the period of the presumed susceptibility and at higher doses than had been tested before. Two or three injections of 100 mg/kg b.w. (subcutaneously) on day 10 of gestation induced typical malformations in rats. The embryos were evaluated on day 11.5 in an identical way as the in vitro cultured embryos (scoring system, protein content, histological evaluation, etc.) and showed the same signs of abnormal development. The in vitro and in vivo exposed rat embryos could not be distinguished macroscopically or histologically. With the dose regime mentioned, malformations of the tail and head can be seen if the fetuses are evaluated conventionally on day 21 of gestation. Some of these abnormalities persist postnatally (brain, eye, tail). The results of the in vitro studies are essential for recognizing the defects as specifically drug induced and to prove that the impairment is not a result of maternal toxicity. This is the first example of the primary recognition of the teratogenic effect with the whole-embryo culture system, which subsequently could be demonstrated under in vivo conditions. The combined in vivo/in vitro approach has been found to be extremely valuable in assessing teratogenic potentials.

Renal Development In Vitro

LAURI SAXÉN AND EERO LEHTONEN
Department of Pathology
University of Helsinki
SF-00290 Helsinki, Finland

OVERVIEW

Early organogenesis of the metanephric kidney can be explored in detail in both organotypic and cell cultures. An inductive action exerted on the metanephric mesenchyme by the branches of the ureter bud leads to a cascade of differentiative events: stimulated proliferation, aggregation of cells, profound changes in the composition of the extracellular matrix, and reorganization of the intermediate filaments. Such changes in whole kidney rudiments and in transfilter-induced mesenchymes in three-dimensional cultures are only partially reproduced in monolayer cultures of dispersed induced cells.

A speculative synthesis is presented of the sequence and role of these changes in early kidney tubulogenesis.

INTRODUCTION

The developing metanephric kidney has proved to be a good model system for analysis of cytodifferentiation and morphogenesis. The formation of its functional units, the nephrons, includes a cascade of events involved in embryogenesis: morphogenetic cell and tissue interactions, guided migration of cells and cell clusters, cell proliferation, changes in the metabolism and chemical composition of the tissue and, ultimately, expression of new cellular phenotypes and assembly of cells into spatially organized structures.

In vitro techniques have been devised to dissect these various events for an experimental analysis. Results of investigations employing such techniques should be used as a deductive approach to understand the complex in vivo development of the nephron and the regulation of this spatially and temporally strictly controlled process (for comprehensive review, see Saxén 1987). Several efforts have been made to apply the kidney model system to teratological and toxicological studies (summarized by Saxén 1983). Table 1 is a recent example of the use of embryonic kidney tissue to test the possibly harmful effect of an exogenous agent.

Banbury Report 26: Developmental Toxicology: Mechanisms and Risk

Table 1
Effect of Interferon on the Incorporation of Thymidine into Transfilter-induced Metanephric Mesenchymes

Medium[a]	Interferon concentration (units/ml)	Thymidine incorporation (dpm/μg DNA)
Serum	None	16.800
	10^3	13.100
	10^5	5.600
Transferrin	None	17.000
	10^3	18.400
	10^5	15.600

[a]Cells were cultivated for 24 hr in medium containing either 10% fetal calf serum or 50 μg/ml of transferrin (modified from Saxén 1985).

RESULTS

Organ Culture

Organ culture of the early rudiments of murine metanephric kidneys allows the exploration of the very first detectable steps in organogenesis. Time-lapse cinematography is very applicable to such an analysis at the level of the whole anlage, and it visualizes the two main events that initiate morphogenesis: invasion of the ureter bud into the mesenchymal blastema, where it branches dichotomously, and the aggregation of the mesenchymal cells around the ureter (Fig. 1). The primary mesenchymal aggregate subsequently splits in

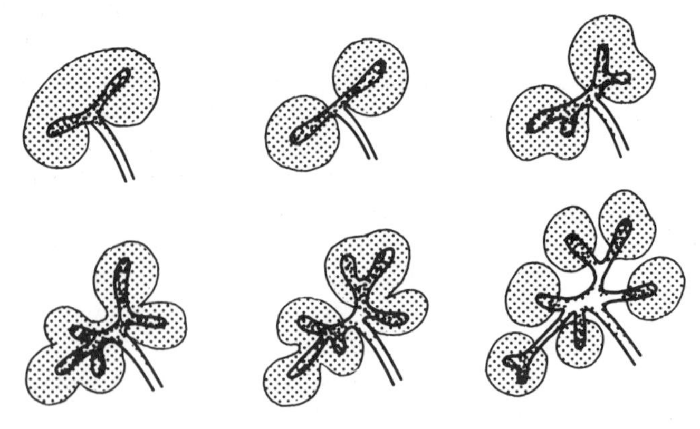

Figure 1
Camera–lucida analysis of a time-lapse motion picture illustrating the early stages of development of a mouse kidney rudiment (modified from Saxén and Wartiovaara 1966).

two along the equatorial plane, determined by the tips of the ureter that gradually move apart. The phenomenon has been attributed to an increasing intercellular adhesiveness of the cells within the aggregate (Saxén and Wartiovaara 1966).

Branching of the ureter and aggregation of the mesenchymal cells are regulated by a reciprocal interaction between the two cell populations (Grobstein 1955). When surgically dissected and cultivated in isolate, both cell lineages fail to differentiate further. The transfilter technique devised by Grobstein (1956) allows further dissociation of the various postinductory events and their temporal correlation. The technique consists of an interposition of porous membrane filters between the interacting tissues, the mesenchyme and its inductor, which is usually a fragment of embryonic central nervous system.

The transfilter technique has been used to study the transmission characteristics of the inductive signals and the kinetics of the process. Two main conclusions have evolved from such experiments.

1. Induction of the metanephric mesenchyme leading to epithelial transformation of the cells and formation of tubules operates via close cell-to-cell contacts rather than being mediated by diffusible signal substances (Lehtonen 1976; Saxén et al. 1976).
2. Induction is a time-consuming event, and a maximal, irreversible determination of the target cells is obtained after 24 hours of transfilter contact (Fig. 2). This inductive "pulse" sets the program for subsequent epithelial transformation of the cells and the expression of several new phenotypes within the early nephron (Ekblom et al. 1981b; Lehtonen et al. 1983).

The mechanism by which the developmental program is subsequently implemented is only partially known. The following direct consequences have been recorded in the mesenchyme.

1. Stimulation of the DNA synthesis beginning at 12 hours and reaching its peak after 24 hours of transfilter contact (Fig. 2).
2. Degradation of certain interstitial proteins (collagen type I and type III, fibronectin) of the extracellular matrix at the aggregating sites (Ekblom et al. 1981a).
3. Enhanced synthesis and redistribution of certain epithelial proteins (collagen type IV, laminin, heparan-sulphate proteoglycan) (Ekblom et al. 1980; Lash et al. 1983). These components subsequently contribute to the basement membrane laid down around the primitive renal vesicle (Wartiovaara 1966; Bonadio et al. 1984).
4. Changes in the cytoskeleton monitored as loss of the vimentin-type intermediate filaments and expression of cytokeratin(s) (Lehtonen et al. 1985).

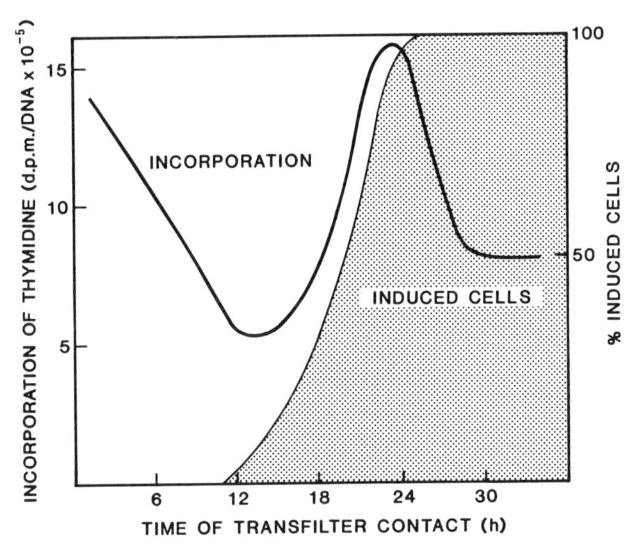

Figure 2

Graph of the "minimum induction time" for the metanephric mesenchyme-exposed transfilter to an inductor for different periods. Number of tubules is given as percentage when compared with similar cultures in which the inductor had not been removed. The curve of the incorporation of tritiated thymidine into induced mesenchymes is superimposed (modified from Saxén and Lehtonen 1978; Saxén et al. 1983).

Interconnections and causal relationships of these events singled out from the process of early epithelial transformation and formation of the pretubular aggregates and nephric vesicles will be discussed later in this review.

Cell Cultures

Cell cultures prepared from the metanephric mesenchyme may allow a dissection of cytodifferentiation and morphogenesis, that is, analysis, separately and in comparison, of the expression of new phenotypes in single cells as opposed to the three-dimensional assemblies. Observations in the transfilter experiments (above) have laid the basis for such comparisons: After 24 hours of transfilter contact with an inductor, the mesenchymal cells are programmed toward an epithelial destiny, yet they show no morphological changes as compared to similarly precultivated, uninduced cells. Consequently, the differentiation of the programmed mesenchymal cells can be followed in parallel in dispersed monolayer cultures and in three-dimensional, organotypic conditions creating aggregates and tubules. Such studies were started only recently, but some preliminary observations can be summarized (Lehtonen et al. 1985; Lehtonen and Saxén 1986 and unpubl.):

1. Cells of the mesenchymal blastema express vimentin-type intermediate filaments but apparently no cytokeratin prior to induction. Upon induction followed by cultivation as monolayers, vimentin is lost from many cells that now express cytokeratin. But small numbers of cells showing cytokeratin-type filaments are obtained in dispersed subcultures of uninduced cells attached to proper substrata.
2. Fibronectin is synthesized and deposited in monolayer cultures of both uninduced and induced mesenchymal cells, whereas in organotypic cultures fibronectin is lost from the induced cells. Moreover, fibronectin and cytokeratin show frequent coexpression.
3. Laminin, already detectable in the starting material, continues to be synthesized and deposited by both uninduced and induced cells. Unlike the three-dimensional aggregates, deposition of laminin shows no signs of polarization.

DISCUSSION

The fragments of information available on the early postinductory changes in the kidney mesenchyme could be arranged to a deductive working hypothesis. The contact-mediated, time-comsuming inductive signal leads to the following primary changes in the target mesenchyme: increased rate of proliferation, increased intercellular adhesiveness, degradation of interstitial proteins, enhanced synthesis and redistribution of epithelial proteins, and changes in the cytoskeleton. These differentiative events in the whole-kidney rudiments are mimicked in three-dimensional culture conditions allowing aggregation and homotypic interactions of induced cells, but they can be reproduced only partially in dispersed cell cultures. Hence, one may conclude with the following speculative sequence of postinductory events: An increased proliferation rate and reduction of the extracellular matrix create an increased cell density (aggregation), allowing close cell-to-cell interactions and recognition of like (induced) cells. Increased synthesis or neosynthesis of a still hypothetic adhesive molecule stabilizes the aggregates. Cells tend to maximize their mutual contact area, which leads to polarization of cells within the aggregate and a peripheral accumulation of the compounds constituting the basement membrane. The basement membrane subsequently acts as a scaffold, maintaining the epithelial arrangement of cells that attach to it. In dispersed cultures, homotypic cell interactions are prevented and cells follow their differentiation program only partially. Because the "epithelial" proteins laminin and cytokeratin can be expressed by uninduced cells as well as by induced ones, the key molecules for tubule formation should be searched for elsewhere, and an "adhesive" molecule seems to be a good candidate (see Edelman 1985). Thus far, conclusive evidence of the role of such compounds in early kidney tubule formation is lacking, although some have been shown to be expressed in the kidney tissue during early tubule formation (Thiery et al. 1984; Vestweber et al. 1985).

REFERENCES

Bonadio, J.F., H. Sage, F. Cheng, J. Bernstein, and G.E. Striker. 1984. Localization of collagen types IV and V, laminin, and heparan sulfate proteoglycan to the basal lamina of kidney epithelial cells in transfilter metanephric culture. *Am. J. Pathol.* **116:** 287.

Edelman, G.M. 1985. Expression of cell adhesion molecules during embryogenesis and regeneration. *Exp. Cell. Res.* **161:** 1.

Ekblom, P., A. Miettinen, and L. Saxén. 1981a. Induction of brush border antigens of the proximal tubule in the developing kidney. *Dev. Biol.* **74:** 263.

Ekblom, P., E. Lehtonen, L. Saxén, and R. Timpl. 1981b. Shift in collagen type as an early response to induction of the metanephric mesenchyme. *J. Cell. Biol.* **89:** 276.

Ekblom, P., K. Alitalo, A. Vaheri, R. Timpl, and L. Saxén. 1980. Induction of a basement membrane glycoprotein in embryonic kidney: Possible role of laminin in morphogenesis. *Proc. Natl. Acad. Sci.* **77:** 485.

Grobstein, C. 1955. Inductive interactions in the development of the mouse metanephros. *J. Exp. Zool.* **130:** 319.

———. 1956. Trans-filter induction of tubules in mouse metanephric mesenchyme. *Exp. Cell Res.* **10:** 424.

Lash, J., L. Saxén, and P. Ekblom. 1983. Biosynthesis of proteoglycans in organ cultures of developing kidney mesenchyme. *Exp. Cell Res.* **147:** 85.

Lehtonen, E. 1976. Transmission of signals in embryonic kidney. *Med. Biol.* **54:** 108.

Lehtonen, E. and L. Saxén. 1986. Cytodifferentiation vs. organogenesis in kidney development. In *Progress in developmental biology*, part A (ed. H.C. Slavkin), p. 411. A.R. Liss, New York.

Lehtonen, E., I. Virtanen, and L. Saxén. 1985. Reorganization of intermediate filament cytoskeleton in induced metanephric mesenchyme cells is independent of tubule morphogenesis. *Dev. Biol.* **108:** 481.

Lehtonen, E., H. Jalanko, L. Laitinen, A. Miettinen, P. Ekblom, and L. Saxén. 1983. Differentiation of metanephric tubules following a short transfilter induction pulse. *Wilhelm Roux's Arch. Dev. Biol.* **192:** 145.

Saxén, L. 1983. In vitro model-systems for chemical teraogenesis. In *In vitro testing of environmental agents: Current and future possibilities*, Part B: *Development of risk assessment guidelines* (ed. A.R. Kolber et al.), p. 173. Plenum Press, New York.

———. 1985. Effect of interferon tested in a model-system for organogenesis. *J. Interferon Res.* **5:** 355.

———. 1987 *Organogenesis of the kidney.* Cambridge University Press, Cambridge, England.

Saxén, L. and E. Lehtonen. 1978. Transfilter induction of kidney tubules as a function of the extent and duration of intercellular contacts. *J. Embryol. Exp. Morphol.* **47:** 97.

Saxén, L. and J. Wartiovaara. 1966. Cell contact and cell adhesion during tissue organization. *Int. J. Cancer* **1:** 271.

Saxén, L., J. Salonen, P. Ekblom, and S. Nordling. 1983. DNA synthesis and cell generation cycle during determination and differentiation of the metanephric mesenchyme. *Dev. Biol.* **98:** 130.

Saxén, L., E. Lehtonen, M. Karkinen-Jääskeläinen, S. Nordling, and J. Wartiovaara. 1976. Are morphogenic tissue interactions mediated by transmissible signal substances or through cell contacts? *Nature* **259:** 662.

Thiery, J.P., A. Delouvée, W. Gallin, B.A. Cunningham, and G.M. Edelman. 1984. Ontogenetic expression of cell adhesion molecules: L-CAM is found in epithelia derived from the three primary germ layers. *Dev. Biol.* **102:** 61.

Wartiovaara, J. 1966. Studies on kidney tubulogenesis. V. Electron microscopy of basement membrane formation in vitro. *Ann. Med. Exp. Biol. Fenn.* **44:** 140.

Vestweber, K., R. Kemler, and P. Ekblom. 1985. Cell adhesion molecule uvomorulin during kidney development. *Dev. Biol.* **112:** 213.

COMMENTS

Shepard: That was very interesting. I was wondering whether you have used the Danforth short (Sd) mice in your system. It is a genetic syndrome, where the heterozygotes have unilateral absent kidney and the homozygotes have both kidneys absent.

Saxén: I think that is a very good suggestion, but we have not examined these mice.

Pratt: Have you ever examined the effect of any well-known kidney teratogens on your recombined epithelium in culture?

Saxén: I wouldn't say kidney-specific teratogens. We used a great variety of antimetabolites, protein synthesis inhibitors, DON, and many others. Most of them affect the process during the first 24 hours, during the induction.

Brent: You used thalidomide, didn't you?

Saxén: Yes, and we used a human embryonic kidney rudiment in culture.

Pratt: TCDD, or dioxin, is a very potent inducer of certain malformations, cleft palate being one, but this is observed mainly in the mouse. However, the kidney defect can be induced by TCDD in the mouse and rat, and there is some evidence that hydronephrosis appears in the human. According to several labs, including mine, some of the extracellular matrix components may be disturbed by dioxin. This seems to be an excellent teratogen to use with your sophisticated techniques. Have you ever explored this possibility?

Saxén: No. So far we have been rather busy with the basic mechanisms.

McLachlan: The question that you started to address was whether or not the program of differentiation, transfilter as well as in organ culture, is inherent in the genetic program of the different cell types. If not, is there some external signal, for example, a hormone, that initiates the first step of differentiation, either in the mesenchyme or in the epithelium, and is the rest of it then a cascade?

Saxén: I think we are now dealing with one of the basic questions in the whole area of developmental biology. The only thing I can say here is that in eight-cell-stage embryos, the cells are still totipotent, and you can remove one of them without affecting development; you can even get the whole embryo out of one of the cells. But by the 16-cell-stage embryo, cells have been committed toward a certain direction. It is not what you would call an inherited genetic system, because it still requires triggers from the outside, as in this kidney system. I don't think that the trigger is informative in my system. For instance, the epithelial bias is already there; you just have to trigger it somehow.

McLachlan: So you are saying that the trigger would be more stimulatory rather than directive?

Saxén: Yes.

Kochhar: Ever sine Grobstein's time, we have known that extracellular matrix and collagen, in particular, is very important in epithelial–mesenchymal interactions of the kind you study in the kidney. In recent times, a mouse mutant has been generated that lacks collagen type I. I wonder if kidney development has been investigated in that system.

Saxén: Yes, Dr. Klaus Kratochwil has done it, and collagen type I is genetically missing from those mice. The problem is that you cannot study embryogenesis in vivo because the embryos die at mid-gestation, but you can take the organs out and grow them in vitro. The kidney does beautifully without collagen type I. Nobody knows the explanation.

Kochhar: I think this is a good way to explore epithelial–mesenchymal interactions. For a long time, we have depended on vague explanations for the mechanistic role of extracellular matrix. Perhaps such mutations can eliminate these factors separately so that we can pinpoint the specific role of each component.

Saxén: Yes, but there is a paradox here. We and others have shown that there are profound changes in collagen type I in the matrix at the beginning of kidney development, and yet the total lack of type-I collagen does not affect that.

Kochhar: There may be some other type of collagen that replaces type I.

Pratt: Is type-III collagen known to be present in that system?

Saxén: Yes, type III is present in those mutants.

Welsch: If one were to use your system to do experiments, as Dr. Pratt proposed, and examine agents that have no known teratogenic effects in the embryo, would you think that thymidine incorporation is an appropriate endpoint? What endpoints would you suggest?

Saxén: I think labeled thymidine incorporation is one because the induction is followed by a rapid increase in the incorporation, which is an early response. The second would be to then follow the maturation of the three segments of the nephron.

Welsch: Would you propose morphologic criteria to examine this event?

Saxén: Yes, and immunohistologic criteria as well, where you could follow its development. There is also a rough method of quantitating the number of tubules that develop after the trigger.

Brent: But you couldn't use it for screening?

Saxén: No.

Welsch: If you know from in vivo studies, or you suspect (TCDD in mice might be a candidate chemical) that an agent does something to kidney development, could your system detect this and give clues as to what factors are affected?

Shepard: As a pediatrician, I think that we need to know why kidneys don't develop in certain children. If you study the parents of children who are stillborn without any kidneys, I think 10–15% of them have unilateral missing kidneys or some marker suggesting that it is a genetic condition. So rather than use agents in your system, I wonder if you could use serum from these patients to examine genetic differences in a population.

Saxén: It's not going to be easy. I might add that the whole kidney system started from some very early observations of Professor Gruenwald. He showed that there were mutant chick embryos where the kidneys were lacking from homozygotes because the bud never grew all the way into the mesenchymal blastema. So, there are animal models for this kind of genetic defect.

Brent: Hypoplastic or absent kidney is probably not one disease, and there are likely to be many etiologies for it. In some of our radiation studies, we did serial sections of the embryo. On the twelfth day, we had a very high incidence of unilateral kidneys, and they were due to destruction of the mesenchyme between the two blastemas, causing them to fuse. In other words, it wasn't the destruction of one kidney; due to the fact that the two blastemas lost the mesenchyme in between, they fused and went to one side or the other; usually the left was missing.

Saxén: How many types of polycystic diseases are there in the kidney?

Brent: There are a lot of polycystics in an adult, and the hypoplastic kidneys are very difficult to evaluate. So, there are multiple causes for these diseases, and a serum factor is a possibility.

Oakley: The rates of renal agenesis have been increased threefold over the last 15 years. We have just finished reviewing the charts from a large group of cases and do not have any clue as to why it's happening. We do not think that it is a change in diagnostic process. There could be environmental agents responsible.

Molecular and Experimental Embryology

Analysis of Cell Lineage During Early Mouse Embryogenesis

ROGER A. PEDERSEN
Laboratory of Radiobiology and Environmental Health
and Department of Anatomy, University of California
San Francisco, California 94143

OVERVIEW

The analysis of cell lineage in early mouse embryogenesis has been undertaken using microinjected lineage tracers. This approach reveals the fate of individual or pairs of sister progenitor cells in the intact preimplantation or early postimplantation embryo. Our results indicate that the inner cell mass of the mouse embryo is established during the fourth and fifth cleavage divisions and that all remaining outer cells contribute to the trophectoderm. Our results do not, however, support the idea that trophectoderm and inner cell mass are mutually exclusive clonal entities at the blastocyst stage. Rather, we find that inner cells contribute to the polar trophectoderm during blastocyst growth. An analysis of cell fate in the early postimplantation endoderm, using this approach, indicates that individual embryonic endoderm cells labeled at mid- or late-primitive-streak stages contribute descendants either to extraembryonic visceral endoderm or to endoderm lining the embryonic gut, but rarely to both tissue populations. We conclude that the endoderm layer of the mid- to late-streak-stage embryo is a mixed population, with progenitors arising from primitive endoderm contributing extraembryonic descendants and other progenitors contributing descendants to the gut endoderm of the embryo proper. This interpretation is consistent with the view, derived from analysis of injection chimeras, that the gut endoderm progenitors arise from the embryonic ectoderm. In summary, an analysis using lineage tracers to determine the fate of individual progenitor cells in early embryogenesis indicates two instances in which descendants eventually occupy different cell layers than their progenitors do. The mechanisms for these and other cellular movements in the early embryo are not yet understood but have important consequences for morphogenesis.

INTRODUCTION

Morphogenesis in vertebrate embryos involves cell proliferation, differentiation, migration, interaction, and death as formative forces for generating the phenotype of the offspring. Although these processes have been studied for

Banbury Report 26: Developmental Toxicology: Mechanisms and Risk

years in model embryonic systems, the inaccessibility of the mammalian embryo has restrained the experimental analysis of morphogenesis, particularly at the early postimplantation stages when the axis of the embryo forms, gastrulation occurs, and the organ rudiments begin to differentiate. An understanding of this period is crucial from the standpoint of teratology, because this is when the embryo is most sensitive to environmentally induced birth defects.

The objective of our studies is to describe the foundation of the tissue lineages that differentiate in the early mouse embryo. We have studied the preimplantation embryo as a conceptual model for cell allocation and for growth, bearing in mind that the early tissue lineages formed in mammalian embryos are destined to develop into extraembryonic structures. To study allocation processes during organ rudiment formation, we have addressed similar issues at postimplantation stages. The results of a cell lineage analysis consist of the tissue localization of descendants; they can be used to infer the fate and, in some cases, the potency of the marked progenitors. We have used intracellular microinjection of lineage tracers to mark single progenitor cells in intact embryos. This has the advantage of providing clonal information about fate and avoiding the disruption of the embryo caused by the chimera approach (Gardner 1978) or by deletion and extirpation procedures (Beddington 1986). The disadvantages of our approach are the short effective lifetime of the microinjected lineage tracers (a maximum of six cell cycles) and the requirement for carrying out the procedures in vitro, which limits the period of development that can be studied to that of successful culture (4 days preimplantation and 1–2 days postimplantation). Despite these limitations, microinjection of lineage tracers into mouse embryo cells has provided novel insights into the allocation of descendants to the inner cell mass and trophectoderm populations and into the origin and fate of endoderm in the gastrula-stage embryo. The results presented here are based on work published previously (Cruz and Pedersen 1985; Lawson et al. 1986; Pedersen et al. 1986) and, to a limited extent, on unpublished observations (G.K. Winkel and R.A. Pedersen; K.A. Lawson and R.A. Pedersen; both in prep.).

Progenitor cells were labeled with a mixture of horseradish peroxidase (Weisblat et al. 1978) and rhodamine-conjugated dextran (Gimlich and Braun 1985) by iontophoretic injection (Balakier and Pedersen 1982). The fate of the descendants was determined by staining for peroxidase activity after 1–2 days of culture and scoring whole mounts or serially sectioned, reconstructed embryos. The presence of the fluorescent tracer in microinjected cells provided an indication of the presence and location of peroxidase activity. The accuracy of injection could be verified, and horseradish-peroxidase-labeled cells scored in a control series stained without culture. This approach thus enabled us to conduct an authentic clonal analysis of early morphogenetic events in a mammalian embryo.

RESULTS

Preimplantation Stages

We analyzed the origin of the mouse inner cell mass by microinjecting lineage tracers into progenitor cells at the 8-, 16-, 32-, and 64-cell stages. In one approach (prospective), we cultured the embryos until the labeled cells had undergone a cell division. Injected embryos had either one or two labeled cells at the outset. The second labeled blastomere was shown to be a sister of the cell injected with tracers, labeled by passive diffusion through the cytoplasmic bridges that join recently divided cells (Lo and Gilula 1979; Pedersen et al. 1986). Because pairs of cells in unincubated embryos were sisters, their location within the embryo also could be used to infer the fate of a progenitor cell from the previous cleavage stage. In this approach (retrospective), embryos were injected with horseradish peroxidase and then stained and scored for the positions of labeled sister cells without further culture.

The results of injecting outer cells at the 8-, 16-, and 32-cell stages are summarized in Table 1. The probability of a division generating either an outer and an inner cell or two outer cells was calculated from the distribution of embryos with outer-inner or outer-outer pairs, respectively. These probabilities, together with the estimated number of outer cells at each stage, could be used to calculate the average number of cells allocated to the inner population at each cleavage division (Fig. 1; Table 1). The results showed that the inner cell mass of the mouse blastocyst is generated from approximately three to four cells that acquire their internal positions at the fourth cleavage division and an additional two to four cells that are internalized in the fifth cleavage division. Cells remaining outside at subsequent stages (32 and 64 cells) have descendants exclusively in the trophectoderm population.

Table 1
Estimated Number of Cells Allocated to the Inner Cell Mass

Cleavage division	Approach (no. of labeled cells)	Probability of outer/inner division	No. of inner cells allocated
Fourth	prospective (1)	0.500	4.0
	prospective (2)	0.366	2.9
	retrospective	0.494	3.9
Fifth	prospective (1)	0.200	2.4
	prospective (2)	0.325	3.9
	retrospective	0.201	2.4

The estimated number of cells allocated at each division is the product of the probability of an outer/inner division and the total number of outer cells. In the fourth cleavage division, all cells are assumed to be outer cells. In the fifth cleavage division, 12 cells are assumed to be outer cells. Data from Pedersen et al. (1986).

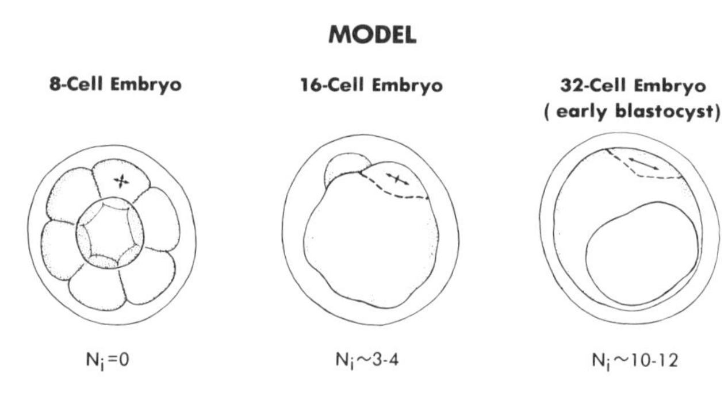

Figure 1

A model for the origin of the inner cell mass in mouse blastocysts. (↔) The probable fates at the succeeding cleavage division; the relative lengths of the arrows indicate the probabilities of outer/inner vs. outer/outer divisions (probabilities from Table 1). (N_i) Number of inner cells. All the progeny of allocated inner cells are assumed to remain inside at the fifth cleavage. (Reprinted, with permission, from Pedersen et al. 1986.)

The fate of trophectoderm cells in blastocysts was analyzed by microinjecting them with horseradish peroxidase alone or combined with fluorescent dextran (Fig. 2) and culturing them for 1 or 2 days to determine the location of descendants (Cruz and Pedersen 1985). Mural trophectoderm cells showed no net displacement from their original abembryonic position after 24-hr culture, compared with controls (Table 2). In contrast, the labeled central polar trophectoderm cell and its descendants became displaced from the embryonic pole in 24 or 48 hr of culture, as compared with controls (Fig. 2; Table 2). The distance traversed by the descendants of a labeled polar trophectoderm cell in 48 hr was approximately two cell diameters. In addition, the distance between the labeled descendants and the abembryonic pole increased, presumably because of cell proliferation or enlargement of mural trophectoderm cells during these stages (Table 2). The descendants of labeled

Table 2

Movement of Trophectoderm Cells Labeled with Horseradish Peroxidase and Rhodamine-conjugated Dextran

Site of injection	Incubation time (hr)	Displacement in controls	Displacement in incubated embryos	Distance traversed (μm)	d_2-d_1
Polar	48	6.9 ± 1.5	94.6 ± 3.1	87.7	25.5
Mural	24	15.7 ± 1.9	17.8 ± 2.0	2.1	–

Distances were determined as described in Fig. 2; average displacement from embryonic or abembryonic pole was determined for control and incubated embryos of each series. Data from Cruz and Pedersen (1985).

trophectoderm cells remained together as a coherent patch. Thus, a marked translocation of labeled cells occurred within the polar trophectoderm during blastocyst expansion, such that descendants of the embryonic pole came to occupy positions within the mural trophectoderm, and labeled cells in the central polar trophectoderm location were replaced by unlabeled cells. Because of the nonrandom movement at the embryonic pole and the lack of translocation at the abembryonic pole, we hypothesized that the inner cell mass contributed the unlabeled descendants to the polar trophectoderm during blastocyst growth (Cruz and Pedersen 1985).

Recent results showing that descendants of single labeled inner cell mass cells appear not only in the inner cell mass but also in the polar trophectoderm and primitive endoderm confirm the major prediction of this hypothesis (G.K. Winkel and R.A. Pedersen, in prep.). Among blastocysts injected in a single inner cell mass cell by penetrating through the mural trophectoderm, 23% had labeled descendants in polar trophectoderm after 24-hr culture, whereas none had labeled polar trophectoderm in controls; 35% had descendants in the endoderm, and 58% in the inner cell mass itself. The labeled descendants in the primitive endoderm population had the morphology of migratory cells, probably partial endoderm. We conclude from this distribution, along with the normal rates of cell proliferation and the apparently normal integration of the labeled descendants into differentiating tissues, that the inner cell mass serves as a stem cell population for the entire embryo during blastocyst growth (G.K. Winkel and R.A. Pedersen, in prep.). Therefore, not only are labeled descendants in the polar trophectoderm and primitive endoderm capable of translocating a substantial distance within a single tissue layer, but their intermediate progenitors are also capable of moving away from the inner cell mass, by mechanisms as yet undefined.

Postimplantation Stages

The fate of endoderm cells between mid- to late-primitive-streak and early-somite stages was studied by microinjecting horseradish peroxidase into single cells located in four zones along the midline of the anterior-posterior axis and culturing the embryos for 24 hr before staining (Lawson et al. 1986). Control (injected but not incubated) embryos were generally labeled in either one or two endoderm cells; a small fraction of the control embryos had label in ectoderm cells (6%). Cultured embryos had 0–13 cells; there was no evidence for a detrimental effect of the labeling procedure on either the incidence of cells in S phase or the incidence of mitotic cells. A comparison of the fraction of embryos with labeled cells in controls (71%) and in cultured embryos (42%) was interpreted as indicating extensive cell death in the

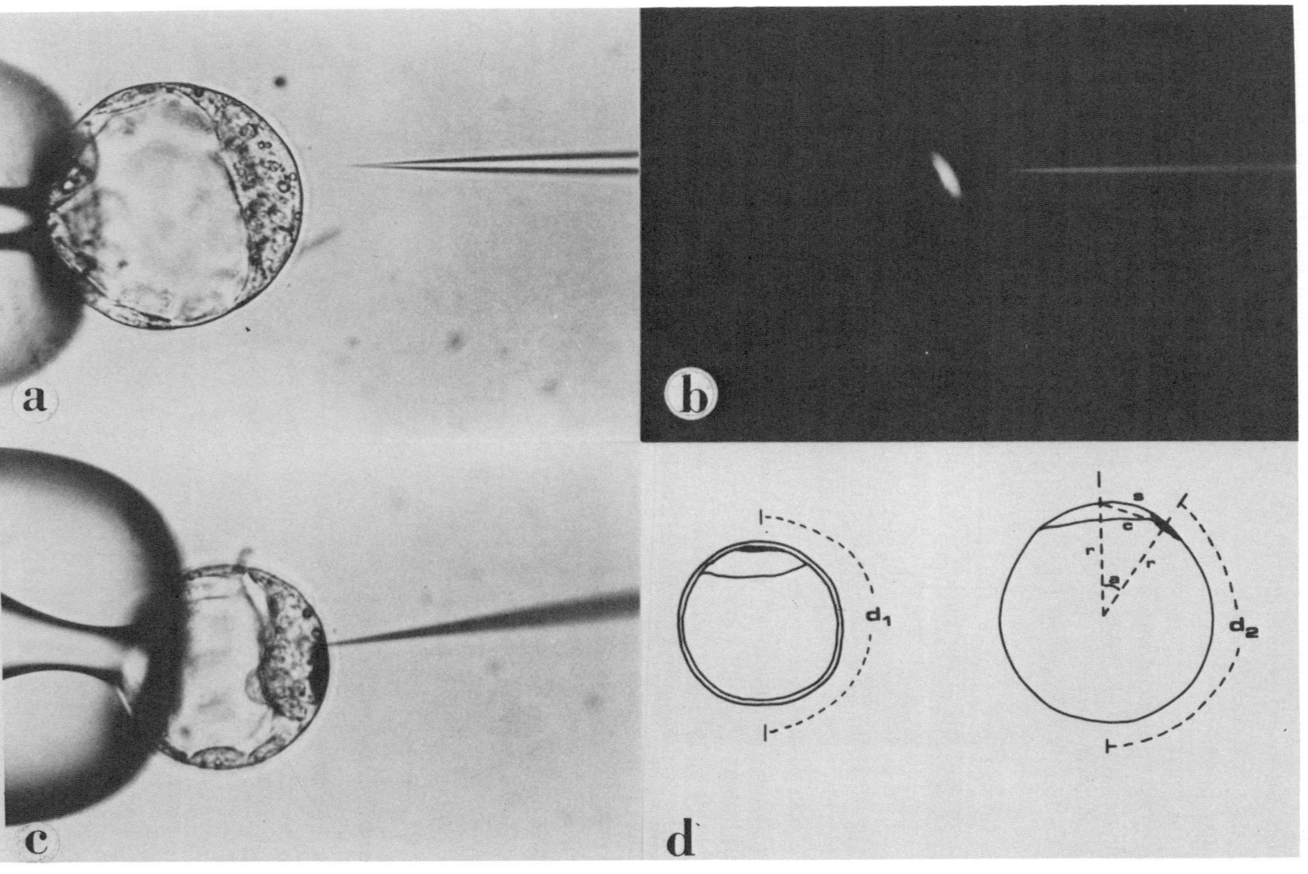

Figure 2 *(See facing page for legend.)*

endoderm population. Despite this cell death, the population increase was 2.01-fold in 24 hr, implying a population doubling time of 23.9 hr. The population doubling time of the surviving labeled cells was 13.7 hr.

Embryos injected in anterior embryonic endoderm had labeled descendants in the anterior visceral yolk sac endoderm. Those injected in the more distal regions of the egg cylinder had descendants predominantly overlying the heart, in the anterior intestinal portal, and lining the foregut. Head process cells of late streak embryos had descendants spread along the trunk endoderm or notochord; some were separated by as much as 500 μm from each other. Cells descended from posterior endoderm remained in the posterior part of the embryo, mainly overlying the primitive streak, but occasionally in the posterior part of the trunk endoderm.

These results show that embryonic endoderm of the primitive-streak-stage mouse embryo is a heterogeneous population composed of cells that give rise to visceral yolk sac endoderm or to embryonic gut endoderm. These contributions are made by separate progenitor cells, because with rare exception (2/53), the marked cells contributed descendants only to visceral yolk sac and cells overlying the heart, or to cells lining the embryonic gut. These rare exceptions may be attributable to a low incidence of injection into adjacent nonsister cells or to other inaccuracies in injection.

These results indicate that there is extensive morphogenetic movement of endoderm cells during gastrulation in the mouse embryo. Interestingly, the descendants in the visceral yolk sac tended to remain coherent and aligned perpendicular to the embryonic axis, whereas endoderm cells along the trunk of the embryo were aligned along the embryonic axis and tended to be dispersed, separated from each other by several or more cell diameters. The extensive anterior translocation of cells in the anterior region of the embryo contrasts with the fate of cells in the distal and posterior regions. Descendants of anterior progenitors were predominantly extraembryonic, whereas those of distal progenitors were spread throughout the embryo, and those of posterior progenitors generally retained their posterior embryonic position.

Figure 2

Injection of polar trophectoderm cells with horseradish peroxidase and rhodamine-conjugated dextran. Central polar cell was injected with lineage tracers (*a*), viewed for localization of fluorescence (*b*), and stained for peroxidase activity (*c*). The distance from the injected polar cell to the abembryonic pole in controls (d_1) was determined from the radius of the injected blastocysts; the comparable distance after incubation (d_2) was determined from embryonic radius and from *s*, the distance traversed by the labeled cells. This arc was estimated by constructing the triangle *rcr*, where *r* is the radius, *a* is the central angle, and *c* is the measured chord between the pole and the labeled cells. The procedure was facilitated by the fact that the labeled descendants remained together as a coherent clone, or patch of cells. (See Table 2 for estimates of distance traversed and d_2-d_1.) (Reprinted, with permission, from Cruz and Pedersen 1985.)

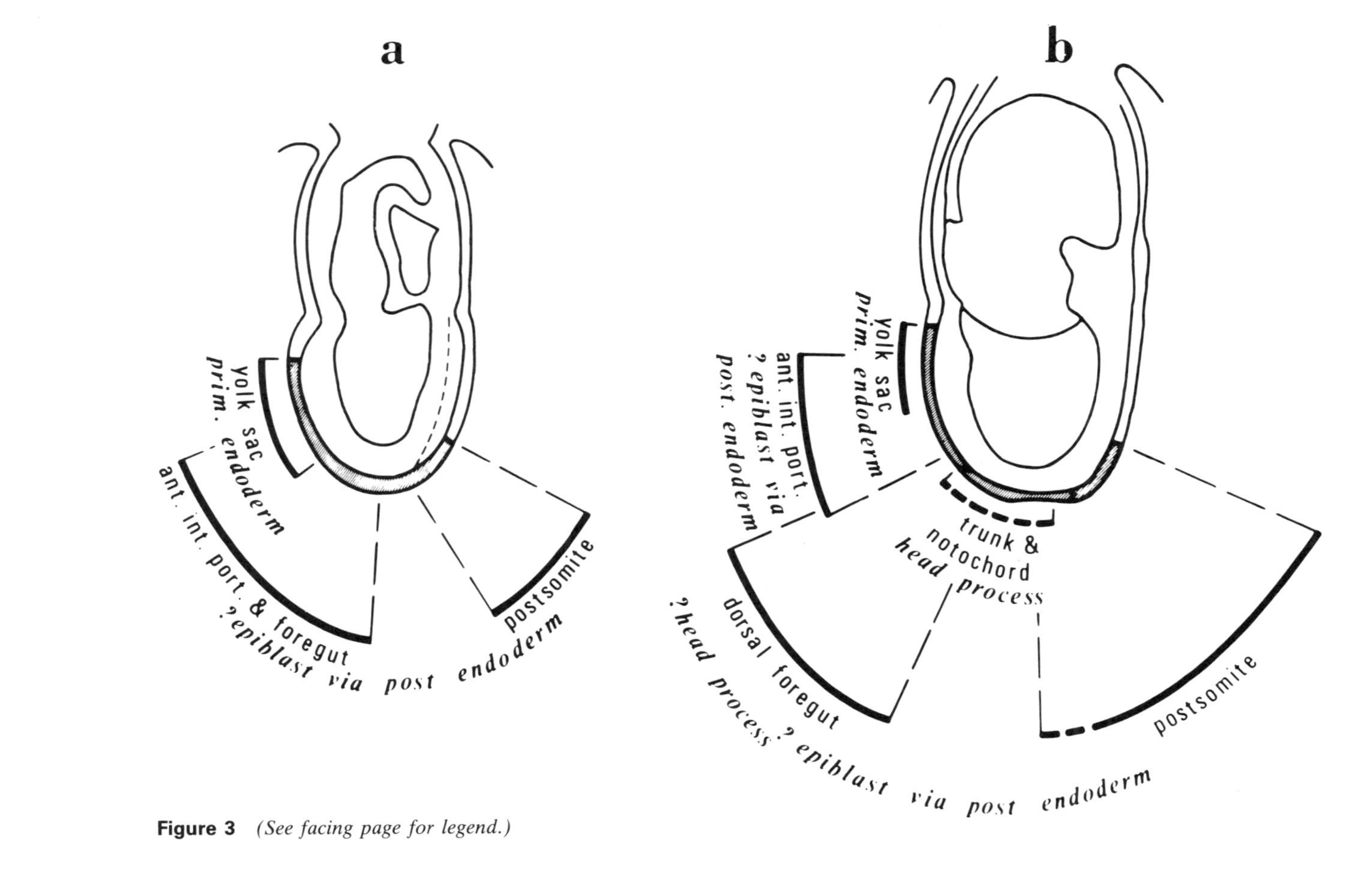

Figure 3 *(See facing page for legend.)*

The differences between mid- and late-streak stages indicate a progressive, anterior movement of cells that give rise to endoderm lining the embryonic foregut and a posterior extension of the primitive streak. The fate of axial endoderm cells inferred from this analysis is summarized in Figure 3 and discussed below.

DISCUSSION

During preimplantation development of mammalian embryos, epithelial layers differentiate that give rise to the trophectoderm and primitive endoderm populations. These extraembryonic cell lineages appear to emerge by a succession of binary decisions based on the position of pluripotent cells within the embryo, outer cells forming trophectoderm, and cells lining the blastocoel cavity forming primitive endoderm (reviewed by Pedersen 1986; Rossant 1986).

Our results show that the trophectoderm is derived from cells that have outer surfaces as early as the 8-cell stage; indeed, any 8-cell blastomere and any outer cell at the 16-cell stage have at least one descendant in the trophectoderm in subsequent divisions. After the division to the 32-cell stage, all outer cells have trophectoderm descendants exclusively. Our findings support the hypothesis of Johnson and Ziomek (1983) that the polarized phenotype of late cleavage stage mouse embryo blastomeres constrains polarized cells to have at least one trophectoderm descendant. Their conclusions, which were based on the behavior of isolated 8- and 16-cell blastomeres (Johnson and Ziomek 1981, 1983; Ziomek and Johnson 1982), thus apply as well to the intact embryo. The origin of the inner cell mass occurs gradually, requiring two cleavages to establish the population of cells that will give rise to the fetus.

Our conclusion from a cell lineage analysis of inner cell mass and trophectoderm fate in the intact mouse embryo is that the inner cell mass serves as a stem cell population for the entire embryo, having descendants in trophectoderm, in primitive endoderm, and in the core of the inner cell mass, presumptive primitive ectoderm. This hypothesis, which accounts for the movement of labeled polar trophectoderm cells and their descendants out of

Figure 3

Origin and prospective fate of axial endoderm cells in mid-streak stage (*a*) and in late-streak/neural-plate-stage embryos (*b*). The zones studied occupy the hatched area (shown in diagrammatic sagittal sections). Italic type indicates probable origin (i.e., ancestral cell lineage), with uncertainties indicated by question marks; roman type indicates cell fate (i.e., location of descendants after 24 hr). Anterior of embryonic region is to the left. (Reprinted, with permission, from Lawson et al. 1986.)

the embryonic pole (Copp 1979) and for the presence of labeled inner cell descendants in trophectoderm, is a novel view of the mammalian blastocyst (Cruz and Pedersen 1985). It was initially suggested by Handyside (1978) as an explanation for the decreasing proportion of total embryo cells in the inner cell mass during blastocyst growth. Such a contribution of inner cell mass descendants to trophectoderm could not occur unless the inner cells were totipotent. Indeed, numerous studies have demonstrated inner cell mass differentiation into trophectoderm after isolation by immunosurgery (Handyside 1978; Hogan and Tilly 1978; Spindle 1978; Rossant and Lis 1979; Nichols and Gardner 1984; Chisholm et al. 1985).

What remains to be determined is how long an inner cell mass contribution to trophectoderm persists in the intact mouse embryo. The restriction of inner cells to a strictly inner fate (with descendants only in primitive ectoderm and primitive endoderm) probably occurs during expansion of the blastocyst cavity and before the time of implantation, if we are to account for the results of injection chimera and blastocyst reconstruction studies. Those studies show that with few exceptions, the extraembryonic ectoderm and ectoplacental cone of the trophectoderm lineage are descendants of the trophectoderm overlying the inner cell mass (Gardner et al. 1973; Papaioannou 1982; Rossant and Croy 1985). At issue, then, is the time of formation of the definitive polar trophectoderm in the intact mouse embryo.

Although the origin and fate of the cell types formed in preimplantation development is an inherently interesting aspect of mammalian embryology, we cannot conclude a priori that mechanisms involved in their differentiation apply to the definitive germ layers. Thus, an understanding of decision-making processes during allocation to the definitive germ layers requires a direct lineage analysis of postimplantation stages, when gastrulation occurs.

The clonal analysis of cell fate described here indicates that embryonic endoderm of the mid- to late-streak mouse embryo is a mixed population of cells that contribute either to extraembryonic or embryonic cells, but not both. Extensive analyses of embryonic ectoderm cell fate following injection into blastocysts (Gardner and Rossant 1979) or into gastrula-stage embryos (Beddington 1981) demonstrate its capacity to form embryonic gut in the mid-gestation (day 16) embryo (Gardner and Rossant 1979) or after culture to mid-somite stages (Beddington 1981). Primitive endoderm cells, in contrast, contribute only to visceral and parietal yolk sac in blastocyst injection chimeras (reviewed by Rossant 1986). Taken together, the results of clonal analysis described here and of lineage analysis in chimeras can be interpreted as follows: During early gastrulation, cells emerge from the primitive streak into the embryonic endoderm layer, displacing or replacing resident cells in that position. During further gastrulation, the early emerging cells move anteriorly, where they differentiate into the endoderm at the anterior intestinal portal and in the ventral foregut while the displaced endoderm cells (derived from the primitive endoderm) move onto the yolk sac; later emerging cells also move anteriorly, occupying positions in the dorsal foregut and along the embryonic trunk, the latter being descendants of the head process.

Further evidence for this view has been obtained by marking endoderm cells of prestreak and early-streak embryos (K.A. Lawson and R.A. Pedersen, in prep.). In this case, prestreak endoderm cells contribute only to extraembryonic visceral endoderm, whereas endoderm cells overlying the tip of the early primitive streak migrate into anterior positions after 1 day of culture and into the anterior intestinal portal, ventral and dorsal foregut, and trunk endoderm after 2-day culture. Moreover, injection into embryonic ectoderm of prestreak and early-streak stages (K.A. Lawson et al., in prep.) directly reveals ectoderm descendants in the endoderm layer after 1-day culture. Although we cannot prove that cells lining the embryonic gut at early somite stages are destined to form the definitive endoderm, there is no indication that they are less viable or proliferate more slowly than cells in other embryonic regions. It therefore seems reasonable to conclude that the gastrula-stage mouse embryo is undergoing dynamic rearrangements of gut endoderm cells that have their origins in the ectoderm layer. This interpretation is also consistent with the fate of embryonic endoderm cells in cultured embryos deprived of their visceral extraembryonic endoderm (Hogan and Tilly 1981) and of visceral endoderm cells in blastocyst injection chimeras (Gardner 1982). In this view, the origin of embryonic gut endoderm in the mouse bears a striking resemblance to that in the chick embryo (Nicolet 1970; Vakaet 1970; Rosenquist 1972).

The similarity between emerging inner cell mass cells contributing to trophectoderm and emerging ectoderm cells contributing to embryonic endoderm is one of the few indications that similar morphogenetic mechanisms are involved at pre- and postimplantation stages of mouse development. Although the cellular mechanisms involved in these tissue rearrangements have not been determined yet, they would most likely require extensive junctional remodeling and cell motility. In conclusion, the further lineage analysis of gastrula-stage mouse embryos, with the objective of obtaining a complete fate map for the organ rudiments, should be a high priority as a basis for understanding teratogenesis in mammals.

ACKNOWLEDGMENTS

This work was supported by the Office of Health and Environmental Research, U.S. Department of Energy (contract no. DE-AC03-76-SF01012). I am grateful to my associates Hanna Balakier, Yolanda Cruz, Kirstie Lawson, Juanito Meneses, Glen Winkel, and Kitty Wu for their vital roles in bringing these approaches to fruition. I thank Kirstie Lawson and Glen Winkel for their comments on the manuscript and for letting me cite their unpublished observations.

REFERENCES

Balakier, H. and R.A. Pedersen. 1982. Allocation of cells to inner cell mass and trophectoderm lineages in preimplantation mouse embryos. *Dev. Biol.* **90:** 352.

Beddington, R.S.P. 1981. An autoradiographic analysis of the potency of embryonic ectoderm in the 8th day postimplantation mouse embryo. *J. Embryol. Exp. Morphol.* **64:** 87.

———. 1986. Analysis of tissue fate and prospective potency in the egg cylinder. In *Experimental approaches to mammalian embryonic development* (ed. J. Rossant and R.A. Pedersen), p. 121. Cambridge University Press, Cambridge, England.

Chisholm, J.C., M.H. Johnson, P.D. Warren, T.P. Fleming, and S.J. Pickering. 1985. Developmental variability within and between mouse expanding blastocysts and their ICMs. *J. Embryol. Exp. Morphol.* **86:** 311.

Copp, A.J. 1979. Interaction between inner cell mass and trophectoderm of the mouse blastocyst. II. Fate of the polar trophectoderm. *J. Embryol. Exp. Morphol.* **51:** 109.

Cruz, Y.P. and R.A. Pedersen. 1985. Cell fate in the polar trophectoderm of mouse blastocysts as studied by microinjection of cell lineage tracers. *Dev. Biol.* **112:** 73.

Gardner, R.L. 1978. The relationship between cell lineage and differentiation in the early mouse embryo. In *Genetic mosaics and cell differentiation* (ed. W.J. Gehring), p. 205. Springer-Verlag, Berlin.

———. 1982. Investigation of cell lineage and differentiation in the extraembryonic endoderm of the mouse embryo. *J. Embryol. Exp. Morphol.* **68:** 175.

Gardner, R.L. and J. Rossant. 1979. Investigation of the fate of 4.5 day *post-coitum* mouse inner cell mass cells by blastocyst injection. *J. Embryol. Exp. Morphol.* **52:** 141.

Gardner, R.L., V.E. Papaioannou, and S.C. Barton. 1973. Origin of the ectoplacental cone and secondary giant cells in mouse blastocysts reconstituted from isolated trophoblast and inner cell mass. *J. Embryol. Exp. Morphol.* **30:** 561.

Gimlich, R.L. and J. Braun. 1985. Improved fluorescent compounds for tracing cell lineage. *Dev. Biol.* **109:** 509.

Handyside, A.H. 1978. Time of commitment of inside cells isolated from preimplantation mouse embryos. *J. Embryol. Exp. Morphol.* **45:** 37.

Hogan, B. and R. Tilly. 1978. In vitro development of inner cell masses isolated immunosurgically from mouse blastocyst. II. Inner cell masses from 3.5- to 4.0-days p.c. blastocysts. *J. Embryol. Exp. Morphol.* **45:** 107.

———. 1981. Cell interactions and endoderm differentiation in cultured mouse embryos. *J. Embryol. Exp. Morphol.* **62:** 379.

Johnson, M.H. and C.A. Ziomek. 1981. The foundation of two distinct cell lineages within the mouse morula. *Cell* **24:** 71.

———. 1983. Cell interactions influence the fate of mouse blastomeres undergoing the transition from the 16- to the 32-cell stage. *Dev. Biol.* **95:** 211.

Lawson, K.A., J.J. Meneses, and R.A. Pedersen. 1986. Cell fate and cell lineage in the endoderm of the presomite mouse embryo, studied with an intracellular tracer. *Dev. Biol.* **115:** 325.

Lo, C.W. and N.B. Gilula. 1979. Gap junctional communication in the preimplantation mouse embryo. *Cell* **18:** 399.

Nichols, J. and R.L. Gardner. 1984. Heterogeneous differentiation of external cells in individual isolated early mouse inner cell masses in culture. *J. Embryol. Exp. Morphol.* **80:** 225.

Nicolet, G. 1970. Analyse autoradiographique de la localisation des differentes ebauches presomptives dans la ligne primitive de l'embryon de poulet. *J. Embryol. Exp. Morphol.* **23:** 79.

Papaioannou, V.E. 1982. Lineage analysis of inner cell mass and trophectoderm using microsurgically reconstituted mouse blastocysts. *J. Embryol. Exp. Morphol.* **68:** 199.

Pedersen, R.A. 1986. Potency, lineage, and allocation in preimplantation mouse embryos. In *Experimental approaches to mammalian embryonic development* (ed. J. Rossant and R.A. Pedersen), p. 3. Cambridge University Press, Cambridge, England.

Pedersen, R.A., K. Wu, and H. Balakier. 1986. Origin of the inner cell mass in mouse embryos: Cell lineage analysis by microinjection. *Dev. Biol.* **117:** 581.

Rosenquist, G.C. 1972. Endoderm movements in the chick embryo between the early short streak and head process stages. *J. Exp. Zool.* **180:** 95.

Rossant, J. 1986. Development of extraembryonic cell lineages in the mouse embryo. In *Experimental approaches to mammalian embryonic development* (ed. J. Rossant and R.A. Pedersen), p. 97. Cambridge University Press, Cambridge, England.

Rossant, J. and B.A. Croy. 1985. Genetic identification of tissue of origin of cellular populations within the mouse placenta. *J. Embryol. Exp. Morphol.* **86:** 177.

Rossant, J. and W.T. Lis. 1979. Potential of isolated mouse inner cell masses to form trophectoderm derivatives in vivo. *Dev. Biol.* **70:** 255.

Spindle, A.I. 1978. Trophoblast regeneration by inner cell masses isolated from cultured mouse embryos. *J. Exp. Zool.* **203:** 483.

Vakaet, L. 1970. Cinephotomicrographic investigations of gastrulation in the chick blastoderm. *Arch. Biol.* **81:** 387.

Weisblast, D.A., R.T. Sawyer, and G.S. Stent. 1978. Cell lineage analysis by intracellular injection of a tracer enzyme. *Science* **202:** 1295.

Ziomek, C.A. and M.H. Johnson. 1982. The roles of phenotype and position in guiding the fate of 16-cell mouse blastomeres. *Dev. Biol.* **91:** 440.

COMMENTS

Scott: Some years ago, Mike Snow found a high proliferative rate in the early mouse embryo. Have you located that zone, and do you have any ideas about how those cells that come from that proliferative zone are allocating?

Pedersen: We have not located the proliferative zone, but it is also possible that we would have missed it because the peroxidase in those cells would have become diluted. We see cells with an average cell-cycle time of about 12 hours in the endoderm. There is also a lot of cell death in the endoderm labeled at 7 1/2 days and cultured 24 hours. This is consistent with the idea of primitive endoderm cell layer being displaced, or replaced, in that position with descendants from the epiblast. But, we have not seen the very rapidly cycling population that Snow described (1977). His conclusion was based on cell counts and mitotic indices.

Johnson: Have you tried any experiments explanting these and moving them to different places where they should have different fates? Perhaps, one could find out how long they maintain their multipotency.

Pedersen: No, but Beddington has done some experiments like that, which I think bear very much on this issue. Her results show (Beddington 1986) that if epiblast cells are taken from different regions, they all have a full range of developmental potential in ectopic sites. Alternatively, when labeled fragments of epiblast are transferred as orthotopic grafts, only one region (distal epiblast) contributed to the gut.

So, it would seem that in the intact chimeric embryo, there are other constraints besides potency that determine what a cell is going to differentiate into. The obvious candidate is cell interactions.

It would be interesting to see what the constraints are in the intact embryo on allocation of cells to the different layers that comprise an organ; for example, do the cells of the mesoderm start to become coherent and descend in a targeted way to given organs, which would imply that mesoderm is directing the formation of that particular organ from a totipotent embryonic shield.

Brent: Let me paraphrase what you said with regard to the development of the yolk sac and the primitive endoderm that goes on to form the GI tract. Your experiments indicate that this is a very early divergence; in other words, very few of the yolk sac cells would end up in the gut endoderm.

Pedersen: We think that all of the primitive endoderm that covers the embryo before gastrulation begins is displaced to yolk sac regions and has strictly extraembryonic descendants. After the beginning of gastrulation, the endoderm is a mixed population; we sometimes label primitive endoderm cells and sometimes hit embryonic endoderm cells. But, with very few exceptions, all the descendants are either in the yolk sac or in the embryo, so it is unlikely that a single endoderm cell is having descendants in both embryonic and extraembryonic regions.

Brent: The reason why I paraphrased that is because it was of personal interest related to our own work, which has to do with the production of congenital malformations with an antibody against the yolk sac. In the early days, we used to think of the yolk sac and the primitive gut as having great similarities—nutritive organs, one nourishing the animal after it is born and one nourishing the embryo when it is in utero—but, as Tom Shepard indicated, their methods of nourishment are very, very different. One transports small molecules after they have been digested, and the other transports macromolecules in order for the yolk sac to digest them. So, it is not surprising that they are very different.

But, the first thing we did in looking to see whether there was a tissue that had a surface antigen similar to the yolk sac was to go into the embryonic endoderm, and we were completely surprised to find that the antigen from the yolk sac was not represented at all in the primitive gut.

Pedersen: But your studies show that it is in the kidney.

Brent: Yes, it is in the kidney, which I still do not understand.

Pedersen: Well, it is not necessary to have a common lineage in order to have a common function. It would be nice, but that just does not always seem to be the case. For example, the yolk sac endoderm has a great similarity to liver in many of its functions, including making α-feto-protein during early embryogenesis. There are many other examples too—P-450 metabolism is very high in the yolk sac. But I think that these are examples of convergent development rather than lineage-derived similarity.

Shepard: Could you estimate what percentage of cells must be destroyed in order to destroy the embryo at 8-, 16-, 32-, and 64-cell stages?

Pedersen: In the mouse, you can get a whole embryo out of half of the embryo, that is, you can destroy one blastomere at the two-cell stage, as Tarkowski (1959) showed, and get a whole embryo. Seidel (1952) demonstrated the same thing in the rabbit. In the sheep, you can get a whole fetus out of a one-eighth blastomere (Willadson 1981), but you cannot do that in the mouse. Anything less than half the embryo in the mouse develops into a trophectoderm vesicle, which implants normally but does not proliferate, because, as Gardner (1972) showed, the inner cell mass induces the proliferation of the trophectoderm.

Shepard: We see a lot of those in the human spontaneous abortuses too, where there is only placenta.

Pedersen: There is reason to think that there is a relationship between the time of blastocoel formation, in terms of cell numbers, and how late you can get a whole embryo out of a single blastomere, because the sheep, for example, forms its blastocyst at a much later stage, the 128-cell stage (Papaioannou and Ebert 1986).

Brent: We looked at a dose of radiation that kills two thirds of the embryo, irradiating on the first day, and we determined the number of cells that died during that period of gestation, using the eosin exclusion technique. We were very surprised to find that very few cells died in those first 4 days. However, there was a delay in blastochyle formation; in other words, that was the predominant event in the first 4 days. So, those cells may die later, but they probably lose a lot of chromosome material and are able to go through those first four divisions with impunity. But, as I say, our one finding was a delayed blastocoel development.

Pedersen: An interesting perspective on the history of toxicological applications of cultured embryos is that for a long time, we have thought that the formation of the blastocyst was a good endpoint for toxicological investigations, and yet, from recent genetic work, it is clear now that an embryo can develop handily through the blastocyst stage, missing a chromosome, that is, as monosomies. Therefore, the embryo can sustain quite a bit of genetic damage and get to the blastocyst. The only thing that it cannot be missing is the X chromosome; in this case, the mouse embryo dies at the two-cell stage.

So, despite the early onset of the embryonic genome activity, development of the blastocyst is probably not very taxing to the embryo genetically, but it is the development beyond the blastocyst stage that requires a total chromosome complement.

Models of Organogenesis Based on Mosaic Pattern Analysis in the Rat

PHILIP IANNACCONE,* JONATHAN HOWARD,† WENDY WEINBERG,* LELAND BERKWITS,* AND FRANK DEAMANT*

*Department of Pathology
Northwestern University Medical School and
Northwestern University Cancer Center
Chicago, Illinois 60611
†Agricultural and Food Research Council
Institute of Animal Physiology
Babraham, Cambridge, England

OVERVIEW

Chimeras were produced by combining embryos of two congenic strains of rat. The tissues of these strains are distinguishable in histological sections. The direct visualization of mosaic patterns in various organs has allowed several models of organogenesis to be discussed. The liver of these animals represents nonclonal development without obvious placement of patches of tissue in association with major anatomical structures of the organ. The adrenal gland, on the other hand, demonstrates clear patterns of clonal development in the cortex. The thymus gland shows considerable stem cell proliferation in the cortex. Focal proliferations and areas of altered enzyme expression in carcinogen-treated liver were shown to be clonal expansions of either one of the two lineages present within the chimeras.

INTRODUCTION

The word chimera has conveyed many meanings since its first appearance in the Iliad. Since Homer's description of the Chimera as a she-beast—part lion, goat, and serpent—whose fate was to be slain by Bellerophon, the word has been used to name any unnatural mixture of organisms. In a biological sense, it was first applied to a class of fish by Linnaeus. The name was also given to a group of grafted plants constructed in the early 1900s; the grafts were thought to have attributes of both the stock and grafted plants. The designation chimera in this paper will refer to multizygotic (usually dizygotic) mosaic animals experimentally produced by amalgamating the early embryos of distinguishable strains by one of a number of techniques.

The morula-stage embryos of mammalia are very sticky tissues once the zona pellucida has been removed. Embryos from different strains of animal

Banbury Report 26: Developmental Toxicology: Mechanisms and Risk

or even different species can then be pushed together, and the individual blastomeres will intermingle. This combined embryo can develop normally in utero in a surrogate mother. Embryonic material can also be combined microsurgically. The inner cell mass, which is fated to form the fetus, can be removed from the blastocyst-stage embryo of one strain of animal and placed in the blastocoel of a blastocyst from another strain. In both instances, the tissues of the resulting adult animals are comprised of two populations of cells, one lineage from each of the two combined genetic strains. This process is different from hybridization, where different strains are mated with each other and offspring have only a single, heterozygous cell population.

The population of amalgamated embryonic blastomeres establishes cell lineages, which, if they are distinguishable in some way, can demonstrate mosaicism. The cell lineages are usually discernible at the biochemical level as a result of polymorphisms in electrophoretically demonstrable enzymes within the combined strains. These differences are isoenzymic and have traditionally been dimorphic, that is, of two forms. Some of the enzymes utilized as markers are dimeric as well. As a result, the isoenzymes are present as a dimer of two subunits. In heterozygous individuals, the dimers can be formed from homotypic forms or heterotypic forms. The probability that the holoenzyme will be comprised of different subunits is twice that of a holoenzyme being formed with homotypic subunits. As a result, the heteropolymeric form of the enzyme is present in levels roughly twice as concentrated as the levels of the two homopolymeric forms and has an intermediate electrophoretic mobility (Fig. 1). The classic example of this type of marker is glucose phosphate isomerase, which is a dimeric, dimorphic autosomal enzyme important in the glycolytic pathway (DeLorenzo and Ruddle 1969). In chimeric animals, where there are two distinct populations of cells, only the two homopolymeric forms of the enzyme can be seen in most, but not all, tissues (Iannaccone et al. 1978; Gardner 1982).

From the earliest use of mammalian chimeras in developmental studies, the importance of a marker system allowing direct visualization of the patterns of mosaicism in such animals has been recognized. McLaren has stated that such a marker should be ubiquitously expressed in the tissues of the chimera, located on the cell surface, and should not translocate to the inside of the blastomere during embryonic differentiation. Furthermore, such a marker should be unmodified by embryonic development itself, or by the experimentation (McLaren 1975). Currently, there are a number of such systems available that have been utilized in a variety of studies (Ponder et al. 1983; Rossant et al. 1983; Wareham et al. 1983; Weinberg et al. 1985a). These include the use of a DNA probe that distinguishes nuclei of *Mus musculus* from nuclei of *Mus caroli* by in situ hybridization in interspecific chimeras (Rossant et al. 1983), the use of a histocytochemical determination of or-

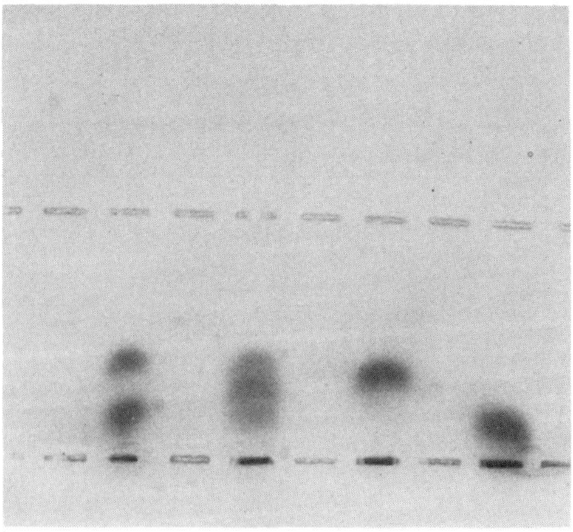

Figure 1

Horizontal starch gel electrophoretigram of glucose phosphate isomerase (GPI-1). (*Right to left*) Lysates of tissue from a homozygous *Gpi-1*^{aa}, from a homozygous *Gpi-1*^{bb}, and from a heterozygous *Gpi-1*^{ab} animal showing a heteropolymeric band of intermediate mobility and from a chimeric animal with two populations of cells.

nithine carbamoyl transferase in chimeric mouse liver (Wareham et al. 1983), and the use of antibodies that recognize differences in the expression of H-2 molecules in congenic strains of mouse (Ponder et al. 1983) and RT1.A molecules in the rat (Weinberg et al. 1985a).

RESULTS

We have reported the direct visualization of the patterns of mosaicism that develop within chimeric animals as a result of the random mixture of embryonic progenitors at the cleavage stage of development (Weinberg et al. 1985b). The progeny of these early embryonic cells assort themselves in a manner that reflects the mechanisms of organ formation and renewal. The marker system is based on the congenic strains of rat described in Table 1. There are a variety of monoclonal antibodies directed against class-I differences between these strains. The monoclonal antibodies are capable of distinguishing tissues of these strains of rat autoradiographically. By directly iodinating the antibodies and applying them to fresh frozen cryostat sections, it is possible to radioactively label cells of one or the other lineage. The labeling is uniform and intense in most visceral tissues of the animals (Fig. 2). One exception to this is the brain, which has a very low class-I major histocompatibility complex determinant density. This is apparent in Figure 3, an unstained autoradiogram of a cryostat section of the frontal lobe. Intensely labeled vascular structures can be distinguished from the unlabeled cells of the substance of the brain. If the antibody is preabsorbed with spleen cell suspensions, the endothelial labeling does not occur.

Chimeras were produced between strains of rat by the method of morula

Table 1
Congenic Strains and Monoclonal Antibodies Used in the Construction and Analysis of Chimeric Rats

Strain designation	Monoclonal antibody	Reactivity[a]
PVG-*RT1*[a]	R2/15S	+
	R3/13	+
	R2/10P	+
	R4/18	−
	NR3/31	−
PVG	R2/15S	−
	R3/13	−
	R4/18	+
	NR3/31	−
PVG-*RT1*[a]	R2/15S	−
	R3/13	−
	R4/18	−
	NR3/31	+

[a]Reactivity determined by indirect hemagglutination as described previously (Weinberg et al. 1985a). Indirect hemagglutination was performed using 10×10^6 red blood cells from the designated strain of rat and incubating for 1 hr in the presence of the designated monoclonal antibody. Cells were washed and then incubated with goat anti-rat IgG. Agglutination in the appropriate strain–antibody combinations occurred at high titers. (Reprinted, with permission, from Weinberg et al. 1985a.)

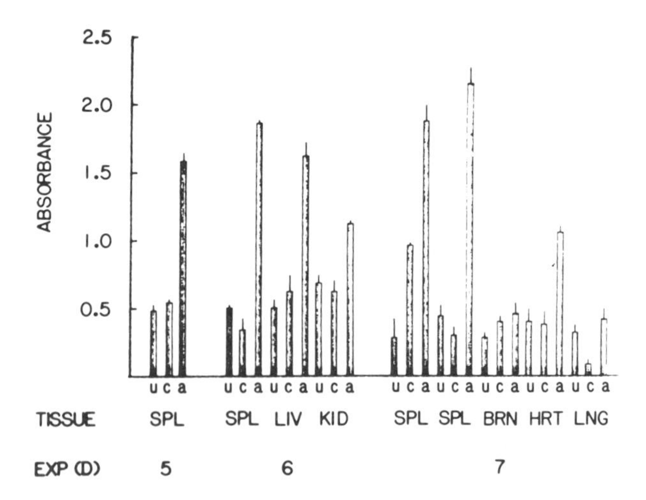

Figure 2
Histogram demonstrating the intensity (absorbance) of grain accumulation in unstained autoradiograms of various tissues prepared as frozen sections and incubated with a monoclonal antibody that recognizes a class-I molecule of PVG-*RT1*[a] haplotype. The tissues are spleen (SPL), liver (LIV), kidney (KID), brain (BRN), heart (HRT), and lung (LNG), examined after 5, 6, or 7 days of exposure (EXP) of the autoradiograms. Absorbance was determined with a scanning densitometer on tissues from PVG-*RT1*[a] (a), PVG-*RT1*[u] (u), and PVG (c) cells. (Reprinted, with permission, from Weinberg et al. 1985a.)

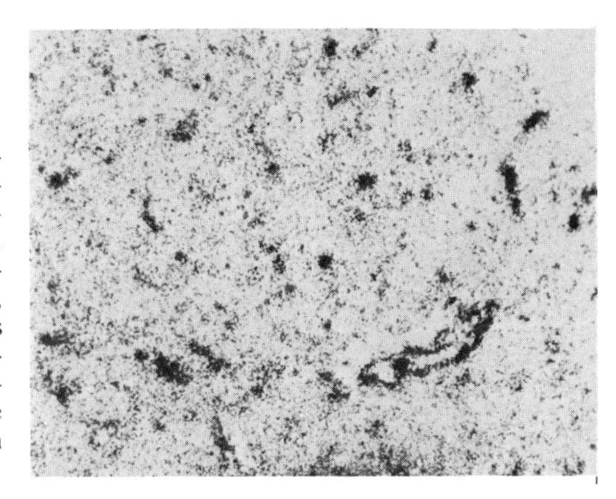

Figure 3
Unstained autoradiogram of frozen section of brain from an animal of the PVG-$RT1^a$ haplotype, incubated with an antibody that recognizes that haplotype. Notice that the endothelium of blood vessels has accumulated many silver grains, whereas the substance of the brain has not, indicating its low reactivity presumably due to low class-I major histocompatibility determinant density in these cells. (Reprinted, with permission, from Weinberg et al. 1985a.)

aggregation, as shown in Figure 4. First, eight-cell-stage embryos are removed from the oviduct of pregnant rats that were mated to yield homozygous embryos of the distinguishable strains. The zonae pellucidae are removed by brief exposure to acidified tyrode solution. Homozygous embryos from two different strains are then physically apposed and cultured until the individual morulae have adhered. These combined embryos are then removed to the uterus of a pseudopregnant surrogate mother and allowed to come to term. The tissues of the adult animals are then removed and rapidly frozen by immersion in liquid nitrogen. Sections are prepared as described above. The autoradiograms demonstrate patterns of mosaicism, as shown in Figures 5, 7, and 9.

The liver, for example, shows a random pattern of mosaicism, with respect to the anatomic structure of the organ. When serial sections, stained alternately with the haplotype marker or with hematoxylin and eosin, are compared by pin registration, there is no apparent association between the patch distribution and either acinar zones or liver lobules. Acinar zones and liver lobules represent the major current concepts of the organization of the liver, yet neither shows any association with patches within the chimeric liver (Iannaccone et al. 1987a). The liver in mammalia has a number of fetal functions that are both vital and distinct from adult liver functions. It seemed reasonable to assume, therefore, that the method of formation of the organ did not require considerations of the anatomical organization of the adult organ. Moreover, the organ forms very rapidly in the rat and in other mammals as well. These observations lead us to speculate that the decisions required to place the hepatoblasts in the right positions at the right time might be very simple, but highly reiterated, probabilistic decisions.

We set out to test this hypothesis by creating a computer program to create a simulated mosaic tissue that could be analyzed quantitatively in the same

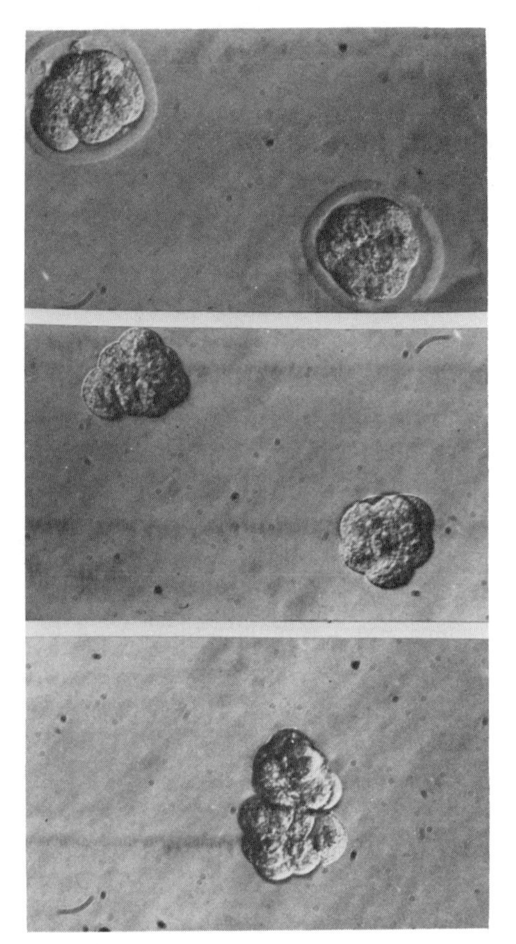

Figure 4

Photomicrographs of eight-cell-stage embryos recovered from the oviducts of donor female rats of two different PVG congenic strains 3 days after mating. (*Top*) The embryos are surrounded by zonae pellucidae, a structure that can be quickly removed in an acid medium, as seen in the *middle*. (*Bottom*) The embryos may then be pushed together and, in the presence of phytohemagglutinin, will begin to merge into a single large morula. The amalgamated embryo is then transferred to a surrogate mother. The embryos are ~100 μm in diameter.

way as the actual chimeric tissue. The program created an "anlage" of 100 "cells" of two types, with a proportion that was requested approximately but known precisely. The cells in the anlage were randomly placed. Each cell was chosen at random to "divide," and the daughter was placed in an adjacent position (one of eight available) that was chosen randomly. When each of the cells had divided, that generation was complete. The simulation ran until the "tissue" had filled the available image-analyzable matrix. The results of this simulation at any given starting proportion in the anlage were computationally indeterminate. The set of rules that represent highly reiterated but simple probabilistic decisions represent a cellular automata. The relationship between the proportion of the two cell types in this mosaic and the number of

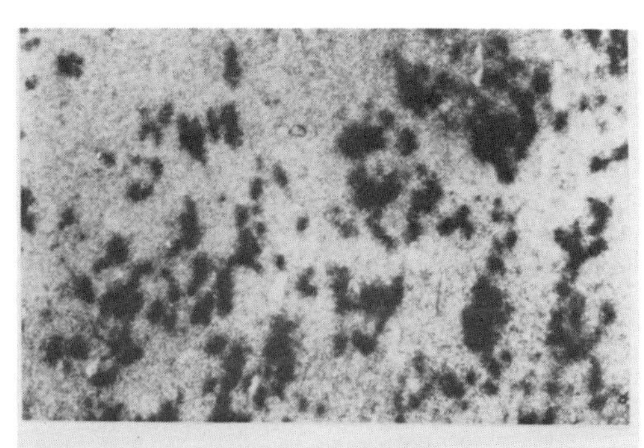

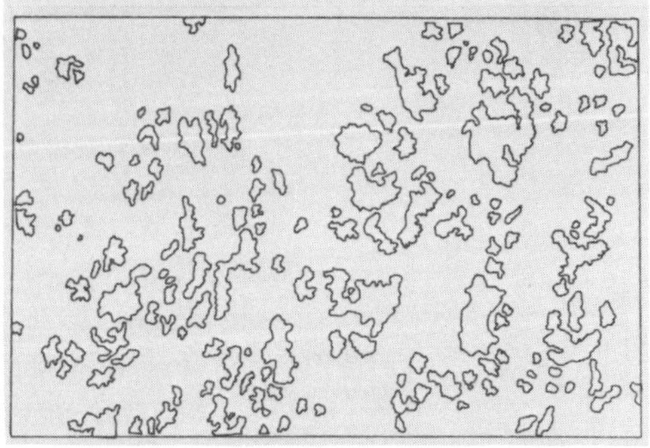

Figure 5
(*Top*) An unstained autoradiogram of section 18 from the liver of animal 0912. In this section, patches of the PVG-*RT1*[a] type are evident. There is no evidence in any of the sections of liver that there is an association of patch distribution with major anatomical units within the tissue. (*Bottom*) Photograph of a computer-generated plot of digitized data from the photomicrograph. These digitized data were used for a quantitative analysis of the patches in the liver. Bar represents 0.25 mm. (Reprinted, with permission, from Iannaccone et al. 1987a.)

patches per unit area and the mean patch area for both the simulated tissue and the actual chimeric liver tissue is shown in Figure 6. Simply stated, as the proportion of the major cell type increases, the mean area of the patches of that type also increases while the number of patches of the major cell type decreases. There is a critical proportion at which the major cell type forms a background lattice. Patches of the minor cell type then form islands in this sea of the major cell type. Thus, the number of patches of the minor cell type per unit area increases as the proportion of this minor cell type decreases. A proportion is reached, after which actual number of patches of the minor type begins to decrease. The area of these patches decreases as the proportion of the minor cell type decreases. The analysis of the patches in the simulated tissue was performed by computer-assisted image analysis and represents the results of feature extraction. These results are comparable to those from the actual chimeric tissue and suggest that as a first approximation, the highly reiterated simple probabilistic decisions would be sufficient to form the type of mosaic tissue seen in actual chimeric rat liver (Iannaccone et al. 1987a).

Different results might be expected in other organs. In the adrenal cortex, for example, the pattern of mosaicism suggests nonrandom placement of the epithelium of the organ in the formation and maintenance of the tissue.

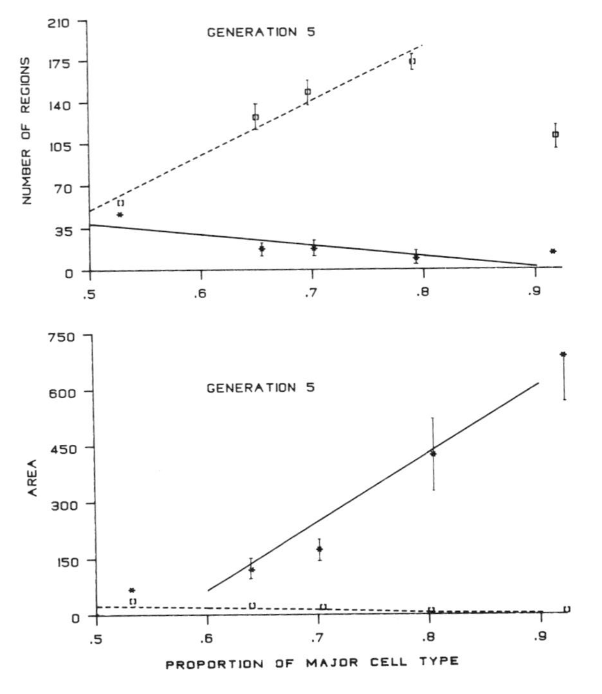

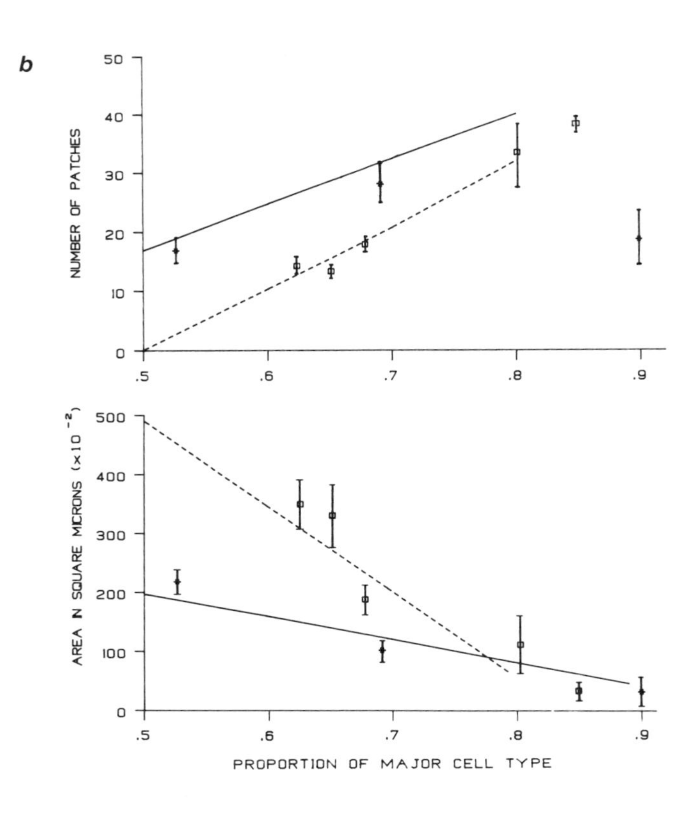

Figure 6 *(See facing page for legend.)*

Figure 7 reveals the radial striped pattern of the two cell populations that extend from the medulla of the adrenal gland to the capsule across all three histogenic zones. The relationship between the number of cords and the proportion of the two cell types is not linear as in the liver, and there is no reciprocal relationship between the major cell type and the minor cell type with respect to proportion of cell types and the number of cords. Moreover, there is no apparent relationship between the proportion of the two cell types and the patch area (Table 2). The cords of both cell types are equal in number (Iannaccone and Weinberg 1987).

From the time of Gottschau's elegant description of the development of the adrenal cortex, there has been controversy over the mechanism by which the organ is developed and maintained. Gottschau originally contended that the organ is formed by the inward migration of primordial cells at the capsule of the organ (Gottschau 1883). The organ can be extirpated from the center and will regenerate completely, presumably from the outer rim of cells attached to the capsule (Greep and Deane 1949; Nickerson et al. 1969). However, radiolabeled thymidine incorporation determined autoradiographically shows

Figure 6

(*a*) (*Top*) The correlation between the number of regions (patches) per unit area in the computer-simulated mosaic tissue and the proportion of type-1 cells (major type) in the tissue at generation 5. Since the computer uses algorithms that can analyze the background lattice as well as the minor cell type, this correlation is presented for both cell types as a function of type-1 proportion. The dashed line represents type-2 (minor type) regions, and the solid line represents type-1 (major type) regions. The number of type-2 regions (minor type) increases rapidly as the probability of the type-1 cells (major type) increases up to a proportion of ~0.8. (*Bottom*) The correlation between the area (in arbitrary units) of regions in the computer-simulated mosaic tissue and the proportion of type-1 cells (major type) in the tissue at generation 5. Since the computer uses algorithms that can analyze the background lattice as well as the minor cell type, this correlation is presented for both cell types as a function of type-1 proportion. The dashed line represents the area of type-2 regions (minor type), and the solid line represents the area of type-1 regions (major type). S.E.M.S are presented as a vertical bar, unless the bar is smaller than the symbol representing the mean. (*b*) (*Top*) The relationship between the number of patches of the minor type per unit area of chimeric tissue and the proportion of the major cell type. The dashed line represents the number of PVG patches, with increasing proportion of PVG-$RT1^a$ cells in the tissue, and the solid line represents the number of PVG-$RT1^a$ patches, with increasing proportion of PVG cells in the tissue. The data are presented in this manner since the major cell type formed a background lattice and, hence, it is the minor cell component that was digitized, as described in Results. Lines represent linear regression of all data between proportion 0.5 and 0.8; points represent the mean values of all sections from a given liver lobe of each chimera. (*Bottom*) The relationship between the area of patches of the minor type in chimeric liver and the proportion of the major cell type. The dashed line represents the area of PVG patches, with increasing proportion of PVG-$RT1^a$ cells in the tissue; the solid line represents the area of PVG-$RT1^a$ patches, with increasing proportion of PVG cells in the tissue. The data are presented in this manner since the major cell type formed a background lattice and, hence, it is the minor cell component that was digitized, as described in Results. S.E.M.S are presented as vertical bars. Lines represent linear regression of all data; points represent the mean values of all sections from a given liver lobe of each chimera. (Reprinted, with permission, from Iannaccone et al. 1987a.)

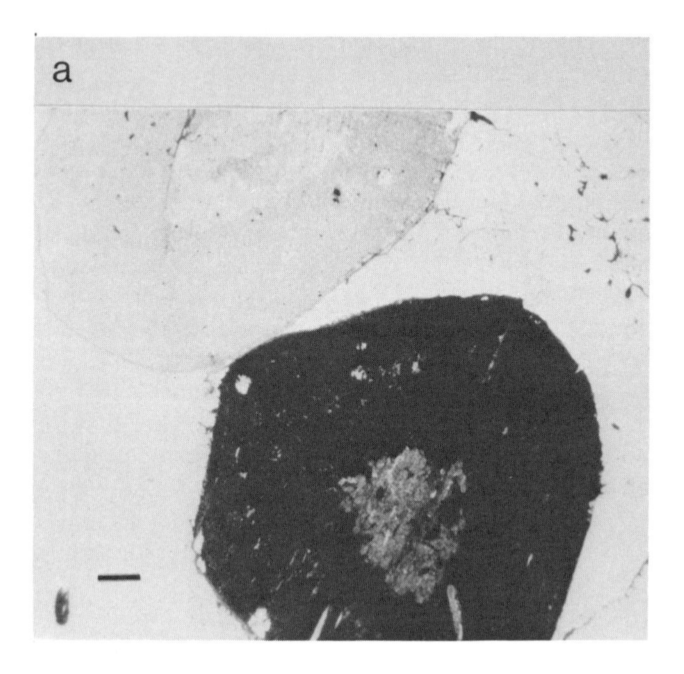

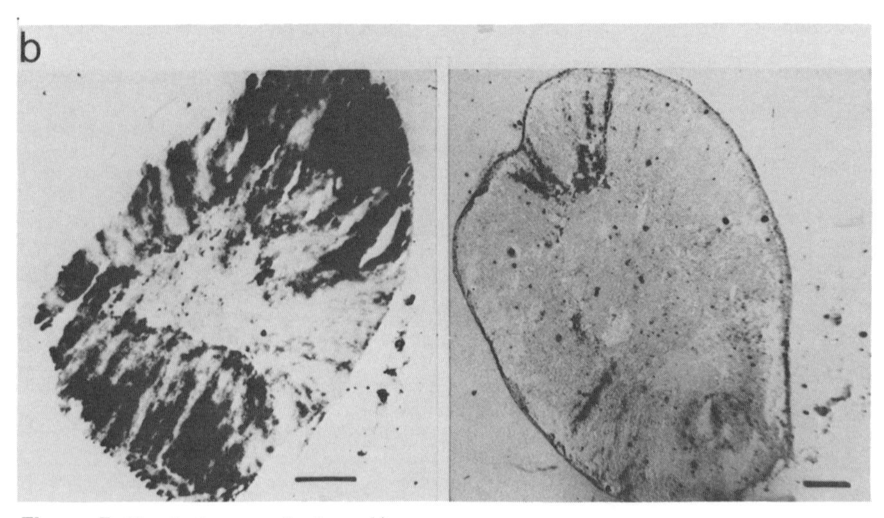

Figure 7 *(See facing page for legend.)*

Table 2
Analysis of Adrenal Cortex Mosaicism

Animal	PVG-*RT1* [a] cells (%)	No. of PVG-*RT1* [a] cords	Mean patch area (mm^2)
0912 left	5[a]	6.1 ± 0.4	0.133[a]
0906 right	85	12.5 ± 1.5	0.086
left	82	15.2 ± 1.6	0.059
0909 left	78	13.3 ± 2.1	0.086
0911 right	62	34.3 ± 0.3	0.226

[a]Percentage of tissue, derived from the PVG-*RT1* [a] lineage, and patch area was determined by computer-based image analysis. (Reprinted, with permission, from Iannaccone and Weinberg 1987.)

that division can occur in any area of the cortex (Diderholm and Hellman 1960). It may be possible to resolve whether cells divide from the inside to the outside or from the outside to the inside by mosaic pattern analysis in informative corners of cross sections through the center of this pyramidoid-shaped organ. The available data fit the concept of inward clonal proliferation of stem cells, with cords meeting and joining toward the medulla. This results in a fragmented patch pattern in the corners (Fig. 8). This pattern is particularly evident in the chimeras that have approximately equal proportions of the two cell types, that is, in balanced chimeras. The radial pattern of patch distribution is maintained even in highly unbalanced chimeras. A central cross section of an animal with 95% PVG cells demonstrates the same radial pattern comprised of six cords of PVG-*RT1* [a] cells. These results imply that during the maintenance of the normal unstressed organ, the cells of the adrenal cortex are regulated in a way that prevents the random migration of the cells throughout the organ.

The thymus gland presents yet a third pattern of mosaicism in these animals. As shown in Figure 9, the thymus gland is a bilobed organ with

Figure 7

(*a*) Unstained autoradiogram of tissue from the adrenal gland of PVG-*RT1* [a] animal, demonstrating uniform, heavy-grain accumulation. The frozen cryostat-sectioned tissue was incubated with iodinated R2/15S monoclonal antibody directed to the class-I alloantigens of this strain of animals. The unreacted tissue in the upper left corner with little or no grain accumulation is cryostat-sectioned adrenal from a PVG animal, treated in an identical manner simultaneously. Bar represents 0.5 mm. (*b*) Unstained autoradiograms of adrenal glands from two different chimeras of widely different proportions of cell types. (*Left*) chimera 0911, which has 38% PVG cells; (*right*) chimera 0912, which has 95% PVG cells. The parallel array of cells in the cortex is conserved regardless of the proportion of parental cell lineages present in the animal. Bar represents 0.5 mm. (Reprinted, with permission, from Iannaccone and Weinberg 1987.)

Figure 8

Summary of the results obtained from all of the glands studied at various levels and planes of section. Since the organ is a pyramidoid in the rat, sectioning from any surface resulted in the same progression of patterns toward the center of the organ. As can be seen, the results are very suggestive that the cortical lineages are arrayed as regular columns of cells from the capsular surface to the medullary interface. (Reprinted, with permission, from Iannaccone and Weinberg 1987.)

several lobules in each lobe. The mosaic pattern of this organ is consistent with the concept of substantial stem cell proliferation of the cortical thymocytes. In the absence of additional data, however, it is not possible to know whether the pattern observed is due to reaction of the antibody with cortical lymphocytes or with epithelial cells. Resolving this issue may be of some

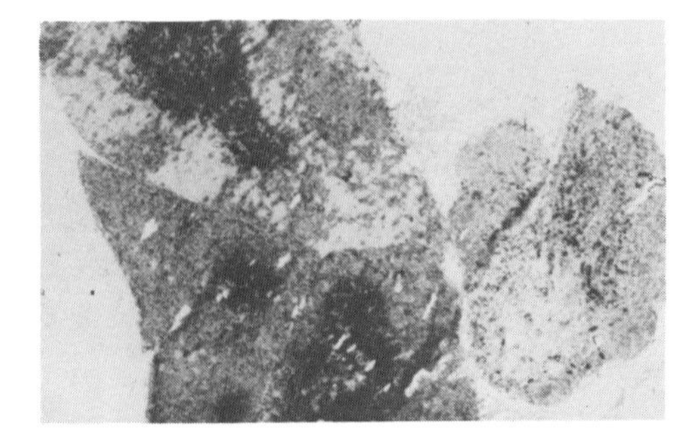

Figure 9

An unstained autoradiogram of a frozen section of the thymus gland incubated with a monoclonal antibody that recognizes the PVG-$RT1^{a}$ haplotype. Several large patches of cells of the PVG lineage are evident within the cortex of this gland. The scalloped interface between the cortex and the medulla indicates that cortical cells may move to the medullary regions on broad fronts. The oligoclonality of the cortex may indicate stem cell growth within the cortex. Magnificiation, 17.6 ×.

importance since the cortical cells may progress to the medulla by direct invasion, as evidenced by the scalloped margin between the cortex and the medulla. The medullary regions are comprised of cells of the same lineage within a lobe. This suggests that there is a single medulla that extends throughout each of the lobules. The data, although preliminary, suggest that medullary areas might be clonal in origin.

This system has been utilized to examine the possible clonal origin of preneoplastic lesions. Abundant evidence has accumulated over the past several years to support the contention that tumors arise by the clonal expansion of single affected cells (Iannaccone et al. 1978; Tanooka and Tanaka 1982; Deamant and Iannaccone 1985). This observation is consistent with the concept that tumors arise as a result of rare critical events. However, it is also clear that tumor formation is the end result of a multistep process. If any of these steps was a clonal event, the resulting tumor would appear to be clonal. To determine whether tumors are clonal from their earliest inception, one must explore lesions that, although not neoplastic in themselves, will lead to the formation of tumors. The best known of these lesions have been described in the carcinogen-treated rat liver. In these animals, a number of morphological alterations and areas of altered expression of enzymes are known to be associated with carcinogenic treatment. These lesions arise following a single administration of a carcinogenic chemical when the animals are further treated with a chemical that, by itself, cannot cause cancer (a promoter, e.g., low concentrations of phenobarbital). Although they do not all inevitably lead to cancer, the lesions are invariably associated with it. Hence, these altered areas of the carcinogen-treated liver are taken to be preneoplastic lesions (Becker 1978; Pitot et al. 1978; Farber 1984; Goldsworthy et al. 1984).

In mosaic animals, the presence of a single lineage of cells within a lesion may be taken to mean that the lesions are clonally derived. Conversely, the presence of several lineages within the lesions may be taken to mean that they

Table 3
Cellular Lineage of Foci of γ-Glutamyltranspeptidase Expression in Carcinogen-treated Chimeric and Control Liver

	PVG-*RT1*[a] lineage	PVG lineage	Mixed lineage
Chimeras	151	323	25
PVG-*RT1*[a] parental animals	367	0	0
PVG parental animals	0	249	0

are not clonally derived. Constraints on these conclusions have been discussed previously (Iannaccone et al. 1987b). Serial sections were prepared from frozen tissues of these animals following a year of promotion of their tumors. The tissues analyzed had independent landmarks to allow correct

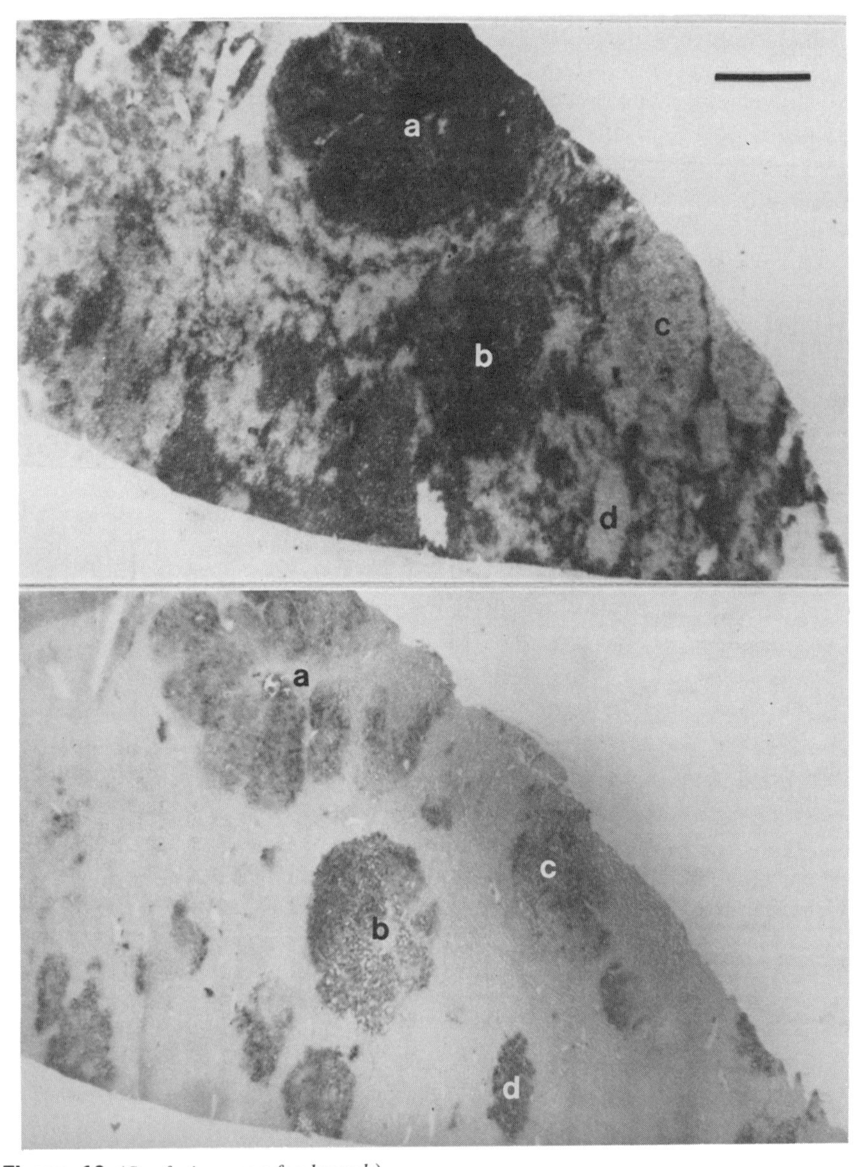

Figure 10 (*See facing page for legend.*)

registration of the sections. The patches and the lesions from separate sections were digitized and stacked on one another in the correct order. In this way, we were able to determine the precise relationships between the distribution of cell lineages within the treated tissues and the lesions that resulted from that treatment. Table 3, for example, shows that of 499 lesions characterized by the expression of γ-glutamyltranspeptidase, 151 were comprised exclusively of cells of the PVG-*RT1*[a] lineage, and 323 were comprised entirely of PVG lineage. The appearance of the patches in surrounding uninvolved tissue strongly suggests that these are expanding lesions. The same results were obtained for other preneoplastic lesions of the carcinogen-treated rat liver. A number of these are illustrated in Figure 10. Figure 11 demonstrates a computer reconstruction of registered serial sections stained for either the haplotype marker or for the enzyme alteration under consideration. The data clearly indicate that irrespective of various degrees of microheterogeneity in areas of altered enzyme expression, the lesions nearly always consist exclusively of one or the other of the two cell lineages present in normal tissue. The exceptions to this are very small lesions and may represent minor misregistration of sections. Thus, small patches and small lesions studied in serial sections separated by 6–18 μm may not line up precisely. Alternatively, they may indicate that these lesions are derived from more than one cell.

In conclusion, the data support the contention that areas of altered enzyme expression, which are generally held to be precursor lesions of hepatocellular carcinoma, are clonal expanding growths and may indicate that liver tumors are clonal from their earliest inception (Weinberg et al. 1987).

DISCUSSION

We have shown, as have others, that mammalian embryonic systems can be manipulated to create histologically demonstrable mosaicism. We have attemped to apply mosaic pattern analysis to the description of potential

Figure 10

Serial frozen sections of liver from chimera 0906 after 1 yr of phenobarbital promotion, begun after partial hepatectomy and treatment with a single dose of the carcinogen *N*-nitrosodiethylamine. (*Top*) Photomicrograph of an unstained autoradiogram of the treated liver stained with an antibody that recognizes cells of the PVG-*RT1*[a] lineage. (*Bottom*) The serial frozen section of treated liver stained for the enzyme γ-glutamyltranspeptidase (E.C. 2.3.2.2). This enzyme is normally expressed only in fetal hepatocytes, but areas of expression are associated with hepatocarcinogenesis. Several nodular lesions that are γ-glutamyltranspeptidase positive are evident as clones of PVG-*RT1*[a] origin (e.g., *a* and *b*). These lesions were histologically evident as hyperplastic nodules in serial sections stained with hematoxylin and eosin. Lesions *c* and *d*, which were not evident as morphologically altered areas, are clearly enzyme-altered areas and are composed entirely of cells of the PVG-*RT1*[c] lineage. Bar represents 1 mm.

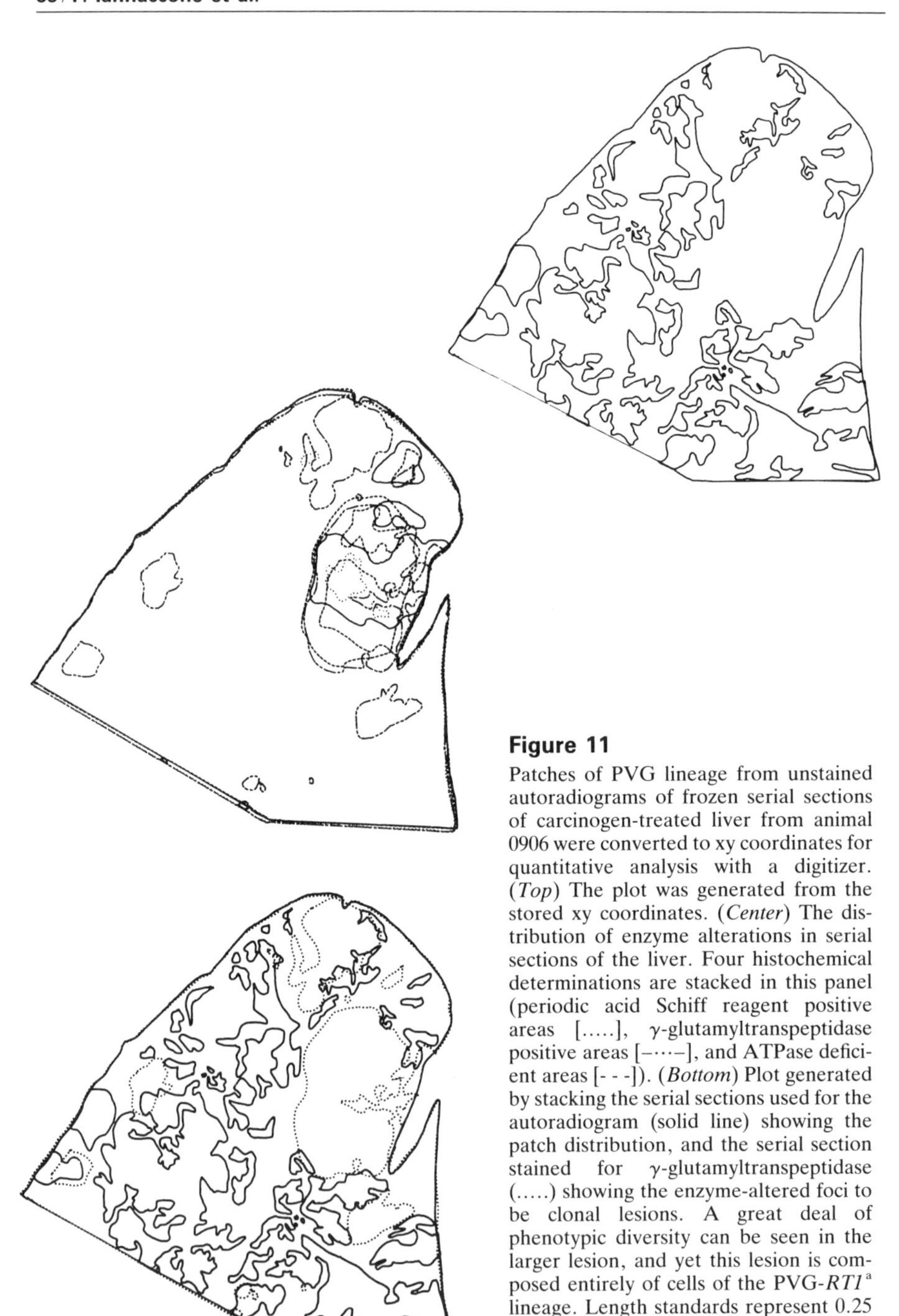

Figure 11

Patches of PVG lineage from unstained autoradiograms of frozen serial sections of carcinogen-treated liver from animal 0906 were converted to xy coordinates for quantitative analysis with a digitizer. (*Top*) The plot was generated from the stored xy coordinates. (*Center*) The distribution of enzyme alterations in serial sections of the liver. Four histochemical determinations are stacked in this panel (periodic acid Schiff reagent positive areas [.....], γ-glutamyltranspeptidase positive areas [–····–], and ATPase deficient areas [- - -]). (*Bottom*) Plot generated by stacking the serial sections used for the autoradiogram (solid line) showing the patch distribution, and the serial section stained for γ-glutamyltranspeptidase (.....) showing the enzyme-altered foci to be clonal lesions. A great deal of phenotypic diversity can be seen in the larger lesion, and yet this lesion is composed entirely of cells of the PVG-*RT1*[a] lineage. Length standards represent 0.25 mm.

mechanisms of organ construction and maintenance. This approach has yielded preliminary evidence in support of three models of organogenesis in different organs. Clearly, a great deal more work is required to describe these processes more precisely. Many approaches will be taken, and it is our hope that mosaic pattern analysis will remain useful among them. Our work with carcinogen-induced precursor lesions has shown that this form of analysis can also be applied to pathologic processes, and we will continue in this manner as well. This form of experimental manipulation of mammalian systems may help point the way toward an explanation of organ formation in normal and diseased states.

ACKNOWLEDGMENTS

This work was supported, in part, by grants CA29078 and ES03498 from the U.S. Public Health Service, Department of Health and Human Services; and by MOD Birth Defects grant 15-49 (P.M.I.). J.C.H. is supported, in part, by grant CA34913 from the U.S. Public Health Service, Department of Health and Human Services.

REFERENCES

Becker, F.F. 1978. Characterization of hepatic nodules. In *Rat hepatic neoplasia* (ed. P.M. Newberne and W.H. Butler), p. 42. MIT Press, Cambridge, Massachusetts.

Deamant, F.D. and P.M. Iannaccone. 1985. Evidence concerning the clonal nature of chemically induced tumors: Phosphoglycerate kinase-1 isozyme patterns in chemically induced fibrosarcomas. *J. Natl. Cancer Inst.* **74:** 145.

DeLorenzo, R.J. and F.H. Ruddle. 1969. Genetic control of two electrophoretic variants of glucosephosphate isomerase in the mouse (*Mus musculus*). *Biochem. Genet.* **3:** 151.

Diderholm, H. and B. Hellman. 1960. The cell renewal in the rat adrenals studied with tritiated thymidine. *Acta Pathol. Microbiol. Scand.* **99:** 82.

Farber, E. 1984. Chemical carcinogenesis: A current biological perspective. *Carcinogenesis* **5:** 1.

Gardner, R.L. 1982. Manipulation of development. In *Embryonic and fetal development* (ed. C.R. Austin and R.V. Short), p. 159. Cambridge University Press, Cambridge, England.

Goldsworthy, T., H.A. Campbell, and H.C. Pitot. 1984. The natural history and dose-response characteristics of enzyme-altered foci in rat liver following phenobarbital and diethylnitrosamine administration. *Carcinogenesis* **5:** 67.

Gottschau, M. 1883. Struktur und Embryonale Entwickelung der Nebennieren bei Saugethieren. *Arch. fur Anatomie und Physiologie, Anat. Abth.* p. 412.

Greep, R.O. and H.W. Deane. 1949. Histological, cytochemical, and physiological observations on the regeneration of the rat's adrenal gland following enucleation. *Endocrinology* **45:** 42.

Iannaccone, P.M. and W.C. Weinberg. 1987. The histogenesis of the rat adrenal cortex: A study based on histologic analysis of mosaic pattern in chimeras. *J. Exp. Zool.* (in press).

Iannaccone, P.M., R.L. Gardner, and H. Harris. 1978. The cellular origins of chemically induced tumours. *J. Cell Sci.* **29:** 249.

Iannaccone, P.M., W.C. Weinberg, and L. Berkwits. 1987a. A probabilistic model of mosaicism based on the histological analysis of chimeric rat liver. *Development* **99:** 187.

Iannaccone, P.M., W.C. Weinberg, and F.D. Deamant. 1987b. On the clonal origin of tumors: A review of experimental models. *Int. J. Cancer* **39:** 778.

McLaren, A. 1975. Sex chimaerism and germ cell distribution in a series of chimaeric mice. *J. Embryol. Exp. Morphol.* **33:** 205.

Nickerson, P.A., A.C. Brownie, and F.R. Skelton. 1969. An electron microscopic study of the regenerating adrenal gland during the development of adrenal regeneration hypertension. *Am. J. Pathol.* **57:** 335.

Pitot, H.C., L. Barsness, T. Goldsworthy, and T. Kitagawa. 1978. Biochemical characterisation of stages of hepatocarcinogenesis after a single dose of diethylnitrosamine. *Nature* **271:** 456.

Ponder, B.A.J., M.M. Wilkinson, and M. Wood. 1983. H-2 antigens as markers of cellular genotype in chimaeric mice. *J. Embryol. Exp. Morphol.* **76:** 83.

Rossant, J., M. Vigh, L.D. Siracusa, and V.M. Chapman. 1983. Identification of embryonic cell lineages in histological sections of *M. musculus*<−>*M. caroli* chimaeras. *J. Embryol. Exp. Morphol.* **73:** 179.

Tanooka, H. and K. Tanaka. 1982. Evidence for single-cell origin of 3-methylcholanthrene-induced fibrosarcomas in mice with cellular mosaicism. *Cancer Res.* **42:** 1856.

Weinberg, W.C., L. Berkwits, and P.M. Iannaccone. 1987. The clonal nature of carcinogen-induced altered foci of γ-glutamyl transpeptidase expression in rat liver. *Carcinogenesis* **8:** 565.

Weinberg, W.C., F.D. Deamant, and P.M. Iannaccone. 1985a. Patterns of expression of class I antigens in the tissues of congenic strains of rat. *Hybridoma* **4:** 27.

Weinberg, W.C., J.C. Howard, and P.M. Iannaccone. 1985b. Histological demonstration of mosaicism in a series of chimeric rats produced between congenic strains. *Science* **227:** 524.

COMMENTS

Johnson: At what stage during development are these clones, or these equivalence groups, established?

Iannaccone: What we do know with respect to the liver is that the process is not altered by regeneration, as it occurs following partial hepatectomy, so our feeling is that this is the type of patch fragmentation, which we presume is leading to the mosaic pattern that may have occurred from the beginning.

Now, our assumptions are based on divisions that would begin in the hepatic blastema or in the hepatoblast as a small nub in the true gut endoderm. We assume that embryonic mixing is absolutely vital up to the point at which the organ begins.

Our arguments of patch fragmentation state that once the primordium begins to divide, it is no longer necessary to argue that a large amount of cell movement is required to come up with a small, highly variegated, geometrically complicated patch pattern that we see in the adult organ.

The reason that this is important may be a nuance. In fact, a number of investigators have argued that by looking at the size of the patch or the clone, one can work backward to determine when in embryogenesis the cell movement actually stopped. We think that may be an oversimplification, because we don't think that the cell movement is necessary; all that is actually required is adjacent cell positioning.

Markert: During development—and certainly even after that—there is substantial turnover of cells in the body of animals. Do you expect that the original patch pattern will be maintained during that time or will there be differential multiplication of the cells?

Iannaccone: Again, the only data that we have to bear on that question are in the regenerating rat liver. We end up with a quantitatively similar patch pattern in uninvolved areas of liver, as we do in the original adult liver with which we started; so, we presume that it could work the same way embryonically.

Markert: If I may follow up on that, there is evidence, as you know, in the mouse that there is sometimes a change in the relative production of gametes in chimeras so that they start out, for example, 50/50 and end up 75/25, in terms of the gametes. That would suggest some relative difference in survival or multiplication during the adulthood of the organism.

Iannaccone: That may be a phenomenon that is demonstrated more easily in cell populations that are highly dependent on smaller stem cell groups for renewal. I suspect that it is not the case in the liver.

In red blood cell populations, we see exactly the same thing. If we follow the red blood cell populations over time in these animals, we can see that there are major shifts in the proportion of red blood cells. In the liver, that does not occur. Whether it is occurring in the adrenal gland or not, we don't know.

Pedersen: Going back to the hepatocytes, how do you get out of the argument that the small patch size is generated by the mixing of a larger clone?

Iannaccone: I'm not trying to get out of it. All I'm saying is that it isn't necessary. What we did was attempt to create a two-dimensional mosaic pattern without that as a requirement. That does not mean, in any way, that it is not occurring; all it means is that it does not have to occur to come up with a finely variegated pattern.

Pedersen: The reason I bring it up is the consistent observation (Lawson et al. 1986) that cells labeled within the embryonic population—whether embryonic ectoderm, mesoderm, or endoderm—are noncoherent in their expansion; they do not stay close together; they are often widely separated, especially along the trunk. But cells in the extraembryonic regions do tend to be more coherent. Other than your data, I don't know any that directly address the coherence in later phases of the organ itself. It would be interesting to ask when the mixing that goes on during gastrulation stops.

Iannaccone: Yes, there is no question that the mixing occurs in gastrulation, and I cannot argue that it absolutely does not occur during organogenesis. But, at least in the case of the liver, I would argue that it is not absolutely required. I would say that it begins when the anlage gets established as an endodermal bud.

At that point, there is very rapid proliferation of hepatoblasic elements to form a mass of cells, which is then invaded by the vascular structures; these structures eventually are going to impose the adult organization on the organ. If you watch that very carefully through serial sections, you see no evidence of cordlike division in that hepatoblastic mass. This would support the argument that all that is required for the first phases of the formation of this organ is very rapid division of those cells.

There are also some very vital fetal functions in rodent livers that are completely different from the adult functions of that organ. It is definitely in the best interests of the developing fetus to have a functional liver in a very short period of time.

Pedersen: In the case of the adrenal gland, you don't have to argue about it because you see large coherent patches. You could not get them if you had a lot of mixing.

Iannaccone: You are absolutely right. The point is that we are looking in the adult organ, so although this may not point directly at the way the organ forms as a primordial bud, it certainly is a requirement that something is holding those cells in a precise pattern.

Hierarchical Interactions among Pattern-forming Genes in *Drosophila*

CHRISTINE RUSHLOW, KATHERINE HARDING, AND MICHAEL LEVINE
Department of Biological Sciences
Fairchild Center
Columbia University
New York, New York 10027

OVERVIEW

The specification of morphologically diverse body segments in *Drosophila* involves the activities of pair-rule genes and homeotic genes. Pair-rule genes divide the embryo into a repeating series of homologous segment primordia. Homeotic genes establish the diverse pathways by which each embryonic segment develops a distinct adult phenotype. Each of the pair-rule and homeotic genes that has been examined shows a unique pattern of expression during early embryonic development. A central problem in the control of morphogenesis in *Drosophila* is how each of these genes comes to be expressed in specific regions of the developing embryo. There is evidence that the expression of a given pair-rule or homeotic gene can influence the expression patterns of others. In this paper, we describe possible hierarchical relationships governing the regulated expression of several pair-rule and homeotic genes.

INTRODUCTION

In *Drosophila*, embryonic cells select different pathways of morphogenesis based on their physical locations within the developing embryo. Many of the genes that specify this positional information have been identified. Elaboration of positional identity along the anterior-posterior and dorsal-ventral embryonic body axes involves early zygotic gene functions that are expressed in response to maternal cues present in the unfertilized egg. Zygotic genes that are required for the specification of positional identity along the anterior-posterior body axis have been described in detail and include the pair-rule genes (Nüsslein-Volhard and Wieschaus 1980) and homeotic genes (Lewis 1978; Kaufman et al. 1980). Pair-rule genes divide the embryo into a repeating series of homologous segment primordia, whereas homeotic genes establish the diverse pathways by which each embryonic segment develops a distinct adult phenotype. Mutations in any of the nine known pair-rule genes

Banbury Report 26: Developmental Toxicology: Mechanisms and Risk

usually result in embryos that lack pattern elements in alternating segments (Jürgens et al. 1984; Nüsslein-Volhard et al. 1984, 1985; Wieschaus et al. 1984). Mutations in homeotic genes do not alter segment number or polarity but, instead, result in the partial or complete transformation of one segment into the homologous tissues of another.

The expression patterns of pair-rule genes and homeotic genes are stringently regulated. The tissue distributions of transcripts encoded by many of these genes have been determined previously by in situ hybridization (Akam 1983; Levine et al. 1983; Hafen et al. 1984a; McGinnis et al. 1984b; Fjose et al. 1985; Harding et al. 1985, 1986; Ingham et al. 1985a; Kornberg et al. 1985; Kilchherr et al. 1986). Each of the five pair-rule and six homeotic genes that has been examined displays a unique pattern of expression. Disruptions of the embryonic segmentation pattern are often associated with altered distributions of products encoded by one or more pair-rule or homeotic genes (Hafen et al. 1984b; Struhl and White 1985; Wedeen et al. 1986). A central problem in the control of positional information is how these different genes come to be expressed in specific regions along the body axis of the developing embryo.

The mechanisms responsible for the selective patterns of homeotic gene expression have been examined in detail. Many homeotic loci reside within one of two gene clusters in the *Drosophila* genome, the bithorax complex (BX-C) (Lewis 1978; Sanchez-Herrero et al. 1985) or the Antennapedia complex (ANT-C) (Kaufman et al. 1980). Each of the six known ANT-C and BX-C homeotic lethal complementation groups is expressed in a discrete region along the anterior-posterior body axis of the developing embryo, as shown in Figure 1. There is evidence that selective patterns of homeotic gene expression involve hierarchical cross-regulatory interactions among these genes. In general, homeotic genes expressed in posterior regions of the embryo (such as *abd-A* and *Abd-B*) negatively regulate homeotic genes expressed in more anterior regions (such as *Antp* and *Scr*) (Struhl and White 1985; Wedeen et al. 1986). The molecular basis for these interactions is not known; however, it has been shown that each of the six ANT-C/BX-C homeotic loci contains a similar 180-bp protein-coding region designated the homeo box (McGinnis et al. 1984a; Scott and Weiner 1984; Regulski et al. 1985). Homeo box protein domains appear to contain sequence-specific DNA-binding activities. It has been proposed that cross-regulatory interactions among homeo-box-containing homeotic genes occur at the level of transcription and are mediated by the homeo box protein domains encoded by these genes (Desplans et al. 1985; Harding et al. 1985). In addition to homeotic genes, several pair-rule genes also contain homeo boxes, as shown in Table 1. Recent studies suggest that pair-rule gene expression involves cross-regulatory interactions similar to those described for homeotic genes (Carroll and Scott 1986; Harding et al. 1986; Howard and Ingham 1986).

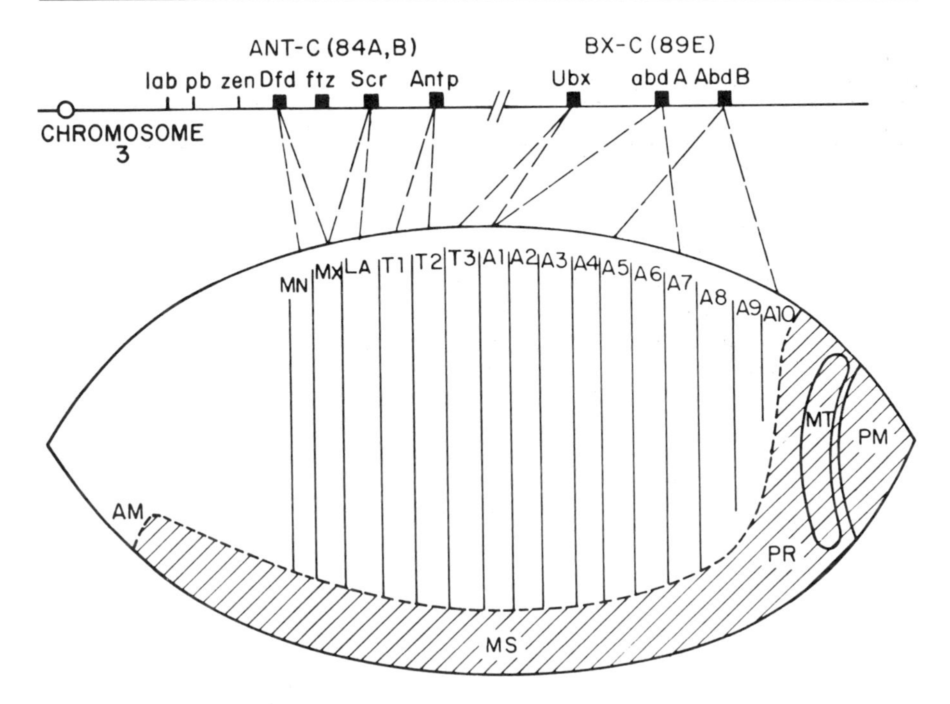

Figure 1

Sites of ANT-C and BX-C homeotic gene expression in an embryonic fate map. The fate map shows the origins of various embryonic tissues at the blastoderm stage of development. The dashed lines show the primary domains of expression for each of the six known ANT-C and BX-C homeotic genes. For example, the primary site of *Antp* expression corresponds to the pro- (T1) and mesothoracic (T2) segment primordia. The top line represents a map of the ANT-C and BX-C. There is a colinear correspondence between the order of the homeotic genes on the right arm of the third chromosome and their domains of expression along the anterior-posterior body axis. The fate map is oriented so that dorsal is up and anterior is to the left. (MS) Mesoderm; (MT) malpighian tubules; (PM) posterior midgut; (PR) proctodeum; (MN) mandibular head segment; (MX) maxillary segment; (LA) labium; (T1–T3) pro-, meso-, and metathorax; (A1–A10) abdominal segments.

Here we describe possible hierarchical relationships for cross-regulatory interactions among pair-rule genes and homeotic genes. In particular, we show that the transcript distribution pattern of the pair-rule gene, *even-skipped* (*eve*), is altered in embryos that are mutant for another pair-rule gene, *hairy* (*h*). There appears to be a reciprocal requirement for eve^+-gene activity in the establishment of the wild-type *h* expression pattern since *h* transcripts show an altered distribution pattern in eve^- embryos. Finally, it is possible that the initiation of homeotic gene expression involves one or more pair-rule genes since transcripts encoded by the homeotic gene *Deformed* (*Dfd*) show an altered distribution pattern in eve^- embryos.

Table 1
Homeo Box Genes in *Drosophila*

Identity of gene	Reference
I. Anterior-posterior genes	
A. Homeotic	
1. *Deformed (Dfd)*	McGinnis et al. (1984b)
2. *Sex combs reduced (Scr)*	Harding et al. (1985); Kuroiwa et al. (1985)
3. *Antennapedia (Antp)*	Garber et al. (1983); Scott et al. (1983)
4. *Ultrabithorax (Ubx)*	Bender et al. (1983)
5. *abdominal-A (abd-A)*	McGinnis et al. (1984b); Karch et al. (1985)
6. *Abdominal-B (Abd-B)*	Karch et al. (1985); Regulski et al. (1985)
B. Possible homeotics	
1. caudal (S67)	Mlodzik et al. (1985); Hoey et al. (1986)
2. F90-2	Hoey et al. (1986)
C. Pair-rule genes	
1. *fushi tarazu (ftz)*	Kuroiwa et al. (1984); Weiner et al. (1984)
2. *engrailed (en)*	Fjose et al. (1985); Poole et al. (1985)
3. *invected (in)*	Poole et al. (1985)
4. *even-skipped (eve)*	Harding et al. (1986); Macdonald et al. (1986); Frasch et al. (1987)
II. Dorsal-ventral genes	
1. z1 (*zerknullt?*)	Doyle et al. (1986)
2. z2 (*zerknullt?*)	C. Rushlow et al. (unpubl.)

RESULTS

There are at least 20 homeo box genes in *Drosophila* (W. McGinnis, pers. comm.). Thus far, 14 of these genes have been isolated and characterized (Table 1). Twelve of the genes are involved in the specification of positional information along the anterior-posterior embryonic axis and include eight homeotic genes, as well as four pair-rule genes. The two remaining genes (z1 and z2) appear to be involved in the differentiation of the dorsal-ventral pattern. Among these homeo box genes, *eve* appears to be particularly important for the overall spatial organization of the *Drosophila* embryo since *eve*$^{-}$ embryos lack all segmental subdivisions in the middle-body region (Nüsslein-Volhard et al. 1985) (Fig. 2). In a previous paper, we have shown that the expression patterns of two other pair-rule genes, *engrailed* (*en*) and *fushi tarazu* (*ftz*), are altered in *eve*$^{-}$ embryos (Harding et al. 1986). Thus, the *eve*$^{-}$ phenotype appears to result not only from the absence of *eve*$^{+}$ products but also as an indirect consequence of altered expression patterns of

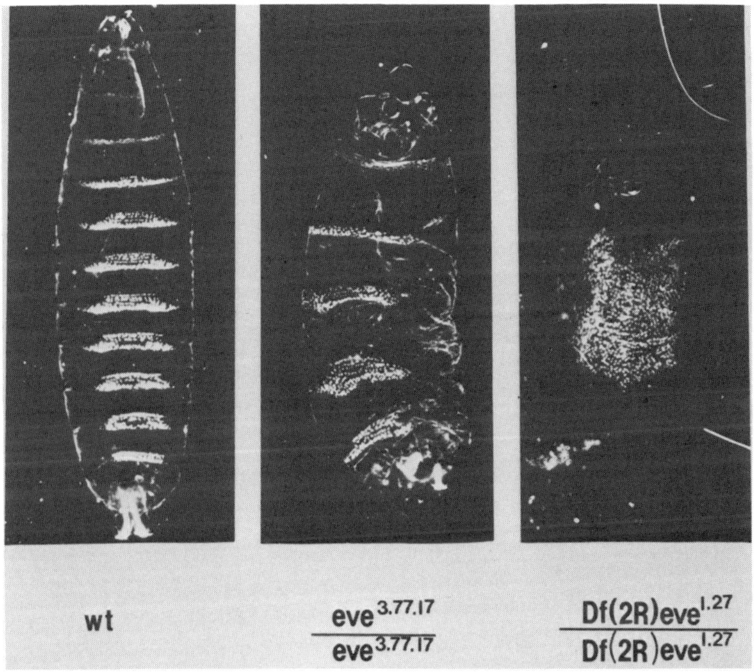

Figure 2

Cuticular phenotypes of wild-type and *eve* mutant embryos. All embryos are oriented so that anterior is up and the ventral surface is displayed. (wt) Wild-type embryo. The denticle belts associated with the anterior portion of each of the three thoracic and eight abdominal segments can be seen. ($eve^{3.77.17}$) An *eve* mutant embryo that displays the "weak" *eve* phenotype. The denticle belts of T2 (T2A) are fused with naked cuticle of T3 (T3P), thereby resulting in a composite T2A-T3P segment. Similar fusions in the abdominal region result in the following composite segments: A1A-A2P, A3A-A4P, A5A-A6P, A7A-A8P. (Df[2R]$eve^{1.27}$) This deficiency uncovers the *eve* locus. Embryos homozygous for the deletion show no overt segmentation of the middle-body region; an uninterrupted lawn of denticle hairs over the thorax and the abdomen can be seen.

other pair-rule genes. These results suggest that *eve* plays a key role in a hierarchy of cross-regulatory interactions among pair-rule genes. To determine whether *eve* influences the expression of other pattern-forming genes, we have localized transcripts encoded by the pair-rule gene *h* and the homeotic gene *Dfd* in tissue sections of eve^- embryos.

Interactions between *eve* and *h*

In wild-type embryos, *h* transcripts show a periodic distribution pattern, similar to those observed for *ftz* and *eve* (Ingham et al. 1985a). Hybridization of an *h* probe to wild-type embryo tissue sections reveals seven stripes of labeling in the middle-body region, as well as a patch of expression near the anterior pole (Fig. 3a,b). In contrast, only six strong stripes of *h* expression are detected in the middle-body region of an *eve*$^-$ embryo of similar age (Fig. 3c,d). In this embryo, the nuclei that comprise the second stripe of expression show substantially reduced levels of hybridization with the *h* probe. Examination of younger *eve*$^-$ embryos suggests that this reduction results from the improper initiation of the *h* expression pattern from approximately 70% to 60% egg length (where 100% corresponds to the distance from the posterior pole to the anterior pole). There is a delay in the initiation of the first expression stripe, and the second stripe never reaches a wild-type level of abundance. Thus, *eve*$^+$ activity appears to be required for the initiation of only the first two of the seven middle-body stripes of *h* expression. The region of *eve*$^-$ embryos, where the *h* pattern is altered, also shows the most severe disruption of the *ftz* expression pattern (Carroll and Scott 1986; Harding et al. 1986).

In a previous study, we have shown that *eve* influences the pattern of *ftz* and *en* expression but that neither of these genes is required for the initiation of the normal *eve* expression pattern (Harding et al. 1986). To determine whether *h*-gene function is required for *eve* expression, we have localized *eve* transcripts within *h*-mutant embryos. The mutant strain used for this analysis (h^{K1}) (Ingham et al. 1985b) does not appear to be a null mutation since an *h* hybridization probe detects a patch of *h* expression near the anterior pole of embryos homozygous for this allele (Fig. 3e; cf. a,b). However, none of the middle-body stripes of *h* expression are observed in these mutants, suggesting that this strain is effectively "null" for the middle-body region, which is where *eve* is normally expressed (from 69% to 19% egg length). Hybridization of an *eve* cDNA probe to a cleavage stage 14 h^{K1} embryo reveals a wild-type pattern of expression (Fig. 3f). However, during subsequent periods of development, h^{K1} homozygotes show abnormally low levels of *eve* expression. Deviation from the wild-type pattern is first observed in gastrulating h^{K1} embryos (Fig. 3g,h) and becomes progressively more pronounced by the onset of germ band elongation. These results suggest that h^+ activity is required for the maintenance of the *eve* expression pattern but not for its initiation.

Interactions between *eve* and *Dfd*

As discussed previously, the maintenance of selective patterns of homeotic gene expression appears to involve cross-regulatory interactions, whereby posteriorly expressed genes negatively regulate those expressed in more

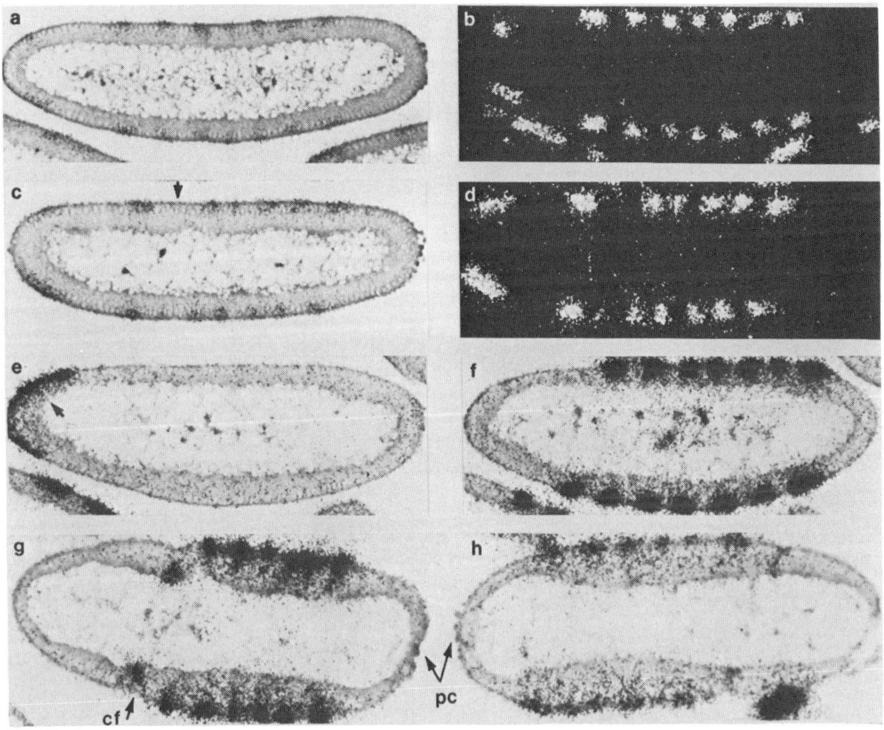

Figure 3

Localization of *h* transcripts in eve^- embryos and *eve* transcripts in h^- embryos. All tissue sections are oriented so that anterior is to the left, with the exception of *h*, where anterior is to the right. The embryos shown in *a–e* were hybridized with an *h* probe, and those shown in *f–h* were hybridized with an *eve* cDNA probe. (*a*) Horizontal section of a wild-type embryo just before cellularization. This embryo was obtained from Df(2R)$eve^{1.27}$ heterozygous adults. Seven strips of *h* expression are observed in the middle-body region. In addition, there is a patch of labeling near the anterior pole. (*b*) Dark-field photomicrograph of the embryo shown in *a*. (*c*) Horizontal section of an eve^- (Df[2R]$eve^{1.27}$ homozygote) embryo just before cellularization. Only six strong middle-body stripes of labeling are detected. The arrow indicates a gap where the second middle-body stripe is normally found. (*d*) Dark-field photomicrograph of the embryo shown in *c*. (*e*) Sagittal section of an h^{K1} homozygote just before cellularization. This embryo was hybridized with an *h* cDNA probe; specific hybridization signals are detected near the anterior pole (arrow). None of the middle-body stripes of hybridization are observed (cf. *a* and *b*, above). (*f*) Serial section through the same h^{K1} embryo shown in *e* after hybridization with the *eve* cDNA probe. A normal expression pattern of seven zebra stripes is observed. Horizontal sections of wild-type (*g*) and h^{K1}/h^{K1} (*h*) embryos from h^{K1} heterozygous adults. Both embryos are at the same stage of development (gastrulation) and occur in the same photographic field. Substantially stronger *eve* hybridization signals are detected in the wild-type embryo, as compared with the h^{K1} embryo. Paraffin tissue sections were prepared as described by Ingham et al. (1985a) and Harding et al. (1986). Single-stranded ^{35}S-labeled RNA probes were used for all hybridizations and were prepared as described by Ingham et al. (1985a).

anterior regions (see Fig. 1). It is unlikely that such interactions are responsible for the initiation of selective patterns of homeotic gene expression during early development. For example, the homeotic gene *Antennapedia* (*Antp*) shows a normal pattern of expression in young embryos that lack all genes of the BX-C (Harding et al. 1985). However, during subsequent stages of BX-C$^-$ development, a sharp alteration of the normal *Antp* expression pattern is observed. To determine whether pair-rule genes might be involved in the initiation of homeotic gene expression, we have localized transcripts encoded by the *Dfd* locus in young *eve*$^-$ embryos.

Dfd is required for the correct morphogenesis of several posterior head segments, including the mandible and maxilla (Hazelrigg and Kaufman 1983; W. McGinnis, pers. comm.). In wild-type embryos, *Dfd* transcripts are first detected during cleavage stage 14 ($\sim$15–20 min prior to the onset of gastrulation), within the presumptive mandibular and maxillary head segments (Fig. 4b). Expression persists in this region throughout embryonic and larval development (e.g., see Fig. 4d,f). Several alterations of this wild-type pattern of expression are observed in *eve*$^-$ embryos. First, the initiation of *Dfd* expression is abnormal such that fewer cells hybridize with the *Dfd* cDNA probe, and those that hybridize show substantially reduced levels of expression (Fig. 4a). Second, *Dfd* expression is not initiated within the mesoderm during gastrulation (Fig. 4c). And third, at more advanced stages of embryogenesis, *Dfd* transcripts are restricted to a narrower portion of the germ band, as compared with wild-type transcripts (Fig. 4e; cf. f). These results suggest that *eve*$^+$ activity is required for the correct initiation of *Dfd* expression.

DISCUSSION

A tentative hierarchy for possible interactions among pair-rule and homeotic genes is shown in Figure 5. In this scheme, pair-rule genes such as *h*, *eve,* and *ftz* affect the embryonic segmentation pattern by regulating, either directly or indirectly, the expression of *en* and several homeotic genes. Thus, mutations in pair-rule genes can lead to altered patterns of *en* and/or homeotic gene expression, which, in turn, affect the wild-type segmentation pattern.

It has been shown that *en* is expressed in the posterior compartment of each segment, where it is required for the initiation or maintenance of segment boundaries (Kornberg 1981). Embryos homozygous for strong *en*-mutant alleles show fusions of adjacent segments. *en* is not expressed in the middle-body region of *eve*$^-$ embryos (Harding et al. 1986). Thus, the absence of segmental subdivisions in *eve*$^-$ (see Fig. 2) might result, at least in part, from the lack of *en* products. We propose that *ftz* and *eve* affect the embryonic segmentation pattern indirectly by regulating the expression of *en*. In this

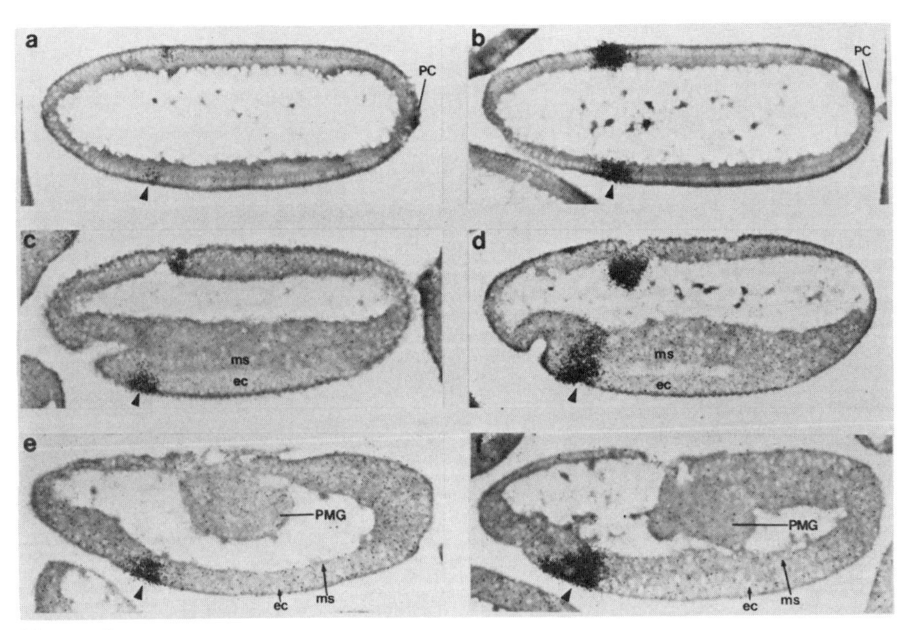

Figure 4

Distribution of *Dfd* transcripts in wild-type and *eve*$^-$ embryos. All tissue sections are oriented so that anterior is to the left. (*a,b*) Horizontal sections; (*c–f*) sagittal sections. (*a*) Cellular blastoderm stage *eve*$^-$ embryo after hybridization with a *Dfd* cDNA probe. Weak autoradiographic signals are detected at ~69% egg length (arrowhead). (*b*) Wild-type embryo at the same stage of development as that shown in *a*. Strong hybridization signals are observed over the presumptive cephalic furrow (~69% egg length; arrowhead). The tissue autoradiograms shown in *a* and *b* were simultaneously hybridized with the same *Dfd* probe and exposed for the same length of time. (*c,d*) Gastrulating embryos after hybridization with the *Dfd* cDNA probe. (*c*) An *eve*$^-$ embryo; (*d*) a wild-type embryo. The wild-type embryo (*d*) shows stronger overall hybridization signals, as compared with *c*. In addition, the *eve*$^-$ embryo (*c*) does not show hybridization above background in the mesoderm. (*e,f*) Embryos undergoing germ band elongation (~4 1/2–5 hr after fertilization). (*e*) An *eve*$^-$ embryo; (*f*) wild-type embryo. The wild-type embryo shows a broader distribution of *Dfd* transcripts than does the *eve*$^-$ embryo. (ec) Ectoderm; (ms) mesoderm; (PC) pole cells; (PMG) posterior midgut.

model, *eve* and *ftz* products positively regulate *en* expression in alternating regions along the anterior-posterior body axis of gastrulating embryos; *eve* is required for the odd-numbered *en* stripes, whereas *ftz* regulates the even-numbered stripes. In *ftz*$^-$ embryos, the even-numbered stripes are absent, whereas in *eve*$^-$ embryos, both the even-numbered and odd-numbered stripes are absent. The total absence of *en* products in *eve*$^-$ embryos might result from the combined lack of *eve*$^+$ products and the premature decay of the *ftz* expression pattern that has been observed in *eve*$^-$ (Harding et al. 1986).

If *eve* and *ftz* are important positive regulators of *en* expression, other

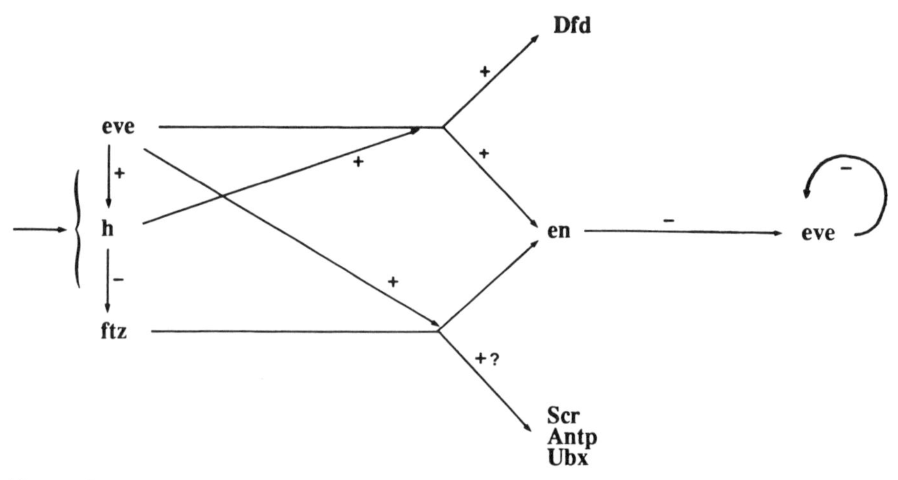

Figure 5

Summary of interactions among pair-rule and homeotic genes. The earliest interactions that are observed involve *eve, h*, and *ftz*. eve^+ activity is required for the correct initiation of the *h* expression pattern, whereas h^+ activity is required for the evolution, but not initiation, of the *ftz* pattern. During later stages of development, h^+ activity is required for the maintenance of the *eve* expression pattern, whereas the maintenance of the *ftz* expression pattern involves eve^+ activity. After cellularization, *en* is expressed as a series of 14 stripes. Both *eve* and *ftz* are required for the establishment of this *en* expression pattern (see Discussion). The absence of *en* transcripts in eve^- might result from the combined effects of removing eve^+ products and altering the *ftz* expression pattern. In wild type, eve^+ products might be required for initiating the odd-numbered *en* expression stripes, whereas *ftz* might be required for the even-numbered *en* stripes. In addition, *eve* might initiate expression of those homeotic genes, such as *Dfd*, that are expressed in odd-numbered parasegments. ftz^+ products might initiate homeotic gene expression in even-numbered parasegments. The *eve* expression pattern gradually diminishes in wild-type embryos undergoing germ band elongation. This decline appears to involve negative regulation by both *eve* and *en* products. It is important to note that the mechanisms responsible for the interactions shown in this diagram are not known. Moreover, it is possible that some or all of these interactions are indirect.

pair-rule genes might then affect segmentation by regulating the expression of *eve* or *ftz*. The results presented in this paper provide an explanation for the h^- pair-rule phenotype within the context of this model. As shown in Figure 3, the *eve* expression pattern prematurely decays in h^- embryos. This premature termination might result in the failure to initiate the odd-numbered *en* stripes of expression. Thus, only one-half the normal number of *en* stripes are established in h^- embryos, which results in the deletion of alternating segment boundaries.

ANT-C and BX-C homeotic gene transcripts are first detected during cleavage stage 14, just prior to cellularization. This corresponds to the time when *ftz* and *eve* proteins reach high steady-state levels in embryonic nuclei.

In this paper, we have shown that the initiation of *Dfd* expression is altered in eve$^-$ embryos. It is possible that this alteration of the *Dfd* expression pattern results directly from the absence of *eve*$^+$ activity since many of the embryonic cells that normally express *Dfd* are contained within the anterior-most band of *eve* expression (K. Harding and M. Levine, unpubl.). The early expression patterns of other homeotic genes, including *Scr*, *Antp*, and *Ubx*, are not obviously altered in *eve*$^-$ (K. Harding and C. Wedeen, unpubl.). However, the domains of *Scr*, *Antp*, and *Ubx* expression do not appear to coincide with any of the seven transverse stripes of *eve* expression. Instead, the expression patterns of these homeotic genes appear to coincide with *ftz*. Recent evidence suggests that the initiation of *Scr*, *Antp*, and *Ubx* expression is altered in *ftz*$^-$ embryos (Ingham and Martinez-Arias 1986). Alternatively, the abnormal *Dfd* expression pattern that is observed in *eve*$^-$ embryos might not result directly from the absence of *eve*$^+$ products. The wild-type domain of *Dfd* expression shows altered activities of several genes in *eve*$^-$ embryos, including *ftz* and *h*, as well as *Dfd*. Thus, it is possible that the absence of *eve*$^+$ activity in this anterior region of the embryo disrupts the localization of a different positional cue which, in turn, results in altered expression of a number of segmentation genes.

The molecular basis for the interactions shown in Figure 5 is not known; however, it is possible that some are mediated by the homeo box. The protein domain encoded by the *en* homeo box (the *en* homeo domain) includes a sequence-specific DNA-binding activity (Laughon and Scott 1984; Desplans et al. 1985). The *en* homeo domain binds to regions upstream from its 5′ terminus and possibly to upstream regions of other homeo-box-containing genes. Moreover, a full-length bacterial *eve* protein has been shown to bind to the same 5′ regions of *en* (T. Hoey and M. Levine, unpubl.). It is therefore possible that homeo box proteins function as specific transcription factors for the regulated expression of pattern-forming genes in *Drosophila*.

ACKNOWLEDGMENTS

We thank Bill McGinnis for sharing unpublished results. This work was funded by grants from the National Institutes of Health, the Searle Scholars Program, and the Alfred P. Sloan Foundation.

REFERENCES

Akam, M. 1983. The location of *Ultrabithorax* transcripts in *Drosophila* tissue sections. *EMBO J.* **2:** 2037.

Bender, W., M. Akam, F. Karch, P.A. Beachy, M. Pfeifer, P. Spierer, E.B. Lewis, and D.S. Hogness. 1983. Molecular genetics of the bithorax complex in *Drosophila melanogaster. Science* **221:** 23.

Carroll, S.B. and M.P. Scott. 1986. Zygotically active genes that affect the spatial expression of the *fushi tarazu* segmentation gene during early *Drosophila* embryogenesis. *Cell* **45:** 113.

Desplans, C., J. Theis, and P.H. O'Farrell. 1985. The *Drosophila* segmentation gene, *engrailed*, encodes a sequence-specific DNA binding activity. *Nature* **318:** 630.

Doyle, H.J., K. Harding, T. Hoey, and M. Levine. 1986. Transcripts encoded by a homeo box gene are restricted to dorsal tissues of *Drosophila* embryos. *Nature* **323:** 76.

Fjose, A., W.J. McGinnis, and W.J. Gehring. 1985. Isolation of a homeo box-containing gene from the *engrailed* region of *Drosophila* and the spatial distribution of its transcripts. *Nature* **313:** 284.

Frasch, M., T. Hoey, C. Rushlow, H. Doyle, and M. Levine. 1987. Characterization and localization of the *even-skipped* protein of *Drosophila. EMBO J.* **6:** 749.

Garber, R.L., A. Kuroiwa, and W.J. Gehring. 1983. Genomic and cDNA clones of the homeotic locus *Antennapedia* in *Drosophila. EMBO J.* **2:** 2027.

Hafen, E., A. Kuroiwa, and W.J. Gehring. 1984a. Spatial distribution of transcripts from the segmentation gene *fushi tarazu* during *Drosophila* embryonic development. *Cell* **37:** 833.

Hafen, E., M. Levine, and W.J. Gehring. 1984b. Regulation of *Antennapedia* transcript distribution by the bithorax complex in *Drosophila. Nature* **307:** 287.

Harding, K., C. Wedeen, W. McGinnis, and M. Levine. 1985. Spatially regulated expression of homeotic genes in *Drosophila. Science* **229:** 1236.

Harding, K., C. Rushlow, H.J. Doyle, T. Hoey, and M. Levine. 1986. Cross-regulatory interactions among pair-rule genes in *Drosophila. Science* **233:** 953.

Hazelrigg, T. and T.C. Kaufman. 1983. Revertants of dominant mutations associated with the *Antennapedia* gene complex of *Drosophila*: Cytology and genetics. *Genetics* **105:** 581.

Hoey, T., H. Doyle, K. Harding, C. Wedeen, and M. Levine. 1986. Homeo box gene expression in anterior and posterior regions of the *Drosophila* embryo. *Proc. Natl. Acad. Sci.* **83:** 4809.

Howard, K.R. and P.W. Ingham. 1986. Regulatory interactions between the segmentation genes *fushi tarazu, hairy,* and *engrailed* in the *Drosophila* blastoderm. *Cell* **44:** 949.

Ingham, P.W. and A. Martinez-Arias. 1986. The correct activation of *Antennapedia* and bithorax complex genes requires the *fushi tarazu* gene. *Nature* **324:** 592.

Ingham, P.W., K.R. Howard, and D. Ish-Horowicz. 1985a. Transcription pattern of the *Drosophila* segmentation gene *hairy*. *Nature* **318:** 439.

Ingham, P.W., S.M. Pinchin, K.R. Howard, and D. Ish-Horowicz. 1985b. Genetic analysis of the *hairy* locus in *Drosophila melanogaster. Genetics* **111:** 463.

Jürgens, G., E. Wieschaus, C. Nüsslein-Volhard, and H. Kluding. 1984. Mutations affecting the pattern of the larval cuticle in *Drosophila melanogaster*. II. Zygotic loci on the third chromosome. *Wilhelm Roux's Arch. Dev. Biol.* **193:** 283.

Karch, F., B. Weiffenbach, M. Pfeifer, W. Bender, I. Duncan, S. Celniker, M. Crosby, and E.B. Lewis. 1985. The abdominal region of the bithorax complex. *Cell* **43:** 81.

Kaufman, T.C., R. Lewis, and B. Wakimoto. 1980. Cytogenetic analysis of chromosome 3 in *Drosophila melanogaster*: The homeotic gene complex in polytene chromosome interval 84A-B. *Genetics* **94:** 115.

Kilchherr, F., S. Baumgartner, D. Bopp, E. Frei, and M. Noll. 1986. Isolation of the *paired* gene of *Drosophila* and its spatial expression during embryogenesis. *Nature* **321:** 493.

Kornberg, T. 1981. Compartments in the abdomen of *Drosophila* and the role of the *engrailed* locus. *Dev. Biol.* **86:** 363.

Kornberg, T., I. Siden, P. O'Farrell, and M. Simon. 1985. The *engrailed* locus of *Drosophila*: In situ localization of transcripts reveals compartment-specific expression. *Cell* **40:** 45.

Kuroiwa, A., E. Hafen, and W.J. Gehring. 1984. Cloning and transcriptional analysis of the segmentation gene *fushi tarazu* of *Drosophila. Cell* **37:** 825.

Kuroiwa, A., U. Kloter, P. Baumgartner, and W.J. Gehring. 1985. Cloning of the homeotic *sex combs reduced* gene in *Drosophila* and *in situ* localization of its transcripts. *EMBO J.* **4:** 3757.

Laughon, A. and M.P. Scott. 1984. Sequence of a *Drosophila* segmentation gene: Protein structure homology with DNA-binding proteins. *Nature* **310:** 25.

Levine, M., E. Hafen, R.L. Garber, and W.J. Gehring. 1983. Spatial distribution of *Antennapedia* transcripts during *Drosophila* development. *EMBO J.* **2:** 2037.

Lewis, E.B. 1978. A gene complex controlling segmentation in *Drosophila*. *Nature* **276:** 565.

Macdonald, P.M., P. Ingham, and G. Struhl. 1986. Isolation, structure, and expression of *even-skipped:* A second pair-rule gene of *Drosophila* containing a homeo box. *Cell* **47:** 721.

McGinnis, W., R.L. Garber, J. Wirz, A. Kuroiwa, and W.J. Gehring. 1984a. A homologous protein-coding sequence in *Drosophila* homeotic genes and its conservation in other metazoans. *Cell* **37:** 403.

McGinnis, W., M.S. Levine, E. Hafen, A. Kuroiwa, and W.J. Gehring. 1984b. A conserved DNA sequence in homeotic genes of the *Drosophila* Antennapedia and bithorax complexes. *Nature* **308:** 428.

Mlodzik, M., A. Fjose, and W.J. Gehring. 1985. Isolation of *caudal,* a *Drosophila* homeo box-containing gene with maternal expression, whose transcripts form a concentration gradient at the pre-blastoderm stage. *EMBO J.* **4:** 2961.

Nüsslein-Volhard, C. and E. Wieschaus. 1980. Mutations affecting segment number and polarity in *Drosophila. Nature* **287:** 795.

Nüsslein-Volhard, C., H. Kluding, and G. Jürgens. 1985. Genes affecting the segmental subdivision of the *Drosophila* embryo. *Cold Spring Harbor Symp. Quant. Biol.* **50:** 145.

Nüsslein-Volhard, C., E. Wieschaus, and H. Kluding. 1984. Mutations affecting the pattern of the larval cuticle in *Drosophila melanogaster*. I. Zygotic loci on the second chromosome. *Wilhelm Roux's Arch. Dev. Biol.* **193:** 267.

Poole, S.J., L.M. Kauvar, B. Drees, and T. Kornberg. 1985. The *engrailed* locus of *Drosophila*: Structural analysis of an embryonic transcript. *Cell* **40:** 37.

Regulski, M., K. Harding, R. Kostriken, F. Karch, M. Levine, and W. McGinnis. 1985. Homeo box genes of the Antennapedia and bithorax complexes of *Drosophila. Cell* **43:** 71.

Sanchez-Herrero, E., I. Vernos, R. Marco, and G. Morata. 1985. Genetic organization of *Drosophila* bithorax complex. *Nature* **313:** 108.

Scott, M.P. and A.J. Weiner. 1984. Structural relationships among genes that control development: Sequence homology between the *Antennapedia*, *Ultrabithorax*, and *fushi tarazu* loci of *Drosophila*. *Proc. Natl. Acad. Sci.* **81:** 4115.

Scott, M.P., A.J. Weiner, T.I. Hazelrigg, B.A. Polisky, V. Pirotta, F. Scalenge, and T.C. Kaufman. 1983. The molecular organization of the *Antennapedia* locus of *Drosophila*. *Cell* **35:** 763.

Struhl, G. and R.A.H. White. 1985. Regulation of the *Ultrabithorax* gene of *Drosophila* by other bithorax complex genes. *Cell* **43:** 507.

Wedeen, C., K. Harding, and M. Levine. 1986. Spatial regulation of Antennapedia and bithorax gene expression by the *Polycomb* locus in *Drosophila*. *Cell* **44:** 739.

Weiner, A.J., M.P. Scott, and T.C. Kaufman. 1984. A molecular analysis of *fushi tarazu*, a gene in *Drosophila melanogaster* that encodes a product affecting embryonic segment number and cell fate. *Cell* **37:** 843.

Wieschaus, E., C. Nüsslein-Volhard, and G. Jürgens. 1984. Mutations affecting the pattern of the larval cuticle in *Drosophila melanogaster*. III. Zygotic loci on the X chromosome and fourth chromosome. *Wilhelm Roux's Arch. Dev. Biol.* **193:** 296.

COMMENTS

Pratt: I would like to extend the last point that you made and ask if homeo-box probes have been used to localize mammalian gene transcripts in tissue sections.

Levine: Yes. This work has been done primarily by Frank Ruddle's and Bill McGinnis' groups at Yale. They find that several homeo box gene transcripts show high concentrations in the spinal cord. Moreover, different homeo box genes appear to show the strongest expression in different regions of the spinal cord.

Scandalios: On one slide you showed the one gene affecting *engrailed* which, in turn, affected another gene. What happens if there is something in an *engrailed* null mutation?

Levine: I showed you that *engrailed* appears to be positively regulated by *even-striped* (*eve*) and that *eve* might be negatively regulated by *engrailed*. If you look at an *engrailed* embryo at relatively late stages of development, *eve* shows 14 stripes of expression (at times when the gene is normally off).

Pedersen: I have a question about this segmental expression and the restriction, in division 14, to a segmental expression. What could possibly be constraining its distribution, as the protein should be able to bind to any nucleus but it is binding only in a segmental pattern; does this mean, therefore, that the specificity of binding lies somewhere in the prepared nucleus?

Levine: Both *eve* and *ftz* expression appear at about the same time, at cleavage stage 10 or 11. At this time, transcripts encoded by both genes are distributed broadly throughout the embryo. During a 30-minute interval of cleavage stage-14 development, there is a selective loss of transcripts in specific regions of the embryo.

If *ftz* and *eve* expression are indeed initiated at the same time, how is it that *eve* influences *ftz* expression and there is no reciprocal requirement of *ftz* activity for *eve* expression? It is possible that this involves a translational control mechanism. We note that soon after the appearance of the *eve* RNA, we detect *eve* protein, but there is at least a 30-minute lag between the first appearance of *ftz* transcript protein. It is therefore possible that *eve* functions earlier than *ftz* which, in turn, allows *eve* to influence *ftz* expression.

Pedersen: The net result is still that they give this patterning, and so there must be some sort of underlying predisposition toward it.

Levine: As best as we can tell, the problem of localized patterns of homeo box gene expression involves de novo transcription. Recent data from Tom Kornberg's and Gerald Schubiger's laboratories suggest that the half-life of the *ftz* transcript is on the order of 10–20 minutes. Thus, the evolution of the *ftz* expression pattern from a general distribution of transcripts to a localized "zebra" pattern might primarily involve de novo transcription.

Kochhar: How much is known about their DNA-binding properties. Are there other peptides from the homeo box that bind to DNA?

Levine: Evidence that homeo box proteins contain sequence-specific DNA-binding activities is rapidly accumulating. For example, a full-length protein, *Ultrabithorax*, has been obtained from *Drosophila* tissue culture cells in Dave Hogness' lab at Stanford. This protein has been shown to contain a sequence-specific DNA-binding activity. However, in this case, the binding activity has not been attributed directly to the homeo box. In the case of *lacZ* fusion proteins, such as those prepared by Pat O'Farrell and Claude Desplan for *engrailed,* it was possible to rapidly establish that the *engrailed* homeo domain is responsible for the DNA-binding activity. However, one must be cautious in interpreting results obtained with fusion proteins. For example, we might be missing some of the nuances of the in vivo binding specificity by using such fusion proteins.

Kochhar: So I presume that we do not yet know what is the least length of the involved peptide sequence.

Levine: No. That kind of data will come fast and furiously from various laboratories; indeed, in vitro mutagenesis experiments are under way. However, it has been proposed that the binding activity within the homeo domain resides in the carboxyterminal third of the domain due to homologies and similarities with known DNA-binding proteins such as yeast α_1/α_2 mating-type proteins, as well as various prokaryotic DNA-binding proteins.

Markert: Considering the nature of positional information, is it absolute in the sense of specifying latitude or longitude, or is it in relationship to the adjacent cells?

Levine: You do not specify the location of a bristle on the wing blade of an adult fly at the cellular blastoderm stage of development. There is some plasticity in the system. There are instances where critical genes that control development show so-called nonautonomy, that is, a mutant patch of tissue in a somatic mosaic can be rescued by surrounding wild-type cells that are presumably secreting a diffusible factor. At the blastoderm stage, the kinds of restrictions of developmental potential you are talking about are basically at the level of the segment.

Bournias-Vardiabasis: It is known that you can produce phenocopies of *bithorax* mutants. Are any of those kinds of experiments being done for *ftz* and *eve*?

Levine: No. This is a very interesting point. You can get very efficient phenocopies of both the *bithorax* and *Antennapedia* mutant phenotypes in a variety of ways, for example, ether treatment and heat shock. This is my one attempt to make this relevant to the current meeting. In particular, *Ubx* phenocopies can be obtained efficiently by such treatments during early embryonic stages. No one really knows why. There has been some speculation on that subject. We imagine that those treatments are somehow resulting in misexpression of homeotic genes in the wrong place, which mimics the basis for the mutations.

Cell Adhesion in the Regulation of Pattern and Structure during Feather Induction

GERALD M. EDELMAN AND WARREN J. GALLIN
Department of Developmental and Molecular Biology
The Rockefeller University
New York, New York 10021

OVERVIEW

Cooperative interactions among the cells of a tissue and the inductive interactions between tissues are pivotal factors in normal development. The cell-adhesion molecules (CAMs) that mediate the binding between cells in a tissue enable those cells to form multicellular complexes that have developmental properties different from a group of isolated cells. The importance of CAM function in forming collectives and borders is exemplified by recent discoveries on the development of feathers in embryonic chicken skin. When cell–cell interactions in the epidermis are disrupted in culture by antibodies to L-CAM, an adhesion molecule that links epidermal cells to each other, an abnormal morphological and histological pattern develops. Instead of forming hexagonal arrays of feather germs, stripes are formed. This perturbation results from a changed pattern of cell accumulation and alteration in the pattern of cell division in the dermis, which does not express L-CAM. In long-term cultures, scalelike structures, rather than normal filamentous structures, are formed in the explanted skin. Thus, alteration of the cooperative interactions among cells of one tissue can alter the developmental pattern that is induced in another tissue.

INTRODUCTION

CAMs are cell-surface glycoproteins that mediate direct cell–cell linkage (Edelman 1983, 1986). The CAMs that have been characterized are widely distributed in the developing embryo (Thiery et al. 1982, 1984; Edelman et al. 1983) and are expressed in many different tissues in characteristic patterns that reflect the early formation of boundaries between mutually inducing cell collectives. Two of these CAMs, the liver CAM (L-CAM), and the neural CAM (N-CAM) appear early in development. They have been isolated, and their molecular structure and expression during development have been characterized extensively (Hoffman et al. 1982; Gallin et al. 1983). L-CAM and N-CAM do not bind to each other, and two adjoining populations of

Banbury Report 26: Developmental Toxicology: Mechanisms and Risk

cells, each expressing one of these CAMs, can form borders based on the different adhesive specificities.

Although analyses of cell sorting in aggregates of mixed cell types have been interpreted to imply that a large number of tissue-specific cell CAMs are responsible for development of organ and tissue specificity (Moscona 1974), molecular analysis of cell-adhesion mechanisms has revealed a completely different picture. CAMs in different tissues undergo a variety of cell-surface modulation events that alter their binding activity. The cell–cell binding mediated by CAMs is nonlinear, with respect to the concentration of CAM in the membrane (Hoffman and Edelman 1983), and small changes in the amount of CAM expressed can have large effects on the adhesive properties of a cell. In addition, N-CAM is structurally modified during development by variations in the amount of sialic acid bound to the molecule (Rothbard et al. 1982), which affect the binding properties of N-CAM, as well as by alternative exon splicing (Hemperly et al. 1986; Murray et al. 1986) to form N-CAMs from the same gene having different forms of association with the cell cortex and membrane. Thus, modulation of the structure and surface distribution of CAMs by cellular mechanisms, rather than the temporally and spatially limited expression of a very large number of tissue-specific CAMs, appears to play a central role in the development of organ and tissue structure.

Disruption of cell–cell contacts by antibodies to the CAMs can alter tissue morphology. For example, antibodies to N-CAM have been demonstrated to alter the normal development of the layered structure of the retina in vitro (Buskirk et al. 1980). Antibodies to L-CAM perturb the structure of histotypic liver cell aggregates in cell culture (Gallin et al. 1983). Conversely, perturbation of morphology can also affect the expression of CAM: when a muscle is denervated, the expression of N-CAM is increased markedly, and the molecule is dispersed on the plasma membrane of the myofibrils. When the muscle is reinnervated, the expression decreases and becomes restricted to the neuromuscular junction again (Covault and Sanes 1985; Rieger et al. 1985; Daniloff et al. 1986).

CAMS AND EMBRYONIC INDUCTION

The development of most organs relies on interactions between two different groups of cells leading to embryonic induction, a form of milieu-dependent differentiation. N-CAM and L-CAM are expressed during development in spatiotemporal patterns that are correlated with pairs of inducing tissues (Edelman et al. 1983; Crossin et al. 1985). These results suggest that different CAMs may function to delineate borders between different inductive cell groups by binding cells into one group or the other. It is also possible that these molecules facilitate communication among cells of a collective and that this is essential to the induction process.

The expression of CAMs in adjacent interacting tissue pairs during induction has been demonstrated in several systems, the best characterized being the developing feather of the embryonic chicken. This is an excellent system for the study of inductive interactions: Transplantation (Cairns and Saunders 1954; Saunders and Gasseling 1957; Mauger 1972), tissue recombination (Rawles 1963), and genetic (Goetinck and Abbott 1963; Sengel and Abbott 1963) studies have demonstrated that the pattern of feather rudiments in the skin and the structural pattern of each individual feather depend on reciprocal interactions between the epidermis and dermis.

When feather rudiments are being formed and when the patterns of individual feathers are arising, the tissues that form each feather rudiment express either L-CAM or N-CAM (Chuong and Edelman 1985a). As the rudiment forms, L-CAM is uniformly present in the epidermis, whereas N-CAM is expressed by the dermal cells that are collecting to form the dermal condensation. Thus, the first inductive interactions that distinguish feather rudiments from the surrounding skin are coordinate with the formation of apposed cell collectives linked by two different CAMs. Similarly, when the feather has developed into a definitive filament, growing from the epidermal collar surrounding the dermal papilla, the collar expresses L-CAM while the papilla expresses N-CAM (Chuong and Edelman 1985a). Finally, as the epidermal cells that will form the feather are undergoing the morphogenetic and histogenetic processes that will form the highly keratinized quill and vanes of the feather, cells that are fated to form the barbs and barbules of the feather express L-CAM, whereas cells that are fated to die, leading to separation of the various branched structures of the feather, express N-CAM (Chuong and Edelman 1985b). The expression of CAMs is therefore not only correlated with morphogenetic processes but also with specific cytodifferentiation pathways.

The repeated correlation of the expression of different CAMs in adjacent cell collectives that undergo inductive interactions and then follow separate fates raises the possibility that different CAMs might play a role in connecting interactive cells within each collective. This would allow such collectives to function as coordinated multicellular units rather than as single cells, and such coordination might be crucial to signaling, leading to the induction of patterns during feather morphogenesis.

To test this hypothesis, antibodies prepared against L-CAM that inhibit L-CAM-mediated cell–cell adhesion in the epidermis were used to perturb the development of feathers in an in vitro organ culture system (Gallin et al. 1986). We exploited the fact that the developing feather germ consists of two tissues, the epidermis and the dermis, each of which expresses one CAM but not the other, resulting in a border between the two functionally and structurally different cell collectives. The experiments were designed to perturb the epidermal cell collective with antibodies to L-CAM and then search for changes of cell pattern in the N-CAM-positive dermal collective, as well as for changes in the fate of both tissues.

RESULTS

Dorsal skin from 6 1/2-day embryos (stage 29-30) was dissected and mounted dermis side down on a collagen-coated 8-μm Nuclepore filter. This organ culture was mounted on stainless-steel grids in standard organ culture dishes and maintained at the liquid-air interface using medium NCTC 135. After 3 days in culture with medium alone or medium containing nonimmune rabbit F_{ab}' fragments, the skin cultures developed a roughly hexagonal pattern of circular feather rudiments, similar in size and pattern to feather rudiments that develop in vivo (Fig. 1A). When 1-mg/ml F_{ab}' fragments of antibodies

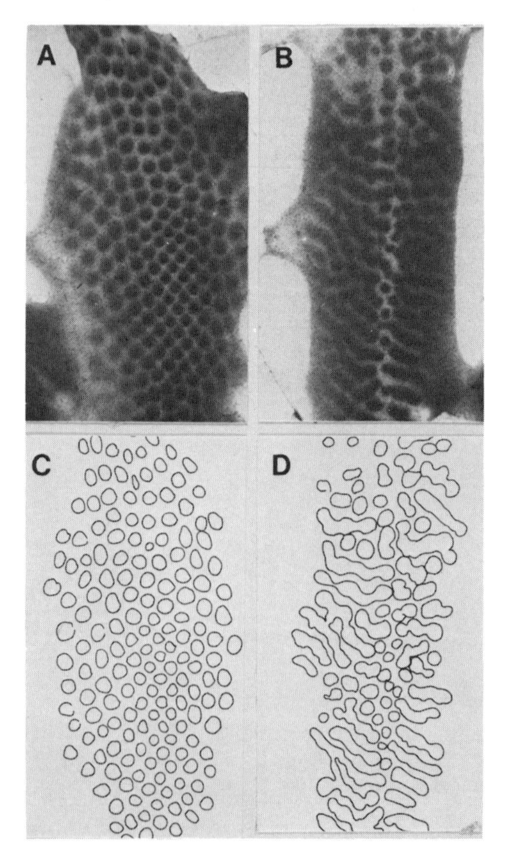

Figure 1

Whole mounts of cultured skin explants. Dorsal skin explants from 6 1/2- to 7-day chicken embryos were maintained in organ culture for 3 days and fixed and stained with borax carmine. (*A*) Explant maintained in 1 mg/ml nonimmune rabbit F_{ab}'. (*B*) Explant maintained in 1 mg/ml anti-L-CAM F_{ab}'. (*C*) Tracing of the pattern of condensations in *A*, showing the roughly hexagonal pattern of circular condensations. (*D*) Tracing of pattern in *B*, showing the meandering stripes of condensation.

against L-CAM were present in the culture medium, the organization of feather rudiments as observed in whole mounts of the explants was radically altered. Instead of forming a hexagonal array of circular feather rudiments, the skins developed meandering striped structures (Fig. 1B). The stripes were approximately as wide (200 μm) as the diameter of a normal condensation. Feather rudiments along the midline of the explant that had clearly formed when the skin was explanted were unaffected by the antibody treatment, indicating that this effect was due to changes in the propagation of the pattern, rather than degeneration of preformed structures.

Histological analysis demonstrated that the altered feather rudiment pattern was due to a striking change in the distribution of cells of the dermis. Sections were cut in the plane of the skin in order to determine the organization of the dermal tissue. In untreated explants, dermal cells formed tight symmetrical cellular condensations surrounded by a loose mesenchyme (Fig. 2A). The dermal cells in the antibody-treated cultures were distributed as stripes of high density surrounded by a mesenchyme of significantly higher cell density than that in unperturbed cultures (Fig. 2B). Moreover, after 3 days in culture, the pattern of mitotic activity in the dermis of perturbed and unperturbed cultures was radically different. Whereas the majority of ^{3}H[TdR]-labeled cells was found in the dermal condensations of 3-day cultures, the majority of labeled cells was in the extracondensation mesenchyme of the perturbed cultures (Gallin et al. 1986).

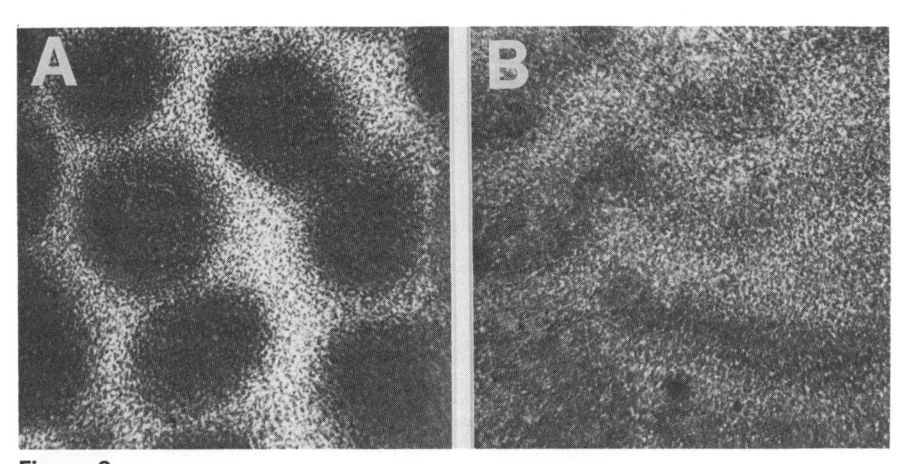

Figure 2

Sections (10 μm) of 3-day skin cultures, cut through the dermis in the plane of the skin and stained with hematoxylin and eosin. (*A*) Unperturbed culture with circular condensations of high density surrounded by loose mesenchyme. (*B*) Culture that was perturbed with 1 mg/ml anti-L-CAM F_{ab}'. The areas of high cell density are stripes, and the mesenchyme between condensations has a significantly higher cell density than that of the unperturbed case.

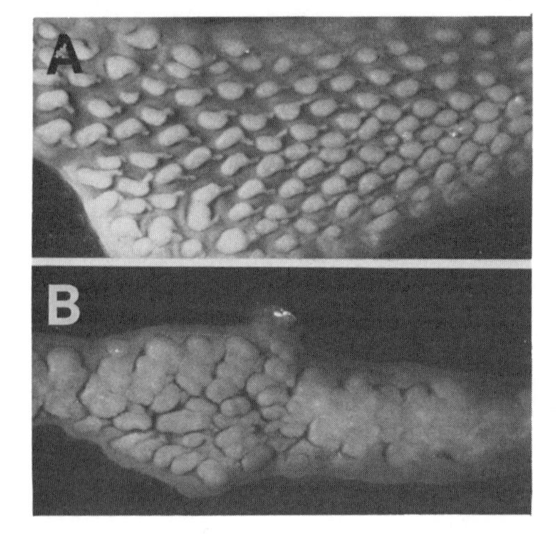

Figure 3

Whole mounts of 6 1/2- to 7-day skin maintained in culture for 10 days. (*A*) Explant maintained in medium alone for 10 days. (*B*) Explant maintained with 1 mg/ml anti-L-CAM F_{ab}' for 2 days, and in medium alone for a further 8 days.

If cultures were maintained with anti-L-CAM F_{ab}', as described above, and then cultured for a further 6–8 days with medium alone (a total of 10 days in culture), the structure of the appendages that develop from the rudiments was also altered. Explants maintained in medium alone for 10 days developed an evenly distributed field of distinct feather filaments (Fig. 3A), each arising from one of the circular rudiments that was observed after 3 days in culture. These structures are similar in appearance to feather filaments that develop in vivo in 9- to 10-day embryos. When the explants were treated with F_{ab}' for 2–4 days and then maintained with medium alone up to 10 days in culture, the appendages that form were either flattened "cobbelstone" structures or were fused scalelike stripes that developed from the striped condensations (Fig. 3B).

The unperturbed 10-day cultures maintained a two-layered epithelial epidermis over both the feather filaments and the surrounding skin (Fig. 4A). The 10-day cultures that were treated with anti-L-CAM for the first 2–4 days in culture formed a multilayered, squamous epithelium overlying the dermis. The dermis was thicker and denser than in the unperturbed culture, and it contained knots of condensed dermal cells that were absent from the unperturbed cultures.

DISCUSSION

The experiments described above were designed to assess the importance of CAM-mediated intercellular interactions within one tissue of an inductive pair of tissues in the development of patterned structures. L-CAM is ex-

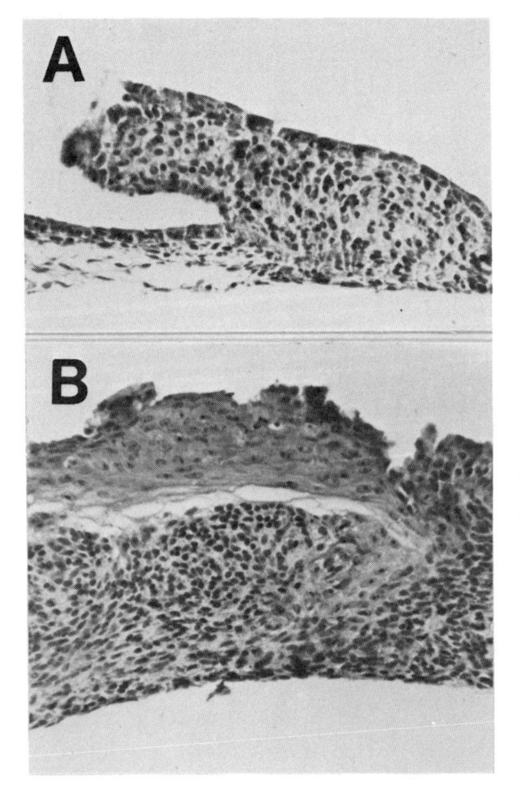

Figure 4
Transverse sections of 10-day cultures, 5 μm thick, stained with hematoxylin and eosin. (*A*) Unperturbed culture develops feather filaments, each with a core of dense dermis covered by cuboidal epidermal epithelium, in a field of loose mesenchymal dermis covered by cuboidal epidermal epithelium. (*B*) Explant perturbed with 1 mg/ml anti-L-CAM F_{ab}' for 4 days, and maintained with medium alone for 6 more days develops a thick multilayered squamous and a thick dense dermis.

pressed by epidermal, but not by dermal, cells (Chuong and Edelman 1985a), and antibodies to L-CAM inhibit epidermal cell aggregation, but not dermal cell aggregation (W.J. Gallin and G.M. Edelman, unpubl.). Nonetheless, when these antibodies are added to cultured embryonic chicken skin, the resulting morphological and histological changes after 3 days are observed in the dermis. The antibody treatment was not severe enough to cause destruction or loss of the epidermis; in fact, after 3 days, the only detectable epidermal change is a thickening of the periderm. The results were therefore not simply due to large-scale removal of the epidermis, which is known to halt normal feather rudiment development in the dermis.

The complexity of the reciprocal interactions between epidermis and dermis during the formation of skin appendages (Dhouailly and Sawyer 1984) suggests that a number of factors are vital for normal regulative development. The results on the effects of anti-L-CAM demonstrate that a major influence is the formation and maintenance of cell collectives linked by CAMs. A coordinated multicellular collective can act and integrate signals over areas that are orders of magnitude larger than single cells. In addition, intercellular coupling, depending initially on CAM linkages, may be a significant element in the way a collective integrates and transmits signals. Although the consequences of the L-CAM linkages for collective integration are unknown, it is significant that a simple computer model (Gallin et al. 1986), involving signaling between N-CAM- and L-CAM-linked collectives, yields hexagonal arrays. Alterations in L-CAM-mediated cooperativity in this model have been shown to result in the formation of striped patterns. Induction may therefore result from multistage signaling events modulated by CAM interactions among the cells in the collectives that are exchanging reciprocal stimulatory and inhibitory signals.

These observations emphasize the fact that morphogenesis is an epigenetic process: Genetically defined factors and cellular processes interact in complex cooperative ways that depend on the milieu and history of the developmental process. Defects in CAM structural genes, in control of CAM expression, or in historegulatory genes (Edelman 1984) can all cause defects in the formation of cell collectives or extracellular milieus that are essential for normal morphogenesis. This picture of dynamic, multilevel control suggests that the relationship between a specific teratological syndrome and the underlying genetic perturbation can be very indirect; conversely, there can be many genetic alterations that may cause changes in form and differentiation.

The analysis of the role of cell–cell interactions in forming functional cell collectives is at an early stage. Nonetheless, it is clear that CAMs can play an important role in the complex web of inductive interactions that lead to pattern. Studies of morphological and histological changes that occur when cell–cell interactions are perturbed can be used to define further the linkage between morphogenetic processes and the expression of differentiation products specified by historegulatory genes; specific keratin types in developing skin provide excellent examples of products under control of such genes. As our understanding of the nature of cell–cell interactions progresses, the signaling mechanisms within cell collectives and between cell collectives during induction will also become more amenable to analysis. The conclusions derived from such studies should have a considerable impact upon our understanding of the complex causes of teratological processes.

ACKNOWLEDGEMENTS

This work was supported by U.S. Public Health Service grants AM-04256 and HD-09635, and a Senator Jacob Javits Center of Excellence in Neuroscience Award (NS-22789).

REFERENCES

Buskirk, D.R., J.-P. Thiery, U. Rutishauser, and G.M. Edelman. 1980. Antibodies to neural cell adhesion molecules disrupt histogenesis in cultured chick retinae. *Nature* **285:** 488.

Cairns, J.M. and J.W. Saunders. 1954. The influence of embryonic mesoderm on the regional specificity of epidermal derivatives in the chick. *J. Exp. Zool.* **127:** 221.

Chuong, C.-M. and G.M. Edelman. 1985a. Expression of cell-adhesion molecules in embryonic induction. I. Morphogenesis of nestling feather. *J. Cell. Biol.* **101:** 1009.

———. 1985b. Expression of cell-adhesion molecules in embryonic induction. II. Morphogenesis of adult feathers. *J. Cell Biol.* **101:** 1027.

Covault, J. and J. Sanes. 1985. Neural cell adhesion molecule (N-CAM) accumulates in denervated and paralyzed skeletal muscles. *Proc. Natl. Acad. Sci.* **82:** 4544.

Crossin, K.L., C.-M. Chuong, and G.M. Edelman. 1985. Expression sequences of cell adhesion molecules. *Proc. Natl. Acad. Sci.* **82:** 6942.

Dhouailly, D. and R. Sawyer. 1984. Avian scale development. XI. Initial appearance of the dermal defect in scaleless skin. *Dev. Biol.* **105:** 343.

Daniloff, J.K., G. Levi, M. Grumet, F. Rieger, and G.M. Edelman. 1986. Altered expression of neuronal cell adhesion molecules induced by nerve injury and repair. *J. Cell Biol.* **103:** 929.

Edelman, G.M. 1983. Cell adhesion molecules. *Science* **219:** 450.

———. 1984. Cell adhesion and morphogenesis: The regulator hypothesis. *Proc. Natl. Acad. Sci.* **81:** 1460.

———. 1986. Cell adhesion molecules in the regulation of animal form and tissue pattern. *Annu. Rev. Cell Biol.* **2:** 81.

Edelman, G.M., W.J. Gallin, A. Delouvée, B.A. Cunningham, and J.-P. Thiery. 1983. Early epochal maps of two different cell adhesion molecules. *Proc. Natl. Acad. Sci.* **80:** 4384.

Gallin, W.J., G.M. Edelman, and B.A. Cunningham. 1983. Characterization of L-CAM, a major cell adhesion molecule from embryonic liver cells. *Proc. Natl. Acad. Sci.* **80:** 1038.

Gallin, W.J., C.-M. Chuong, L.H. Finkel, and G.M. Edelman. 1986. Antibodies to liver cell adhesion molecule perturb inductive interactions and alter feather pattern and structure. *Proc. Natl. Acad. Sci.* **83:** 8235.

Goetinck, P.F. and U.K. Abbott. 1963. Tissue interaction in the scaleless mutant and the use of scaleless as an ectodermal marker in studies of normal limb differentiation. *J. Exp. Zool.* **154:** 7.

Hemperly, J.J., G.M. Edelman, and B.A. Cunningham. 1986. cDNA clones of the neural cell adhesion molecule (N-CAM) lacking a membrane-spanning region with evidence for membrane attachment via a phosphatidylinosital intermediate. *Proc. Natl. Acad. Sci.* **83:** 9822.

Hoffman, S. and G.M. Edelman. 1983. Kinetics of homophilic binding by embryonic and adult forms of the neural cell adhesion molecule. *Proc. Natl. Acad. Sci.* **80:** 5762.

Hoffman, S., B.C. Sorkin, P.C. White, R. Brackenbury, R. Mailhammer, U. Rutishauser, B.A. Cunningham, and G.M. Edelman. 1982. Chemical characterization of a neural cell adhesion molecule purified from embryonic brain membranes. *J. Biol. Chem.* **257:** 7720.

Mauger, A. 1972. Rôle de mésoderme somitique dans le développment du plumage dorsal chez l'embryon de poulet. I. Origine, capacités de régulation et détermination du mésoderm plumigène. *J. Embryol. Exp. Morphol.* **28:** 313.

Moscona, A.A. 1974. Surface specificaton of embryonic cells: Lectin receptors, cell recognition, and specific cell ligands. In *The cell surface in development* (ed. A.A. Moscona), p. 67. Wiley, New York.

Murray, B.A., J.J. Hemperly, E.A. Prediger, G.M. Edelman, and B.A. Cunningham. 1986. Alternatively spliced mRNAs code for different polypeptide chains of the chicken neural cell adhesion molecule (N-CAM). *J. Cell. Biol.* **102:** 189.

Rawles, M.E. 1963. Tissue interactions in scale and feather development as studied in dermal-epidermal recombination. *J. Embryol. Exp. Morphol.* **11:** 765.

Rieger, F., M. Grumet, and G.M. Edelman. 1985. N-CAM at the vertebrate neuromuscular junction. *J. Cell Biol.* **101:** 285.

Rothbard, J.B., R. Brackenbury, B.A. Cunningham, and G.M. Edelman. 1982. Differences in the carbohydrate structures of neural cell-adhesion molecules from adult and embryonic chicken brains. *J. Biol. Chem.* **257:** 11064.

Saunders, J.W. and M.T. Gasseling. 1957. The origin of pattern and feather germ tract specificity. *J. Exp. Zool.* **135:** 503.

Sengel, P. and U.K. Abbott. 1963. In vitro studies with the scaleless mutant: Interactions during feather and scale development. *J. Hered.* **54:** 254.

Thiery, J.-P., J.-L. Duband, U. Rutishauser, and G.M. Edelman. 1982. Cell adhesion molecules in early chicken embryogenesis. *Proc. Natl. Acad. Sci.* **79:** 6737.

Thiery, J.-P., A. Delouvée, W.J. Gallin, B.A. Cunningham, and G.M. Edelman. 1984. Ontogenetic expression of cell adhesion molecules: L-CAM is found in epithelia derived from the three primary germ layers. *Dev. Biol.* **102:** 61.

COMMENTS

Saxen: To me there is no doubt that CAMs and maybe some still unknown molecules are involved. Yet, both in primary induction, which you referred to, and in the dermal and epidermal induction, borders are created where two unlike cells interact.

Edelman: Certainly.

Saxen: You refer to the signals here. Are you willing to state what kind of signals?

Edelman: I can tell you about one that we know of in vitro, but it is not one that I would bank on in any of these experiments. I have become convinced from the experiment on feather induction that although I am a great proponent of cell-surface modulation of a global type for communication, there is one definite signal for a CAM that has been discovered and that is for Ng-CAM. Ng-CAM is the neuron glia CAM, which links neurons to neurons and neurons to glia. It is cell specific in the nervous system and is not found elsewhere. If you take PC-12 cells as models and you add nerve growth factor, that factor will then turn on a great number of biochemical processes and morphological ones, among which is neurite extension. At the same time, one sees up-regulation of Ng-CAM without alteration of N-CAM levels.

We have extended this experiment to dorsal root ganglia in culture. What is seen in that case is a very interesting kind of differential prevalence modulation in which the N-CAM mechanism of side-to-side binding of neurites is immediately switched out by the up-regulation of Ng-CAM, which supervenes. As a result there is an Ng-CAM-modulated mechanism of side-to-side interaction.

Now, that does not mean that nerve growth factor is the actual signal in the real-world case, but it provides one with some feeling that one might regulate things in this way.

I thoroughly agree with you that the essence of this model is not to try to drive to one biochemical solution for morphology. It is to develop asymmetries in early development. Those asymmetries could be arrived at in a great number of convergent ways in evolution. I am asserting that this is a very good candidate for holding together the early boundaries and regulating the expression of genes. Of course, the key question that should be answered is the one about signals.

Pratt: You and others have published some very exciting results, showing the absence of N-CAM when neural crest cells begin to migrate, which I think has some implications for teratology, as well as for normal development.

Edelman: That experiment remains to be really nailed down, in my opinion. Jean-Paul and I are still trying very hard to do that. It is not that easy because of the unavailability of chemically workable amounts of cells, but the basic observation is that neural crest cells follow rule 1. They show at the top of the neural tube and really light up for N-CAM. Just as they begin to move, largely in pathways that seem conditioned by fibronectin, they down-regulate N-CAM and they only up-regulate again when they reach target sites. There is no doubt about that, although the mechanism is not clear at all. It comes back to the nature of the signal.

What we do see additionally is a new protein that seems to follow definite regulatory rules. This is an extracellular matrix protein called cytotactin, which we have discovered recently. It is expressed in three cephalocaudal waves during development, one of which happens to be in the neural crest pathway. That extends the idea of regulation of pattern by expression to the notion that one could also modulate choices involving the highways—not just the choices of how one organizes the bedroom communities.

I suspect that the way this story will be fleshed out is that one will see families of molecules for which various developmental constraints will apply. One example involving cytotactin is to take a cerebellar culture as I have pictured on the board—with the external granule layer and the radial glia going all the way down. This is the surface of the brain. The external granule cells send out side arms that will form parallel fibers in the subjoining molecular layer and then a leading process that interacts with the glia. These granule cells will end up in the internal granule layer after migrating in the glia. If anti-Ng-CAM is added in kinetic experiments, movement will be blocked right at the border of the molecular layer. If anticytotactin is added, which is made by glia and mediates neuron–glia interactions, the cells will all pile up in the molecular layer, but the entry into the molecular layer will not be affected. So borders are being formed as the result of two completely different gene products made by different cells: one, an extracellular matrix protein, and one, a CAM, each with obviously different kinetics of regulation. That suggests what we are up against in trying to analyze the situation.

Nonetheless, it is very exciting to think that one might be able to put the genetic framework and the epigenetic framework together. I would humbly submit that to solve problems of teratology, in most cases, at least you must have that framework. Perhaps this is presumptuous of me because I assume that is why you are having this conference.

Lammer: To expand on Dr. Pratt's remarks, it appears that embryonic mice exposed to retinoic acid prior to the onset of cranial neural crest cell migration have fields of crest cells that lay piled up along the neural tube. They lie plastered together along the neural tube and they presumably do not initiate migration at all (Webster et al. 1986). Perhaps this occurs because the cells did not go through that process of turnover of the CAM molecules that is necessary during crest cell migration. We wondered whether this teratogenic effect could potentially be used to study the control over turnover of CAMs during this phase of embryogenesis.

Edelman: Yes, that is a very good tool. Kathryn Crossin is trying to put that together in my laboratory. I would think that staining with cytotactin would be revelatory in the extreme.

Markert: You spoke at length about the specificity of gene function.

Wouldn't simple changes in the quantity of gene function account for a great deal?

Edelman: Thank you. I am very grateful that you brought this up because I rushed over the point. The fact is, to have bang-bang control is not necessarily needed for gene function.

One case where we know CAM expression is controlled transcriptionally is in a cerebellar tissue line infected with a temperature-sensitive mutant of Rous sarcoma virus. When we down-regulate, we can show that cells lose their CAM and migrate all around; they express nonneuronal *src*-gene product substrates. We can up-regulate to the nonpermissive temperature and get the expression of N-CAM again.

You refer to the fact that you really just may have to change amounts. There is a physicochemical experiment indicating that that would be a very likely way to control things. One can make synthetic lipid vesicles and insert purified N-CAM in the E form or in the A form into them. The formation of superthreshold aggregates of vesicles under defined conditions of shear follows second-order kinetics. If one has exactly the same lipid-to-protein ratio of the vesicles in the A form, one has four times the rate of aggregation seen with the E form. Now, if the form is kept the same and the amount is doubled on the surface of these artificial vesicles, the binding rate goes up 33-fold; in other words, one has a fifth-order dependence upon surface concentration. When the experiment is performed on vesicles made directly from the cerebellum in an exactly parallel mode, the result is the same. That means the graded gene response and the ensuing change in expression at the cell surface could give radical border effects.

Synthesis of Stress Proteins during Normal and Stressed Development of Mouse Embryos

KYU SEONG KIM,* YONG KYU KIM,* FREDERICK NAFTOLIN,[†] AND CLEMENT LAWRENCE MARKERT*[‡]

Yale University Medical School
Department of Obstetrics and Gynecology[†]
and Department of Biology*
New Haven Connecticut 06511

OVERVIEW

Virtually all organisms are equipped with genetic mechanisms for coping with stressful environmental challenges. Faced with acute environmental stresses, organisms reprogram their cellular genomes to enhance the synthesis of stress proteins and to diminish transcription and translation of most proteins present at the time of stress initiation. The genes for the stress proteins are among the most highly conserved of all genes and are clearly fundamental to the genetic makeup of organisms. Moreover, stressful stimuli, including heat, generate teratogenic development in mammals and other vertebrates unless countered by the function of stress proteins. Here, we report the identification and behavior of several stress proteins during the development of mouse embryos. The results show that the capacity to respond to stress develops gradually during the preimplantation stages of mouse development and that in later embryos, the pattern of synthesis of stress proteins is specific for each tissue at each stage of its development. Thus, there is a fine adjustment of the stress response to meet the needs of each type of embryonic and adult cell.

INTRODUCTION

Changing patterns of gene function in response to heat shock and other forms of stress have now been studied extensively in many organisms. Chromosomal responses to stress were first noted in the induction of specific puffs on *Drosophila* chromosomes after brief treatment of larvae at supraoptimal temperatures (Ritossa 1962). Later, the heat shock response in protein synthesis was examined in a wide variety of organisms from bacteria to

[‡]Present address: Department of Animal Science, North Carolina State University, Raleigh, North Carolina 27695-7621.

mammals. Patterns of protein synthesis reflect species specificities, but the heat shock proteins (hsps) show basic similarities, and homologies among hsp genes indicate a remarkable conservation during evolution. Most organisms synthesize hsps with molecular weights ($\times$ 10^3) in the region of 80K–90K and 65K–70K, and most synthesize one or more in the 15K–30K range (Schlesinger et al. 1982a,b). The synthesis of these proteins can be readily induced by physical and chemical stresses such as treatment with arsenite (Bensaude and Morange 1983), ethanol (Dyban and Khoshai 1980), heavy metal ions (Levinson et al. 1980), and ecdysterone (Ireland et al. 1982) and also by mechanically induced stress, such as restricting the aorta (Hammond et al. 1982), slicing and culturing tissue in vitro (Hightower and White 1981; Currie and White 1983), and, most dramatically, by heat shock. Such agents may damage the cell directly or indirectly and may generate teratological development (Webster et al. 1984; Mirkes 1985).

In *Drosophila*, the heat stress brings about developmental delay and/or abnormalities, depending on the developmental stage of the embryo at the time of treatment (Bergh and Arking 1984), and, furthermore, may cause disintegration of nuclei and of cytoplasmic islands, displacement and swelling of nuclei, and blocked mitosis (Grazios et al. 1983). The ubiquity of stress proteins strongly suggests a protective adaptive function in the cell, but the role of these proteins in the structure and function of the cell is still largely unknown.

The synthesis of specific proteins in response to heat has been described for various organs at several developmental stages of *Drosophila* embryos. In *Drosophila,* the hsps (83K, 73K, 68K) were synthesized at all developmental stages, but there was little or no synthesis of hsps 70K and 68K in unfertilized eggs (Grazios et al. 1983). The most apparent quantitative change in hsp synthesis was observed at the beginning of the blastoderm stage in the 70K protein, but no change in the amount of the 68K hsp (Bergh and Arking 1984) was observed. Moreover, 70K mRNA encoding the 70K hsp appears as a heat-inducible mRNA for the first time at gastrulation in *Xenopus*, as identified by cDNA clones for the corresponding mRNA (Bienz 1984). During mouse embryonic development, the 70K hsp is first observed clearly at the morula-to-blastocyst stage of development (Wittig et al. 1983). Although the 70K and 89K hsps were strongly induced by heat at the eight-cell stage, they were abundantly synthesized in the blastocyst without heat stress (Morange et al. 1984). When the heat shock response was analyzed after in vitro differentiation triggered by retinoic acid, the 68K and 105K hsps were also synthesized just as in the blastocyst (Bensaude and Morange 1983). Even in the early two-cell embryo, hsps 68K and 70K are evident, and their synthesis is amanitin-sensitive (Bensaude et al. 1983).

We undertook the present study to investigate the synthetic patterns of gene expression after heat shock in early developmental stages from the one-cell to the blastocyst stage of the mouse, as well as in 15-day embryos and in several different tissues from these 15-day embryos.

MATERIALS AND METHODS

Superovulation

Mice of the CD-1 strain, from Charles River, were used. They were housed under a constant light/dark regimen (dark from 9:00 P.M. to 8:00 A.M.). Adult females from 7 weeks of age can be superovulated by administering 5 IU of PMSG (pregnant mare's serum gonadotropin) in 0.2 ml of normal saline by intraperitoneal injection at the metestrous stage of the estrous cycle, followed by an injection of 5 IU of HCG (human chorionic gonadotropin) 48 hours later (Thadani 1982). Injections were usually given at 5:00 P.M. One female was placed in a cage with a male. The presence of a vaginal plug on the following morning indicated successful mating, and that day was designated day 0.

Preparation of Culture Medium

Mouse embryos were cultured in a medium that was modified from Whitten (1971). Chemicals sufficient for making 100 ml of medium, but without BSA, were weighed out individually and pooled in a Pyrex bottle. Then 100 ml of water (double-distilled in glass) was added, and a gas mixture composed of 90% N_2, 5% O_2, and 5% CO_2 was bubbled through the medium to saturation. Phenol red served as a pH indicator. Subsequently, 0.3 g of bovine serum albumin was added to the medium and allowed to dissolve without agitation. Finally, the medium was sterilized by filtration under positive pressure, using the triple gas mixture, through a 0.45-μ Nalgene filter unit, and stored at 4°C in sterile tissue-culture flasks.

Embryo Collection

The oviducts of pregnant mice were dissected out and placed in medium in a petri dish. The ampullas were then ruptured with fine forceps, and the cumulus cells surrounding the egg mass were dispersed by treatment with 0.1% hyaluronidase (300 U/mg) dissolved in culture medium. Two-cell, four-cell, eight-cell, morula, and blastocyst embryos were flushed from dissected oviducts under the microscope.

Cell Labeling

Twenty embryos from each of the early development stages chosen were transferred into 50 μl of medium at 37°C, containing [^{35}S]methionine at a concentration of 1 mCi/ml. Fifteen-day-old embryos and various organs were isolated, finely minced with scissors, and labeled with [^{35}S]methionine; exposure to the labeled methionine was 3 hours in culture.

Heat Stress Treatment

Cells were subjected to heat stress at 42°C in a water bath for 20 minutes. The volume of culture medium during treatment was brought to 0.5 ml.

Extraction of Protein

After 3 hours labeling, the cells were rinsed three times to remove unincorporated [^{35}S]methionine. For the rinse, a phosphate-buffered saline solution containing 0.5% PVP (polyvinylpyrolidone) was used to facilitate the transfer of tissue. Washed cells were transferred into petri dishes, thoroughly macerated, and centrifuged. The supernatant was mixed with the lysis buffer of O'Farrell (1975).

One- and Two-dimensional Electrophoresis

Visualization of heat stress proteins was achieved by using the methods of O'Farrell (1975). For isoelectric focusing, the upper electrode reservoir contained 0.01 M NaOH and 0.005 M $Ca(OH)_2$. The lower electrode reservoir contained 0.01 M H_3PO_4. After the gels were prerun, the samples were loaded and run for 18 hours at 400 V and for 1 hour at 800 V. After the end of the run, the gels were removed from the tubes and equilibrated in a solution containing 40% (vol/vol) glycerol, 12.5% (vol/vol) β-mercaptoethanol, and 6.25% (wt/vol) $NaDodSO_4$. The gels from the first dimension were then placed on a polyacrylamide gel slab for electrophoresis in the second dimension. The gels were run for 4 hours at a constant current of 2 mÅ/cm. During the electrophoresis, the temperature was maintained at 4°C by means of a circulating water pump. At the completion of electrophoresis, the gels were fixed for a period of 1 hour to overnight with a solution containing 10% trichloroacetic acid (TCA), 10% acetic acid, and 30% methanol. The gels were then immersed in an autoradiographic enhancer solution and processed for autoradiography.

Densitometric Scan of hsps

Densitometric scans of autoradiographs of [^{35}S]methionine-labeled control proteins and hsps from several tissues of the 15-day embryos were performed on a Beckman DU-8 spectrophotometer. The regions of the autoradiograph showing 68K, 70K, 89K, and 105K hsps were cut out, and absorption was measured at 550 nm and compared with the corresponding control region. The total absorption of each area was normalized with reference to a control spot and used as a measure of the synthesis of the protein in that area.

RESULTS

The inducibility and relative abundance of hsps synthesized at developmental stages from the one-cell to the blastocyst stage were determined by two-dimensional electrophoresis. As shown in Figure 1, without heat stress, the 70K protein was synthesized at a higher level at all developmental stages than the 68K hsp. The level of these 70K hsps gradually increased with development even in the absence of heat stress. When subjected to heat stress, the level of 70K was not noticeably increased until the blastocyst stage when its synthesis was increased dramatically by heat treatment. In contrast, the 68K hsp was scarcely evident before the blastocyst stage, when it too became inducible by heat shock (Fig. 1).

The fact that heat stress inducibility appears first at the blastocyst stage demonstrated the developmental dependence of this stress program of gene activation and raised the possibility that stage and tissue-specific patterns of hsp responses to stress would characterize normal mammalian development. We selected the 15-day embryo as sufficiently developed to serve as a model. As shown in Figure 2, [^{35}S]methionine labeling of 15-day embryos showed high incorporation of radioactivity into 68K hsps. In addition, the 89K and 105K hsps were more abundant in embryos after heat treatment of the pregnant mouse at 15 days gestation. Of course, the level of the 70K hsp was also considerably increased as compared to levels of the untreated embryos.

To examine the tissue specificity of response to heat shock, individual organs of the 15-day embryo heat stressed in utero were carefully dissected and labeled for 3 hours in vitro. The proteins from brain, liver, skeletal muscle, eye, kidney, lung, heart, and intestine were then extracted and separated by two-dimensional electrophoresis. The radioactive proteins were visualized by autoradiography (Fig. 3). The 70K hsp was increased in every tissue, as compared with the control (Table 1). We found that two (68A and 68B) or three (68A, 68B, and 68C) electrophoretically distinct forms of the 68K protein were induced by heat stress in a tissue-specific pattern. For instance, hsp 68A and 68B synthesis was detected in eye, kidney, lung, heart, and intestine after heat stress. In general, the 68B was synthesized at the highest level, and detectable amounts of 68B were present in the absence of experimentally applied heat stress. All three forms of 68K (68A, 68B, and

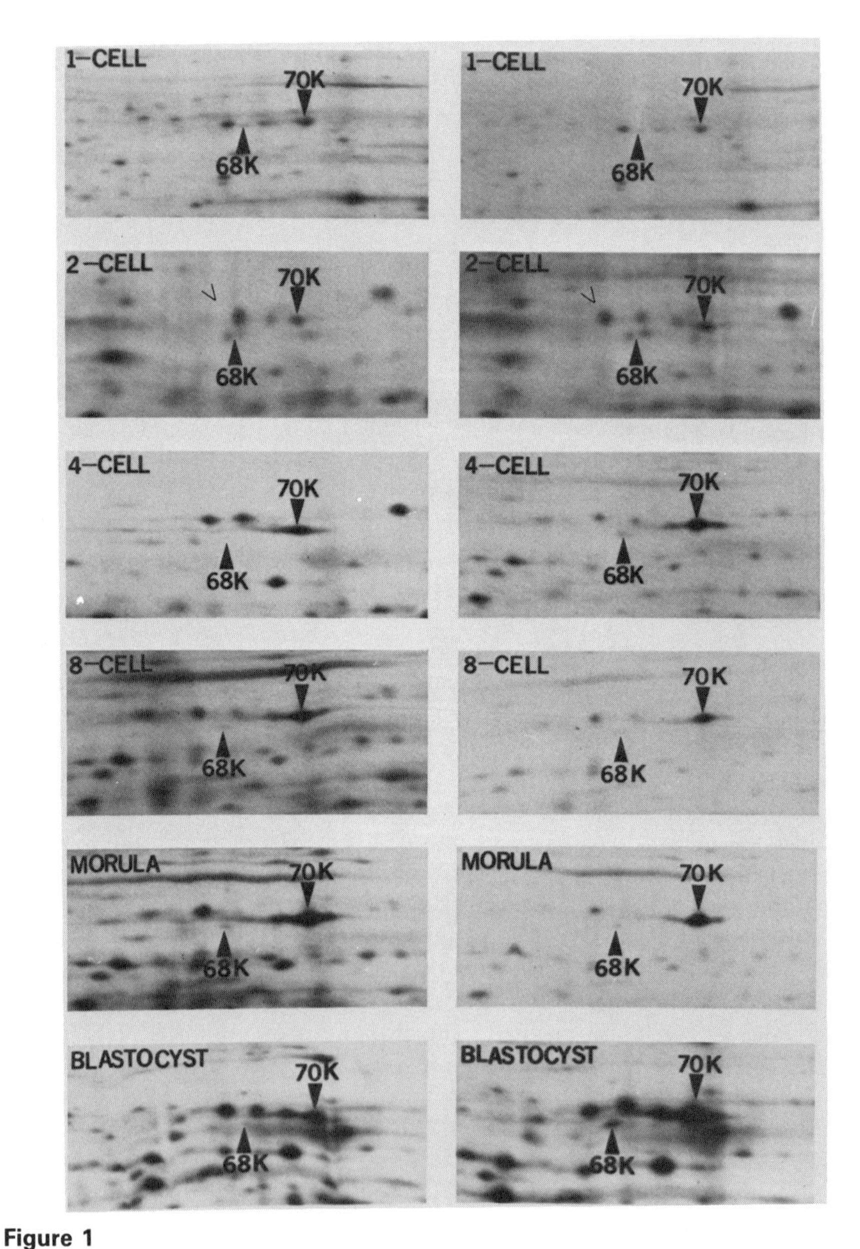

Figure 1

Patterns of stress proteins during early mouse development. Developing in culture through the "two-cell block" is apparently stressful; however, hsps are not readily induced until the blastocyst stage. At the two-cell stage, a new protein (>) does appear after heat shock. The 68K and 70K hsps have been identified to serve as guides to the total patterns. Note that both these hsps are conspicuously increased by heat shock at the blastocyst stage. (*Left*) Control proteins; (*right*) hsps.

68C) were also detectable in heart tissue without the application of heat stress. The 89K hsps were increased in every tissue but not in brain when compared with the control. Heat stress also induced the synthesis of two distinct 105K hsps in every tissue.

DISCUSSION

The work of many laboratories has now demonstrated that the hsps are a family of normal ubiquitous proteins in virtually all organisms. Many of these proteins show large degrees of homology across a wide spectrum of species throughout the evolutionary sequence. The synthesis of these proteins is greatly enhanced in response to a wide variety of stresses and, at the same time, normal proteins are greatly and abruptly diminished in synthesis during the stress. The variety of effective stresses makes it difficult to see any common denominator. Nevertheless, many stresses do produce anoxia in the cell and diminish the available energy for carrying on metabolic activities. During normal embryonic development, metabolic demands on cells fluctuate greatly as, for example, during rapid cell division, migration of cells, and metabolic differentiation. Perhaps these normal demands upon cells fall into the category of stressful requirements and lead to the production of some of the stress proteins. The pattern of the two-cell stage after heat stress can also be generated by cultivating the mouse egg in vitro from the one-cell to the two-cell stage (K.S. Kim et al., unpubl.). In any event, we found some of the stress proteins present at all stages of embryonic development, except the earliest cleavage stages, but their abundance was considerably increased under heat stress. We have also found that experimental stresses that produce congenital abnormalities are also associated with increased formation of stress proteins in the abnormal conceptuses (K.S. Kim et al., unpubl.).

Of particular significance is the fact that the pattern of response to stress in terms of synthesizing stress proteins is cell- and tissue-specific and also specific for each stage of development. This observation implies that the requirements of different cells for recovery or defense, when confronted with stress, are different and may depend upon the particular stage of differentiation of the cell. The different stress proteins may therefore be fulfilling specific functions within the cell, but the nature of these functions is quite unknown. New mRNA has been shown repeatedly to be synthesized on heat shock and, thus, new proteins are synthesized; but some previously present proteins may be post-translationally modified by the heat shock. The fact that new patterns of gene function are induced by stress and that translation of mRNA in already extant polysomes is sharply reduced shows that the responses are truly fundamental and characteristic of virtually all cells at all stages of development. Just as different cells have different patterns of gene function, so do the patterns of gene function involving the synthesis of stress proteins

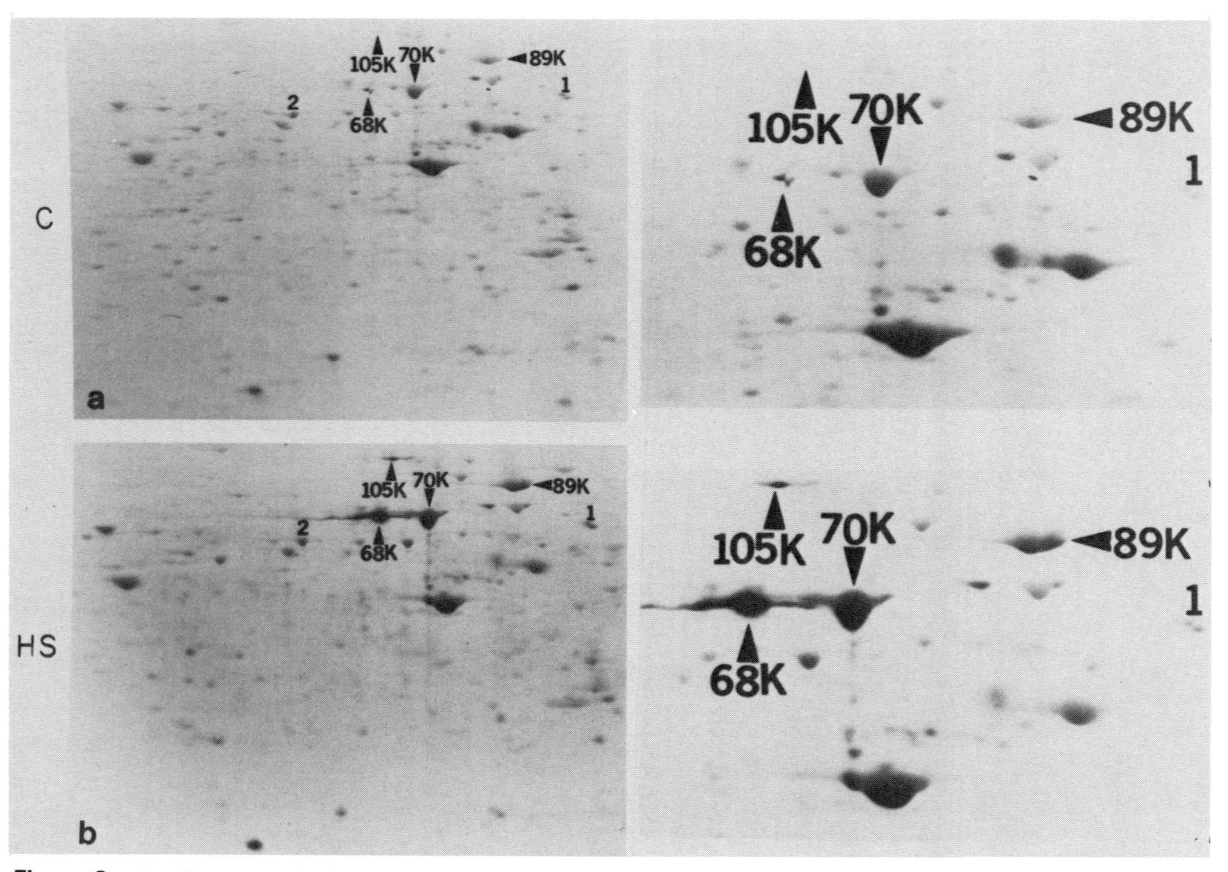

Figure 2 *(See facing page for legend.)*

Table 1
Densitometric Measurements of Four hsp Spots on Autoradiographs of Four Tissues

		M	68 K	70 K	89 K	105 K
Brain	C	1.340	0.525	1.860	3.550	0.071
	E	0.954	1.040	3.110	2.730	0.320
	C′		0.390	1.390	2.640	0.053
	E′		1.090	3.260	2.860	0.330
	R		2.790	2.340	1.080	6.340
Liver	C	1.710	0.048	0.565	0.586	0.025
	E	1.960	3.250	0.714	3.230	0.157
	C′		0.028	0.330	0.343	0.015
	E′		1.650	0.363	1.640	0.080
	R		58.600	1.100	4.990	5.340
Muscle	C	2.580	0.168	0.086	0.854	0.010
	E	2.690	6.210	3.000	6.760	0.162
	C′		0.065	0.330	0.331	0.004
	E′		2.310	1.120	2.510	0.061
	R		35.400	3.350	7.570	15.200
Heart	C	5.910	3.970	3.160	8.060	0.378
	E	4.170	5.690	3.440	8.680	1.960
	C′		0.670	0.535	1.360	0.064
	E′		1.360	0.825	1.530	0.470
	R		2.030	1.540	1.120	7.340

Lines C (control) and E (experimental) list the measured amounts. Lines C′ and E′ list the normalized amounts compared with the standard spots (column M). The ratio of control to experimental is given in line R and shows the relative increase in synthesis that follows heat stress for each of the four hsps. (For autoradiographs of the tissues, see Fig. 3.)

vary with each stage of differentiation. In this respect, these particular genes for hsps are not fundamentally different from other genes in being activated to varying degrees in accord with the stage of differentiation of the cell. The family of genes responsible for hsps offers fruitful material for investigating the organization of the genome and its capacity to respond to stressful and teratogenic environmental stimuli.

Figure 2
Patterns from minced 15-day mouse embryos labeled in culture for 3 hours. (*Left*) Complete patterns are displayed: control (C, *top*); after heat shock (Hs, *bottom*). Enlargements show several of the hsps. Because they did not change with stress, proteins 1 and 2 were used as standards for calculating relative increases in the hsps. Many changes in relative amounts are evident; note the dramatic increases in the 105K and 68K hsps.

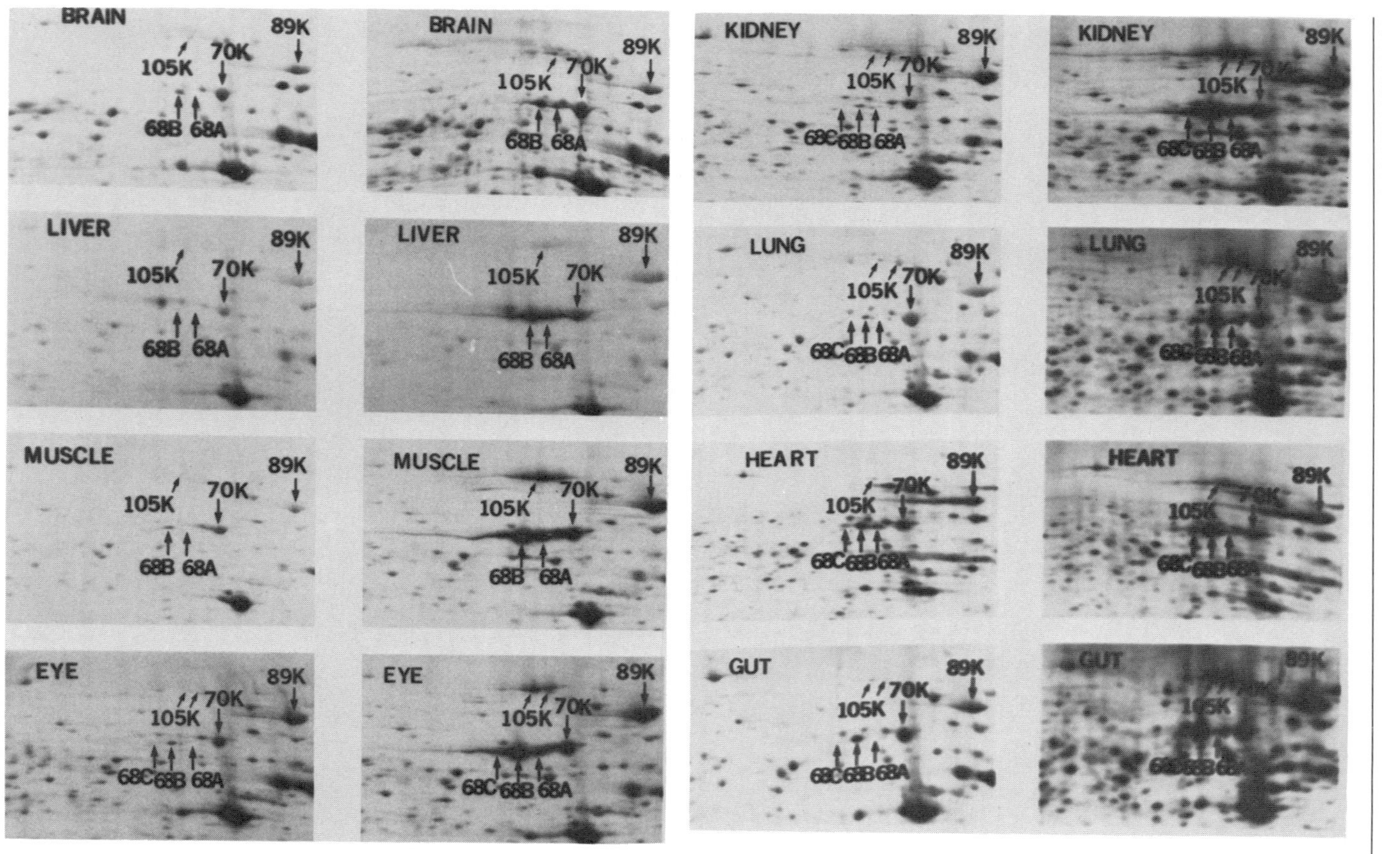

Figure 3 *(See facing page for legend.)*

ACKNOWLEDGMENTS

The original research reported here was supported by the National Science Foundation (grant DAR-7910054) and by the Andrew W. Mellon Foundation.

REFERENCES

Bensaude, O. and M. Morange. 1983. Spontaneous high expression of heat-shock proteins in mouse embryonal carcinoma cells and ectoderm from day 8 embryo. *EMBO J.* **2:** 173.

Bensaude, O., C. Babinet, M. Morange, and F. Jacob. 1983. Heat shock proteins, first major products of zygote gene activity in mouse embryo. *Nature* **305:** 331.

Bergh, S. and R. Arking. 1984. Developmental profile of the heat shock response in early embryos of *Drosophila. J. Exp. Zool.* **231:** 379.

Bienz, M. 1984. Developmental control of the heat shock response in *Xenopus. Proc. Natl. Acad. Sci.* **81:** 3138.

Currie, R.W. and F.P. White. 1983. Characterization of the synthesis and accumulation of a 70-kilodalton protein induced in rat tissues after hypothermia. *Can. J. Biochem. Cell Biol.* **61:** 438.

Dyban, A.P. and L.I. Khoshai. 1980. Parthenogenetic development of ovulated mouse ova under the influence of ethyl alcohol. *Bull. Exp. Biol. Med.* **89:** 528.

Grazios, G., F. DeCristini, A. DiMarcotullio, R. Mazari, F. Micali, and A. Savoi. 1983. Morphological and molecular modifications induced by heat shock in *Drosophila melanogaster* embryos. *J. Embryol. Exp. Morphol.* **77:** 167.

Hammond, G.L., Y.K. Lai, and C.L. Markert. 1982. Diverse forms of stress lead to new patterns of gene expression through a common and essential metabolic pathway. *Proc. Natl. Acad. Sci.* **79:** 3485.

Hightower, L.E. and F.P. White. 1981. Cellular responses to stress: Comparison of a family of 71–73-kilodalton proteins rapidly synthesized in rat tissue slices and Canavalin treated cells in culture. *J. Cell. Physiol.* **108:** 261.

Ireland, R.C., E. Berger, K. Sirotkin, M.A. Yung, D. Osterbur, and J. Fristrom. 1982. Ecdysterone induces the transcription of four heat-shock genes in *Drosophila* 53 cells and imaginal discs. *Dev. Biol.* **98:** 498.

Levinson, W., H. Opperman, and J. Jackson. 1980. Transition series metals and sulfhydryl reagents induce the synthesis of four proteins in eukaryotic cells. *Biochim. Biophys. Acta.* **606:** 170.

Figure 3

Patterns of stress proteins in eight separated organs of the 15-day mouse embryo (see also Table 1). The treated organs (rows 2 and 4) were heated to 42°C for 20 min and incubated at 37°C for 3 hr in medium containing [^{35}S]methionine. Each organ exhibits a specific distinctive pattern of response in the relative increases in the synthesis of hsps after heat shock. Heart and brain appear to be the least responsive to heat shock. (Control proteins, columns 1 and 3.)

Mirkes, P.E. 1985. Effects of acute exposures to elevated temperatures on rat embryo growth and development in vitro. *Teratology* **32:** 259.

Morange, M., A. Diu, O. Bensaude, and C. Babinet. 1984. Altered expression of heat shock proteins in embryonal carcinoma and mouse early embryonic cells. *Mol. Cell. Biol.* **4:** 730.

O'Farrell, P.H. 1975. High resolution two-dimensional electrophoresis of proteins. *J. Biol. Chem.* **250:** 4007.

Ritossa, F. 1962. A new puffing pattern induced by temperature shock and DNP in *Drosophila*. *Experientia* **18:** 571.

Schlesinger, M.J., G. Alperti, and P.M. Kelley. 1982a. The response of cells to heat shock. *Trends Biochem. Sci.* **7:** 222.

Schlesinger, M.J., M. Ashburner, and A. Tissieres, eds. 1982b. *Heat shock from bacteria to man.* Cold Spring Harbor Laboratory, Cold Spring Harbor, New York.

Thadani, V.M. 1982. Mice produced from eggs fertilized *in vitro* at a very low sperm : egg ratio. *J. Exp. Zool.* **219:** 277.

Webster, W.S., M.-A. Germain, and M.J. Edwards. 1984. The induction of microphthalmia, encephalocoele, and other head defects following hyperthermia during the gastrulation process in the rat. *Teratology* **31:** 73.

Whitten, W.K. 1971. Nutrient requirements for the culture of preimplantation embryos in vitro. *Adv. Biosci.* **6:** 129.

Wittig, S., S. Hensse, C. Keitel, C. Elsner, and B. Wittig. 1983. Heat shock gene expression is regulated during teratocarcinoma cell differentiation and early embryonic development. *Dev. Biol.* **96:** 507.

COMMENTS

Miller: In the list of tissues you examined, what came to mind here was the blastocyst and the fact that probably a lot of that blastocyst may be trophectoderm. How specific is the alteration of hsps in placental tissue? Did you examine that at 15 1/2 days, and have you looked at it any further at the blastocyst stage?

Markert: I am slightly embarrassed to admit that we did not look at the extraembryonic tissues. We looked at a lot of other normal tissues in the embryo, but we left the placenta aside for the time being. That certainly is a very important tissue to examine.

Scandalios: You said that there is a residual level in all tissues at all times and it is just a matter of quantitative increases upon the shock treatment. Is it possible, therefore, that as differentiation takes place, these cells have different levels of an inhibitory molecule that the heat shock removes differentially, rather than merely inducing synthesis per se or new transcription?

Markert: There are two things regarding that question. First, in adult tissues it can be shown that heat shock leads to the transcription of the gene; it is not a matter of removing inhibitors; it is completely new transcription of the gene. So, removing an inhibitor would not explain the data for adult tissues, and I doubt that it would explain the data for embryonic tissues. Second, although in the embryonic tissues these heat shock genes were all expressed to some level, in many adult tissues they are not expressed at anything near that level, or even at a detectable level. It is almost as if the development of the embryo was stressful—speaking in very anthropomorphic terms—in that the heat shock genes were all turned on to some degree, and once they had done their job, the adult tissues were formed, and it was not so stressful, these genes were turned off. Embryonic development could be stressful, whatever the biological meaning of stress is, and that may account for the low level of expression of these stress genes all of the time. I don't really know what stress means in biochemical or physiological terms. Stress is a very human kind of word to describe some biological response.

One common consequence that tends to accompany these stresses is a reduction in the available oxygen in the cells. They are all subjected, in general, to a certain degree of anoxia, which means that the ability of the cells to produce energy is relatively reduced in terms of the demand that is being put on them. When the cells are heated, their metabolism tends to increase enormously, and at the same time, the oxygen availability to carry out molecular events diminishes; it is that discordance that seems to underlie many of the responses to stress. But, I cannot translate that observation into biochemistry within the cell either.

Pedersen: In view of the observation that heat shock can induce thermal tolerance, it seems reasonable that having hsps around would be protective or, at least, that tissues which can respond by synthesizing them would be protected to a greater extent than ones that cannot respond. There is a fair amount of literature on heat shock during development by investigators using various forms of radiation. The common denominator of all those effects is the rise in the maternal body temperature. It would be interesting to compare the organs that are affected by heat with the levels of hsp synthesis.

Shepard: What is the function of hsps?

Markert: I think they have a role to play in the normal physiology of the cells, and I think that there is some connection between the synthesis of these proteins and the pattern of gene activation that follows stress.

The hsps are clearly protective, and the evidence for that is quite good. Treatment with heat that is sublethal, usually followed by a lethal treatment, permits the animals or the tissue to survive. There is no question that these proteins are beneficial in terms of confronting the same stress again. Even stress proteins induced by a nonthermal mechanism are protective against thermal stress.

Because these proteins are so ubiquitous in all organisms and so conserved in structure through evolutionary time, they must represent a fundamental biological mechanism that allows cells to confront sharply varying conditions.

But what I cannot understand yet is how to translate this response into biochemistry. I don't know what it really means in the life of the cell to stress it. To interpret stress in molecular terms is not easy.

Mechanisms of Transplacental Carcinogenesis: Mutation, Oncogene Activation, and Tumor Promotion

JERRY M. RICE, BHALCHANDRA A. DIWAN,* PAUL J. DONOVAN, AND ALAN O. PERANTONI

Laboratory of Comparative Carcinogenesis
National Cancer Institute
Frederick Cancer Research Facility
Frederick, Maryland 21701-1013

OVERVIEW

Direct-acting alkylating agents, especially those of the *N*-nitrosoalkylurea family, are both potent teratogens and powerful transplacental carcinogens. In contrast to the teratogenic effects, which appear to be mediated by massive killing of vulnerable cell populations by nonspecific alkylation reactions, evidence is mounting that at least some neoplasms of rodents that are inducible by transplacental or neonatal exposure to alkylating agents characteristically contain activated oncogene sequences that transform NIH-3T3 indicator cells. Renal mesenchymal tumors of the rat have been shown to contain activated Kirsten-*ras* (Ki-*ras*), whereas schwannomas contain activated *neu*, an unrelated oncogene. Both activated genetic sequences differ from their normal cellular homologs (or protooncogenes) by a single-base substitution that results in a protein product differing in sequence by a single amino acid and is consistent with the preferential alkylation by the carcinogenic alkylating agents of thymine and guanine residues at specific sites within specific DNA sequences as initiating events in carcinogenesis in these tissues. This finding suggests that direct assay of cells from embryos exposed to alkylating agents in utero for mutations at other loci that are more easily quantifiable may be a useful quantitative probe for target tissue specificity for carcinogenesis during development. Studies in rats and hamsters using resistance to ouabain or thioguanine, or in hamsters using resistance to diphtheria toxin as markers for transplacental mutagenesis, show a pronounced maximum susceptibility to mutation during early embryogenesis, much earlier than the apparent maximum susceptibility to carcinogenesis late in gestation. A possible explanation for this discrepancy may lie in the capacity of many epithelial cell types to undergo persistent initiation of carcinogenesis by transplacental exposure to alkylating agents or other carcinogens but to

*Present address: Program Resources Incorporated, NCI–Frederick Cancer Research Facility, Frederick, Maryland 21701-1013.

proliferate to form tumors only in response to prolonged postnatal exposure to nongenotoxic promoting agents. Examples include promotion of transplacentally initiated squamous epithelial neoplasms by postnatal exposure to phorbol esters and of thyroid follicular and renal cortical tubular neoplasms by certain barbiturates.

INTRODUCTION

Exposure of gravid mammals to chemical carcinogens, some, but by no means all, of which may also be teratogenic agents, can result in the development of neoplasms in the offspring during postnatal life (Tomatis and Mohr 1973). The anatomic locations, histogenetic origins, and relative latencies of transplacentally induced tumors may vary markedly from species to species in response to an otherwise comparable exposure to a given transplacental carcinogen, and potentially neoplastic cells may persist, quiescent, for a large fraction of the life span, or indeed the entire life span, of the transplacentally exposed individual without undergoing the induction of abnormal genetic expression and acquiring the uncontrolled proliferative invasive and metastatic qualities that characterize the cancer cell. However, exposure to promoting agents during postnatal life may stimulate tumor development from such latent cells, at least those comprising certain epithelia. In the experimental situation, this broadens the apparent range of target tissues and organs that otherwise would be perceived to result from a given transient transplacental exposure to an initiating agent (Burns et al. 1984).

Direct-acting alkylating agents, especially those of the *N*-nitrosoalkylurea family, are potent both as teratogens and transplacental carcinogens. Exposure of fetal rats to high single doses of any of these agents during days 12–14 of gestation results in massive necrosis of the rapidly proliferating cells of the periventricular germinal matrix of the cerebral hemispheres, and fetuses that survive such damage are born markedly microcephalic (Wechsler 1973). At lower doses, measurable microcephaly does not occur, but offspring are at risk for glioma development, especially in the cerebral white matter, as a result of nonlethal carcinogenic damage to the matrix cells. Although the mechanism of the teratogenic damage is clearly massive and nonreparable cell destruction, the mechanism of carcinogenesis has generally been thought to involve a much more specific interaction of the alkylation with intracellular target sites. Since the 1960s, the working hypothesis that initial damage to target cell DNA is a critical early event in the carcinogenic process has been progressively reinforced by the demonstration of specific adducts of chemical carcinogens with cellular DNA (Singer 1975) and the demonstration of a variety of repair mechanisms that can reverse the genetic damage that, in the absence of such systems, can demonstrably result. The initiation step in carcinogenesis thus has been generally considered to involve events related to mutagenesis, but it has not been possible to further refine the definition of the likely target site within the genome or to determine whether unique DNA sequences constitute obligate target sites for chemical carcinogenesis. The explosive advances in recent years in the identification and characterization of dominant transforming genetic sequences, or oncogenes, most of which were originally identified as the genetic sequences responsible for the oncogenic potential of acutely transforming retroviruses, have provided an experimental approach to this problem (Varmus 1984). Demonstration that such sequences occur in an activated form in chemically induced experimental tumors, as well as in a variety of human neoplasms, and that the activation to a genetically dominant transforming gene from a nontransforming precursor sequence, or protooncogene, can occur by way of simple mutagenic events such as single-base transition or transversion mutations has provided a route for exploration of possible specific target sites that may be essential for chemical carcinogenesis in various tissues. In this paper, we discuss recent experiments using transplacental mutagenesis, recently discovered tumor promoters, and oncogene analysis to probe specific events at the genetic level during transplacental and neonatal carcinogenesis in the kidney and the nervous system.

RESULTS

Studies with a wide variety of alkylating agents in our laboratory and elsewhere have led to a consistent picture of transplacental carcinogenesis in the kidney, which reveals a marked divergence between the rat and mouse in the manifestations of carcinogenic events, and both rodent species differ markedly from the human. Renal tumors in children are a major form of pediatric malignancy and are almost invariably of the nephroblastic, or Wilms' tumor, type, which is characterized by undifferentiated tubular epithelial elements and intermingled neoplastic mesenchymal elements that resemble undifferentiated metanephric mesenchyme. Such tumors can be induced by transplacental exposure of nonhuman primates to direct-acting alkylating agents such as *N*-nitrosoethylurea (ENU), but comparable exposures of mice to the same agent give rise to renal tumors that are uniformly of adult epithelial morphology. The corresponding tumor in human beings is characteristically a neoplasm of adulthood. These neoplasms that characteristically appear in the renal cortex late in the life of the mouse result from exposure to ENU even if that exposure occurs so early during fetal development, on day 12 of gestation, that metanephric differentiation has not progressed to fully differentiated tubular epithelium from which these tumors arise (Vesselinovitch et al. 1977). The histogenesis of the tumors has been studied carefully (Lombard et al. 1974), and it has been shown that the precursor cell is the fully mature epithelial lining cell of the proximal convoluted tubule. This suggests that the transforming event in transplacental chemical carcinogenesis in this tissue in mice may well have occurred in a renal blastemal cell which, despite this lesion, has nonetheless succeeded in differentiating to renal tubular epithelium. This suggestion gains credence from the fact that although the rat rarely develops tumors of any kind in the kidney from exposure to ENU, exposure to a wide variety of other chemical carcinogens, including both metabolism-dependent and direct-acting alkylating agents, shows that the predominant neoplasm generated as a result is an undifferentiated mesenchymal tumor, the component cells of which grow extremely rapidly and bear a striking morphologic resemblance to metanephric blastemal cells (Hard 1979). The same tumors can be induced by exposure of neonatal or older juvenile rats to alkylating agents. In addition, however, in the rat, a certain number of fully differentiated adult-type epithelial tumors develop in animals that survive beyond the first year of life, and a few true nephroblastic, or Wilms'-type, tumors may occur in this species. It is tempting to infer, and is in fact a reasonable working hypothesis, that metanephric blastemal cells of the rat escape the differentiation stimulus of the ureteric bud much more easily and are able to express a neoplastic phenotype far more readily than the corresponding cells of the mouse.

Strong evidence that the frequency with which the purely epithelial tumors occur is virtually always underestimated comes from the discovery of promoting agents for renal tubular epithelium among the barbiturates that have been used in human medicine. The most potent such agent is the long-acting hypnotic, barbital (5,5-diethyl barbituric acid), a congener of phenobarbital, the classic promoter for hepatic parenchyma. In a series of investigations, we have found that barbital is the most versatile of promoting agents in this class of substances, effectively promoting progression to neoplasia in at least four distinct types of epithelia in both mice and rats: hepatic parenchymal cells, thyroid follicular cells, urothelium, and renal cortical (proximal tubular) epithelium. Placing mice that were transplacentally exposed to ENU on a diet containing barbital after birth significantly increased the incidence of renal cortical neoplasms, among others (Diwan et al. 1985), and confirms that the differentiated renal cortical epithelial cell is one that can sustain a carcinogenic event at some stage of its development, becoming a tumor precursor cell that can remain latent and unidentified indefinitely until induced to proliferate to generate a tumor by some secondary stimulus.

As an initial step in exploring the role of activated oncogenes in the genesis of renal tumors, we examined renal mesenchymal tumors induced in F344 rats by neonatal (rather than transplacental) administration of a methylating nitrosamine, methyl(methoxymethyl)nitrosamine (Fig. 1). Tumors that developed in rats given a single injection of this compound within 24 hours after

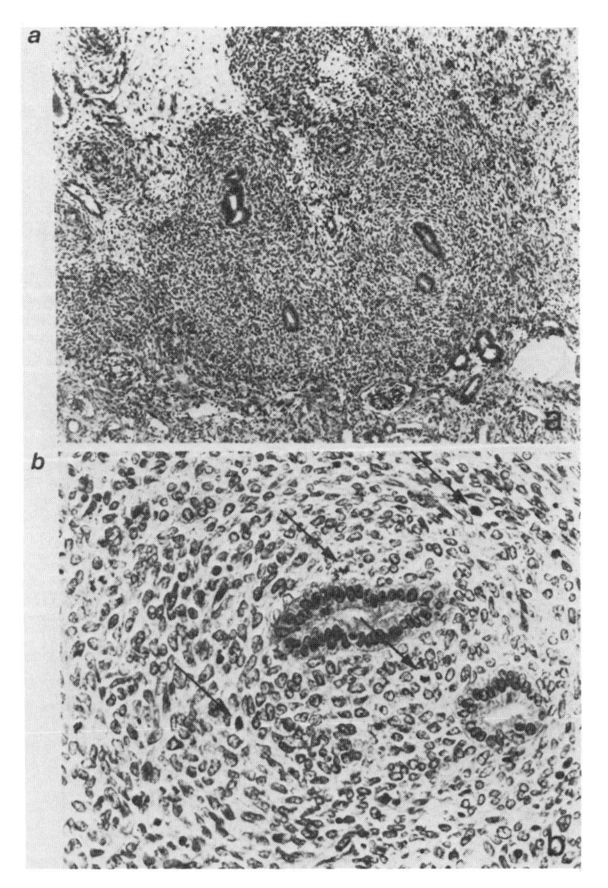

Figure 1

(*a*) Renal mesenchymal tumor (top) surrounding hyperplastic renal cortical tubules as a result of diffuse infiltration of renal parenchyma (bottom). At high magnification (*b*), frequent mitoses are apparent in tumor cells, which resemble the undifferentiated mesenchymal cells of metanephrogenic blastema. This tumor yielded DNA that transformed NIH-3T3 cells and contained an activated K-*ras* oncogene. (Reprinted, with permission, from Sukumar et al. 1986.)

birth were frozen, and tissue samples were extracted to yield high-molecular-weight DNA that was applied to monolayer cultures of NIH-3T3 cells using the calcium phosphate precipitation technique. Transformed colonies were seeded in soft agar, expanded in monolayer culture, and DNA from primary transformants was probed with specific molecular probes for both rat repetitive DNA sequences, as direct evidence for the incorporation of rat DNA into the mouse indicator cells, and for specific activated oncogenes (Fig. 2). Eleven of 25 renal tumors were found to yield transforming DNA at the first attempt, whereas no transforming DNA was isolated from grossly normal

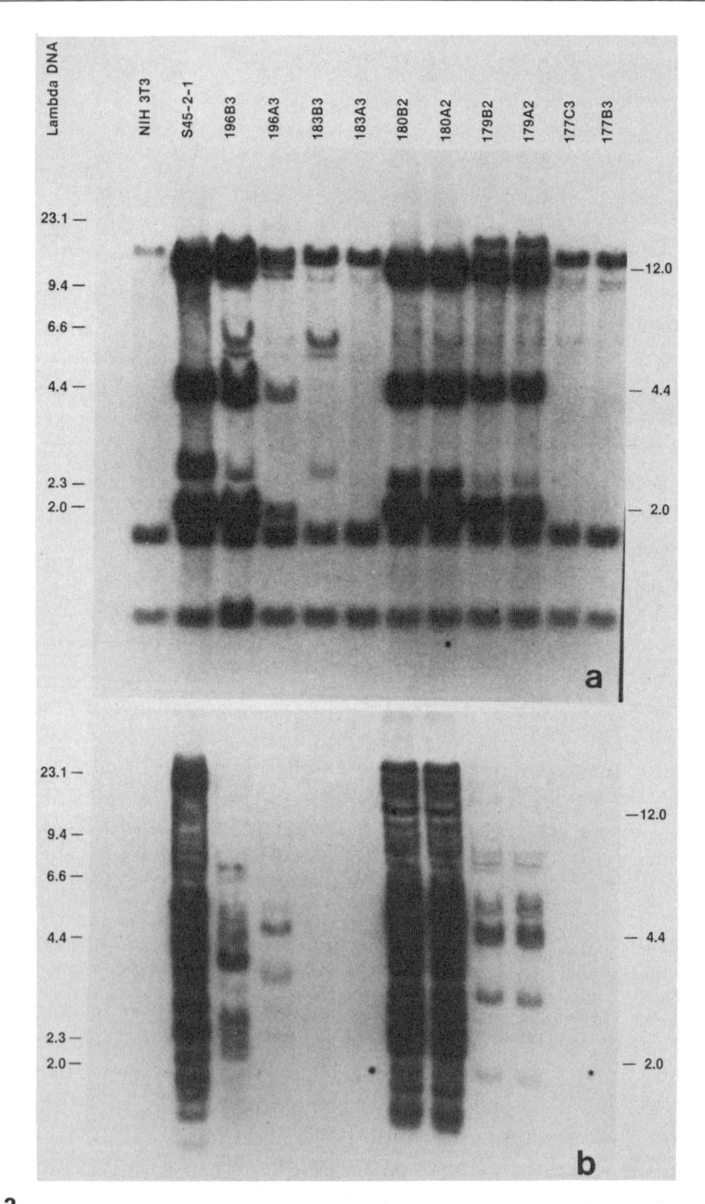

Figure 2

Southern blots of DNA from NIH-3T3 cells digested with *Hin*dIII and probed for K-*ras* (*a*) or rat repetitive sequences (*b*). Mouse cells contain K-*ras* fragments of 2.0, 4.4, and 14.0 kb (*a*), whereas rat K-*ras* fragments of 2.0, 4.4, and 12.0 kb are also seen in NIH-3T3 cells transformed by DNA from rat renal mesenchymal tumors 179 and 180. Rat repetitive DNA sequences (*b*) accompany the transforming sequences. (Reprinted, with permission and modified from Sukumar et al. 1986.)

contralateral kidney, brain, and liver of the tumor-bearing animals. In 10 of the 11 positive specimens, an activated K-*ras* oncogene was demonstrated; a single transformant was demonstrated to contain the smaller N-*ras* oncogene (Sukumar et al. 1986). The rat renal mesenchymal tumor therefore joins the rat mammary tumor (Sukumar et al. 1983) and the mouse epidermal papilloma (Balmain et al. 1984) as examples of chemically induced tumors in which specific activated oncogenes of the *ras* family are consistently demonstrable in the tumor tissues, strongly suggesting that they play a role in the genesis of the tumor.

ras oncogenes have been demonstrated to be activated by single-point mutations (Sukumar et al. 1983). In particular, the activation of these genes by carcinogenic methylating agents in mammary tumors has been shown to involve selective guanine-to-adenine transition mutations, which is perfectly consistent with a direct mutagenic effect of the methylating agent, perhaps by production of O^6-methylguanine residues, which are known to be a promutagenic lesion. The fact that the activated oncogene is demonstrable in the tumor tissue and not in grossly normal tissues of the same animal, which were equally subjected to the carcinogen, is evidence that the activation of the oncogene was the result of a somatic event and was not related to germ-line transmission of an aberrant gene within the tumor bearers.

The presence of an oncogene known to be activated by simple transition mutations in a type of neoplasm readily induced by perinatal exposures of rats to chemical carcinogens provides a basis for inferring that transplacental mutagenesis at other loci could be used as a surrogate probe for the infliction of potentially carcinogenic damage in various tissues of transplacentally exposed fetuses and at different stages of gestation. We have begun to do experiments of this sort (Fig. 3) to determine whether the apparent abrupt onset of susceptibility to transplacental carcinogenesis on approximately day 12 of gestation in the mouse and in the rat and on day 8 in the Syrian hamster (1) reflects the intrinsic vulnerability of different cells and tissues at that stage of gestation, (2) is a reflection of target size, or (3) deviates significantly from the actual pattern of genetic damage in the target cell population, as inferred from quantitative mutational analysis (Donovan 1985). Preliminary studies suggest that the rodent fetus is actually most susceptible to transplacental mutagenesis during the period immediately preceding the onset of apparent susceptibility to carcinogenesis, as measured by actual tumor development in the offspring. We used the Syrian hamster as a model for these studies, because in response to transplacental exposures to ENU, it yields tumors in a wider variety of tissues than the rat and because mutations for ouabain resistance and at the *HPRT* locus, which are commonly used in mammalian mutagenesis studies and which have a significant background, are not the only loci that can be used. Resistance to diphtheria toxin is also applicable to this

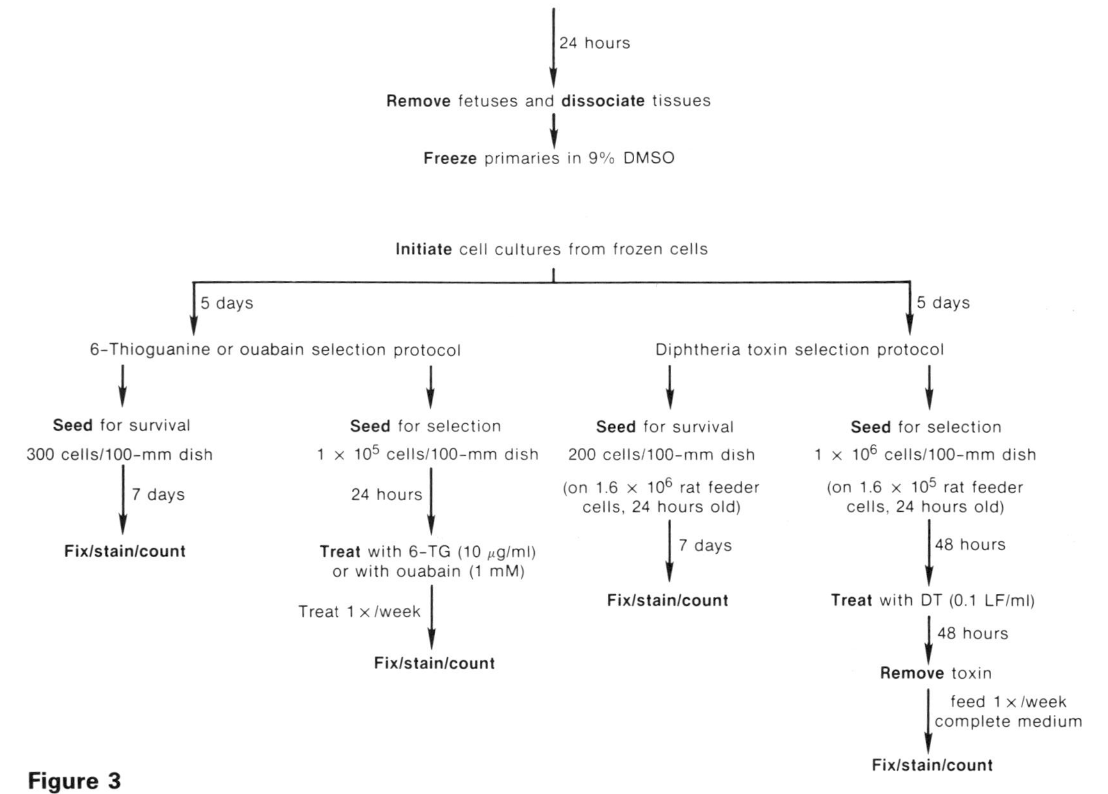

Figure 3
Protocol for transplacental mutagenesis in rodents.

species as a mutagenic marker and is preferable because of its exceedingly low background when one seeks to measure very low induced mutation frequencies. Using this approach, we found a marked maximum in the curve of induced mutants per surviving cell that peaks at approximately day 9, coincident with the apparent onset of susceptibility to transplacental carcinogenesis (B. Diwan et al., unpubl.) in this species (Fig. 4). This mutation curve has not been corrected for the changing rates of cell division in different cellular subpopulations within the embryo over time and thus must be regarded as an approximation. However, the magnitude of the induced mutation rates early in gestation strongly suggests that there is no biologic reason why mutations, including those of the sort required to activate oncogenes, cannot occur during these early stages of prenatal development. The explanation for

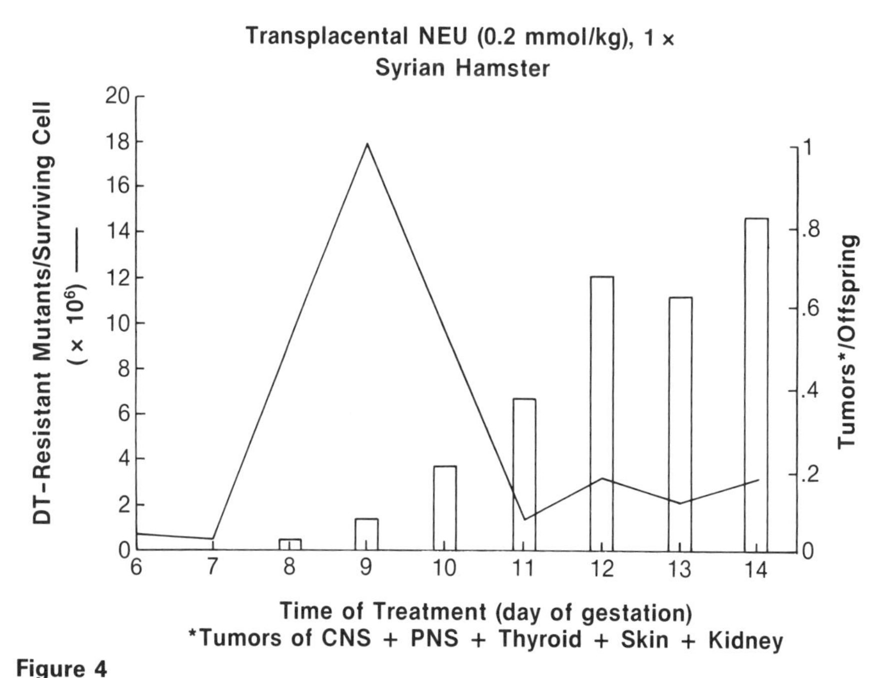

Figure 4

Tumors per offspring in Syrian hamsters (vertical bars), exposed once to ENU (0.2 mmole/kg) on days 6–14 of gestation. Tumors include neoplasms of the CNS and peripheral nervous system, thyroid, skin, and kidney. Diphtheria-toxin-resistant mutants per surviving cell from homogenates of whole hamster fetuses comparably exposed on days 6–12 and cultured on day 13, or exposed on days 13 or 14 and cultured 24 hr later (continuous line) are isolated most readily from fetuses treated on day 9, before maximal apparent susceptibility to carcinogenesis.

the apparent refractoriness of the early mammalian embryo to carcinogenesis probably must be sought elsewhere.

Transplacental carcinogenesis in all three species of rodents, the rat, the mouse, and the hamster, is characterized by marked prenatal susceptibility of the developing nervous system. This is manifested by the appearance of neoplasms of the central nervous system (CNS), nearly all of them gliomas and principally of oligodendroglial origin, and schwannomas of the peripheral nervous system, commonly originating in the ganglia of cranial and spinal nerves. On the one hand, this has confirmed the predictions of human pediatric oncology that prenatal exposure to chemical carcinogens might be a source of pediatric neoplasms, since tumors of the CNS, like Wilms' tumors of the kidney, are a major form of solid tumors of childhood. On the other hand, intracranial tumors of childhood most commonly originate from the astrocytes of the cerebellum and from the small neurons of the internal granular layer, sources of tumors that are among the least common in experimental systems. Furthermore, schwannomas are vanishingly rare in human beings, except in individuals afflicted with von Recklinghausen's disease, which is inherited as an autosomal dominant trait. We have recently completed a major effort to identify and characterize activated oncogenes in schwannomas and in cerebral gliomas of F344 rats that were subjected to a single transplacental exposure to ENU. Subjecting tumor tissue from these categories of neoplasms to the same protocol used for neonatally induced renal mesenchymal tumors in this strain of rat, we have discovered DNA sequences capable of transforming NIH-3T3 cells in nearly 50% of the schwannomas studied but in only 1 of more than 50 primary CNS tumors examined (Perantoni et al. 1986). Unlike the renal mesenchymal tumors, the schwannomas were found to contain an activated *ras* gene in only one case. All six tumors in which an activated oncogene was demonstrated consistently had a very different gene, *neu*, a receptor-type gene originally identified in cultured cells that were originally established from an intracranial tumor thought at the time to be neuroblastic in origin, in a strain BD IX rat that had been exposed transplacentally to ENU (Schubert et al. 1974). The lesion that activates this gene has also recently been shown to be a single-base mutation, involving a thymine-to-adenine transversion at a specific site (Bargmann et al. 1986). This is strikingly consistent with the ability of direct-acting alkylating agents, including ENU, to react with thymine residues to produce the promutagenic O^4-methylthymine, which is a very poor substrate for the alkyl acceptor protein repair system that appears to be most important for repair of O^6-methylguanine residues in vivo (Becker and Montesano 1985). A different oncogene is selectively activated in tumors of the peripheral nervous system but not the CNS, inviting the inference that specific oncogenes may have to be activated in different tissues to initiate carcinogenesis in cells of a particular lineage.

DISCUSSION

Two different oncogenes, K-*ras* and *neu*, both of which are known to be activated by single-point mutations, have been identified to be characteristically present in primary renal mesenchymal tumors and schwannomas of the rat, respectively, after transplacental or neonatal exposure to alkylating carcinogens. Demonstration that different oncogenes, each capable of activation by single-base mutations, are consistently found in different rat tumors that are among those most characteristically induced by transplacental and neonatal exposure to chemical carcinogens is encouraging evidence that the long search for specific molecular targets for chemically reactive chemical carcinogens may, at least in some cases, involve specific genes and specific sites of chemical reaction within the DNA sequences that constitute those genes. It is particularly noteworthy that these activating mutations are not deletions or other events that would prevent synthesis of the protein product of the gene but rather yield protein products that differ from the wild types by a single amino acid residue.

To date, only cutaneous squamous papillomas and carcinomas of the mouse, among known promotable epithelial neoplasms, have been shown characteristically to contain an activated oncogene, in this case, the smallest member of the *ras*-gene family, H-*ras*. Whether oncogene activation generally plays a role in the genesis of neoplasms that derive from cells that ordinarily do not express the neoplastic phenotype unless normal cell–cell interactions are interrupted by the action of tumor promoters such as barbiturates remains to be determined.

The fact that mutations are most efficiently induced in the rodent fetus prior to the stage of prenatal development at which susceptibility to transplacental carcinogenesis first becomes significant is additional and highly intriguing evidence for the generality of the kind of phenomenon studied by Pierce and his associates in which small numbers of neoplastic cells are coopted into differentiated behavior by cellular interactions with normal neighboring cells of the blastocyst and early embryo (Pierce et al. 1982).

The fact that oncogene activation may involve specific genes in different tissues and specific sites within the oncogene itself indicates that precise molecular targets for the initiation stage of the carcinogenic process do exist, may be specific for different tissues and possibly for different carcinogens, and will provide a fruitful direction for further investigation.

ACKNOWLEDGMENTS

We thank George Smith, Carl Reed, and Masahiro Watatani for their contributions to these studies and acknowledge the gifts of probes for rat repetitive sequences from A. Furano and for the rat *neu* gene from C. Bargmann and R.A. Weinberg.

REFERENCES

Balmain, A., M. Ramsden, G.T. Bowden, and J. Smith. 1984. Activation of the mouse cellular Harvey-*ras* gene in chemically induced benign skin papillomas. *Nature* **307:** 658.

Bargmann, C.I., M.-C. Hung, and R.A. Weinberg. 1986. Multiple independent activations of the *neu* oncogene by a point mutation altering the transmembrane domain of p 185. *Cell* **45:** 649.

Becker, R.A. and R. Montesano. 1985. Repair of O^4-methyldeoxythymidine residues in DNA by mammalian liver extracts. *Carcinogenesis* **6:** 313.

Burns, F.J., R.E. Albert, and B. Altshuler. 1984. Cancer progression in mouse skin. In *Mechanisms of tumor promotion: Tumor promotion and skin carcinogenesis* (ed. T.J. Slaga), vol. 2, p. 34. CRC Press, Boca Raton, Florida.

Donovan, P. 1985. "*In vitro* studies of transplacentally induced point mutation in the Syrian hamster." Ph.D. thesis. Catholic University of America, Washington, D.C.

Diwan, B.A., J.M. Rice, M. Ohshima, and M.L. Wenk. 1985. Transplacental carcinogenesis by *N*-nitrosoethylurea in B10.A mice: Effects of postnatal administration of different barbiturates on the incidence and types of tumors. *Proc. Am. Assoc. Cancer Res.* **26:** 137.

Hard, G.C. 1979. Effect of age at treatment on incidence and type of renal neoplasm induced in the rat by a single dose of dimethylnitrosamine. *Cancer Res.* **39:** 4965.

Lombard, L.S., J.M. Rice, and S.D. Vesselinovitch. 1974. Renal tumors in mice: Light microscopic observations of epithelial tumors induced by ethylnitrosourea. *J. Natl. Cancer Inst.* **53:** 1677.

Perantoni, A., C.D. Reed, M. Watatani, and J.M. Rice. 1986. Tissue-specific activation of *erb*B-related oncogene sequences in nitrosoethylurea (ENU)-induced rat neurogenic tumors. *Proc. Am. Assoc. Cancer Res.* **27:** 75.

Pierce, G.B., C.G. Pantazis, J.E. Caldwell, and R.S. Wells. 1982. Specificity of the control of tumor formation by the blastocyst. *Cancer Res.* **42:** 1082.

Schubert, D., S. Heinemann, W. Carlisle, H. Tarikas, B. Kimes, J. Patrick, J.H. Steinback, W. Culp, and B.L. Brandt. 1974. Clonal cell lines from the rat central nervous system. *Nature* **249:** 224.

Singer, B. 1975. The chemical effects of nucleic acid alkylation and their relation to mutagenesis and carcinogenesis. *Prog. Nucleic Acid Res. Mol. Biol.* **15:** 219.

Sukumar, S., V. Notario, D. Martin-Zanca, and M. Barbacid. 1983. Induction of mammary carcinomas in rats by nitrosomethylurea involves malignant activation of H-*ras*-1 locus by single point mutations. *Nature* **306:** 658.

Sukumar, S., A. Perantoni, C. Reed, J.M. Rice, and M.L. Wenk. 1986. Activated K-*ras* and N-*ras* oncogenes in primary renal mesenchymal tumors induced in F344 rats by methyl(methoxymethyl)nitrosamine. *Mol. Cell. Biol.* **6:** 2716.

Tomatis, L. and U. Mohr, eds. 1973. Transplacental carcinogenesis. *IARC Sci. Publ.* **4**.

Varmus, H.E. 1984. The molecular genetics of cellular oncogenes. *Annu. Rev. Genet.* **18:** 553.
Vesselinovitch, S.D., M. Koka, K.V.N. Rao, N. Mihailovich, and J.M. Rice. 1977. Prenatal multicarcinogenesis by ethylnitrosourea in mice. *Cancer Res.* **37:** 1822.
Wechsler, W. 1973. Carcinogenic and teratogenic effects of ethylnitrosourea and methylnitrosourea during pregnancy in experimental rats. *IARC Sci. Publ.* **4:** 127.

COMMENTS

Welsch: Is it correct to assume that a single administration of ENU is not sufficient to induce tumors in the adult animal?

Rice: Not quite. Using very high doses, tumors of the nervous system can be induced in adults, but it takes a much higher dosage to reach a 50% endpoint. The adult is about 50-fold more resistant. Both kinds of tumors can be induced, peripheral and central; presumably, it is just because of the lower proliferative activity, in part, that there is lower vulnerability.

Welsch: Is the high susceptibility of the conceptuses (embryo/fetus) attributable to the fact that they lack repair enzymes at this stage of development? Why can't they cope with the single administration?

Rice: One reason might be that, even if they had alkyl acceptor protein (AAP), this is a terrible substrate for it. And, by and large, although that mechanism—which has not really been studied in fetal tissues—is pretty effective where it exists at reversing O^6 alkylation of guanine, it is barely measurable as a reverser of O^4 alkylation of thymine residues. And if that is, in fact, the critical event, then the possession of a limited amount of AAP activity would do precious little good. So, that too is consistent.

Brent: You mentioned the fact that some people are concerned with the definition of a transplacental carcinogen. From the standpoint of a definition, would it be useful to think of that group of agents in which the same dose produces a higher incidence of tumors in the offspring that are exposed in utero versus the adult? There are many carcinogens that are actually more potent in the adult organism than in the embryo.

Rice: That is very true.

Brent: If you wanted to classify that, for instance, the first transplacental carcinogen, which is urethane, I think (reported by Larsen), is actually not as potent in the embryo as it is in the neonate or the adult organism.

Rice: The choice of that agent set the field back 20 years because it was such a dull result.

Brent: That is correct. So, if you wanted to group them, you could use that as a criterion. It is not anything magic, but it does help.

Rice: Except that usually—and I think the convention is accepted sufficiently—one defines a transplacental carcinogen as one that (to whatever extent) can, on administration to a gravid female, generate tumors in the offspring. And to attempt to define a subclass—those that are more dangerous to the offspring—would probably bog down quickly in arguments over time of exposure because there is such a radical difference in vulnerability of fetal tissues and kinds of tumors.

Brent: The second thing you alluded to is the central nervous system, the dramatic picture of, literally, not a live cell in the brain at a particular stage of gestation. Radiation does the same thing. If one gives 200 rads on the fourteenth day and one looks at the brain 10 or 12 hours later, the ventricle is filled with dead cells and so is the cortex; yet a lower dose will not result in carcinogenesis. In fact, when ENU and radiation are added, the result is protection; the combination actually decreases the incidence of neurogenic tumors.

Rice: Dead cells grow no tumors.

Brent: Well, I don't know. But how about live embryos? Live embryos can have only so many dead cells.

Rice: A very substantial fraction of the damage to DNA induced by ionizing irradiation consists of large deletions; is that correct?

Brent: Yes.

Rice: And again, if that is correct, deletions are important. This is radical for me. I have had difficulty with this for many years, but I think that there is a pattern emerging that simple point mutations may be important for at least some categories of tumors. I am perfectly prepared to turn around and argue against it at some other time.

Brent: Can you go back further in gestation, when ENU or nitrosourea produces any tumors? At what stage does that occur?

Rice: In the rat, there is a problem in conventional sized experiments finding any significant increase in tumors on day 11 or before, counting day 1 as the day on which there was copulation.

Brent: And yet, if one considers amplification of a mutation, it would be greater. Have you worked out the mathematics of why, all of a sudden, the tumors do not develop?

Rice: We have thought a lot about it and have been trying to formulate approaches to it. It is a very important question. Obviously, as target size gets smaller and smaller, there ought to be a point where, in terms of probability or the number of cells at risk (excluding developmental potency and such), larger and larger populations of animals would be needed just to compensate for the smaller number of cells in each unit. It has been argued that falloff to zero represents exactly that.

Manson: Dr. Rice, I think you can explain the time course from a very simplistic point of view, which is that in all your studies prior to day 12, the compounds are primarily cytotoxic.

Rice: That is dose responsive.

Manson: Well, I think it is also time responsive. We have done comparable studies looking at teratogenesis as an endpoint and can really break down the incidence of mutations; they are resistant in 6-cytoguanine, based on whether we have taken tissues from a prenecrotic, a necrotic, or a postnecrotic endpoint.

Rice: But there is no necrosis demonstrable—we have looked very, very closely for it at these levels.

The point I want to make is that there is evidence here that is consistent with the persistence of a lot of viable mutants in various tissues of an animal exposed to one of these agents earlier in gestation than we see tumors. That is also consistent with the fact that if you take these animals and expose them postnatally to a promoting agent, you will see the occurrence of neoplasms in tissues that previously appeared to be refractory to carcinogenesis. That is one of the questions that is unique to transplacental carcinogenesis, as opposed to carcinogenesis in general, namely the question of the relationship between differentiating tissues, transformation efficiency, and the expression of that phenotype.

We don't really have good probes for that. We only have clumsy methods of trying to demonstrate that there are transformed cells in various tissues that resemble what amounts to the bottom nine tenths of an iceberg. But they are there in tissues, such as most of the epithelium. Very possibly, that is an indication that there may be even more of them generated earlier on than there are in the period that appears to be most efficient, late in gestation, for generation of transplacental tumors.

Hanson: Turning to a different agent and a different species, can you comment on what may be known about the molecular mode of action of phenytoin and the claimed carcinogenesis relating to neuroblastoma or neuroblastomalike tumors? Regardless of what your answer is, in view of your last comments, are you suggesting that perhaps those children who have prenatal phenytoin exposures and do not necessarily manifest neuroblastomas in early childhood may be unusually susceptible to other kinds of postnatal promoting agents to develop other tumors?

Rice: That is a difficult problem and I don't have a straight answer to it. One is tempted to speculate that the diphenylhydantoin, metabolized as it is chiefly by oxidation of the phenyl groups, may, as a result of the accumulation of one of the intermediates, actually be a classic chemical carcinogen. It is not effective as a tumor promoter, but it has only been tested superficially.

Miller: You have been working with a number of species, as have others. You spoke specifically about the rat and hamster here. Have you done the same type of analysis with the mouse and the lung tumors that you investigated a number of years ago, and does the oncogene story fit across species in terms of the guanine O^6 ethylation?

Rice: We do not know whether the oncogene story for a particular type of tumor will persist across species, but we are in the process of probing the unique relationship of this gene that is specific for peripheral nerve tumors in the hamster and in the mouse.

There appears to be a marked divergence between mouse and rat in tumors such as the primary liver cell tumors. There is a marked difference in spontaneous incidence and in the biology of these tumors in the mouse versus the rat. In the mouse, they are very common and are easily induced. The rat presents a very different story. They are both promotable by the same categories of agents in most cases, but not all.

In the mouse, an activated oncogene has been reported, and a lot of work has been done on that. In the rat, nobody has been able to find an activated oncogene consistently in hepatic parenchymal tumors. Our own figure is six activated genes for 193 tumors.

In sum, specific activation that is valid across species will be found in some tumors and not in others. A lot more data must be accumulated to identify those systems that behave in one way and to differentiate them from all others.

Kimmel: From the point of view of risk assessment, it is very important to point out that adult carcinogens can also be transplacental carcinogens. I do not think that is very well recognized by the regulatory agencies.

But because there are some transplacental carcinogens that are more potent and others that are less potent to the adult, is there any way to predict what the case will be?

Rice: Not with absolute confidence, because there are special exceptions. There are polynuclear aromatic hydrocarbons that are as potent transplacentally—as a matter of fact, more, if you consider the variety of tissues affected—as the direct-acting alkylating agents. So, it is not valid, for example, to say that those that require no metabolic activation are always going to be better.

Kimmel: I have one other question. Many of these agents are also teratogenic. Are you proposing that there is a common mechanism of action for teratogenesis and transplacental carcinogenesis?

Rice: No. As a matter of fact, I think teratogenesis and transplacental carcinogenesis generally result from very different mechanisms. Within the same organ, for example, the brain, a carcinogen may cause tumors to arise from one cell population, whereas persistent population deficits but no tumors are seen in other, adjacent populations. 7,12-Dimethylbenz(*a*)anthracene, for example, in rats causes gliomas of the cerebral hemispheres, but in the cerebellum of the same animals it wipes out focal populations of small neurons of the internal granular layer, causing a trembling, kurulike syndrome—a teratogenic effect. But no medulloblastomas arise from surviving granular layer cells at the borders of these lesions.

Frankly, as I said before, there are many ways to kill a cell that do not require a specific target. It may well be that for the kinds of teratogenic effects we see, or that have been described with most carcinogens—which tend to be the big anatomic teratogenic effects—there is just no specific comparable target on the molecular level for the two phenomena. I really think that is probably true.

Nonmammalian Models

Developmentally Regulated Genes in Maize

JOHN GEORGE SCANDALIOS
Department of Genetics
North Carolina State University
Raleigh, North Carolina 27695-7614

OVERVIEW

Cells of a multicellular organism, though derived from a single fertilized cell, become the differentiated components of diverse tissues and organs that generate the progressive differentiation leading to the ultimate phenotype of a higher plant or animal. On considering this complex sequence of events, the obvious question arises: How is phenotypic diversity in form and function generated from the initial condition of genotypic constancy? Answers to this basic question remain incomplete, but available data suggest that development is driven by the differential qualitative and quantitative expression of specific genes at specific places and times. In addition, epigenetic modulation of gene activity may serve as a mechanism of fine-tuning and control during development.

The temporal and spatial regulation of the genes encoding the enzyme catalase in the eukaryote *Zea mays* provides an excellent model with which to study the differential expression of genes during development. These are a family of related "protective" genes whose expression is regulated by genetic and environmental factors and whose product serves to protect cells from the toxic effects of activated oxygen species.

INTRODUCTION

The most remarkable property of a fertilized egg is its ability to nearly always develop normally, utilizing only those instructions provided in its composite genome. It is difficult to imagine any singular way of achieving this, except by the differential activity and modulation of genes to express those gene products essential for cell differentiation at the proper time and place. Currently, there is little information available concerning the control of this process in both plants and animals, although significant progress is being made with the availability of the newer and more efficient techniques of molecular biology.

The programmed expression of protein-encoding genes during development involves some of the most complex sequences of biochemical events.

Banbury Report 26: Developmental Toxicology: Mechanisms and Risk

Therefore, understanding the mechanisms by which information encoded in specific genes is expressed at the appropriate time and place in the life of the organism is of critical importance. The vast size of eukaryotic genomes, the occurrence of repetitive nucleic acid sequences, and the interaction of the DNA with chromosomal proteins have presented difficulties, as well as challenges, in the elucidation of the developmental control of gene expression.

The catalase (H_2O_2:H_2O_2 oxidoreductase; E.C. 1.11.1.6; CAT) gene-enzyme system in *Zea mays* (maize) represents a useful model with which to study differential gene expression during eukaryote development. This system comprises a family of structural genes that are highly regulated temporally and spatially; their expression is influenced by specific temporal regulatory genes and by various environmental signals. Additionally, these genes are important since their product, catalase, serves to protect cells from the damaging effects of activated oxygen species.

RESULTS

Cat Structural Gene System

The maize catalase gene-enzyme system is composed of three unlinked structural genes, *Cat1*, *Cat2*, and *Cat3*, which encode the enzymes CAT-1, CAT-2, and CAT-3, respectively (Scandalios 1979). Each isozyme has been purified to homogeneity and found to be a homotetrameric heme protein of approximately 56-kD subunits; monospecific antibodies have been prepared against each isozyme (Chandlee et al. 1983). CAT-1 is expressed predominantly during kernel development and is found in the milky endosperm, aleurone, and scutellum and also in the root, shoot, and scutellum after germination. CAT-2 is expressed primarily in the scutellum during post-germinative development and in green leaves; it is also temporally expressed in the aleurone cells during kernel development. For the most part, CAT-3 is expressed in the shoots of young seedlings (Scandalios et al. 1984).

Catalase Developmental Program

The developmental program of catalase expression in the scutellum of maize was standardized by examining a significant number of different inbred lines. The scutellum was chosen because it is a tissue that persists from early kernel development to the time when the young seedling is fully differentiated. In addition, the scutellum is a stable, virtually nondividing, nondifferentiating diploid tissue (Kiesselbach 1949). Catalase activity in the scutellum surges to a peak at about the fourth day after seed imbibition and then declines. This time course of activity reflects the differential expression of the *Cat1* and *Cat2* genes, resulting in the gradual disappearance of CAT-1 protein and the appearance and accumulation of CAT-2 protein due to their differential turnover rates (Quail and Scandalios 1971).

Variants of the Catalase Developmental Program

A number of inbred lines with altered developmental programs for catalase expression in the scutellum have permitted us to study the genetic bases of the observed activity profiles. One such high activity variant, R6-67, sustains high levels of CAT-2 activity in the scutellum after the fourth day of postgerminative development, as compared to the typical catalase activity program characteristic of most maize inbred lines (e.g., W64A) in which CAT activity declines drastically. Detailed genetic analyses of such variants have demonstrated that the observed scutellar catalase developmental program during the first 10 days of postgerminative development is controlled by specific temporal regulatory genes. The *Car1* gene, distally located to the *Cat2* structural locus on chromosome 1S, controls the rate of synthesis of the CAT-2 protein by regulating the levels of translatable *Cat2* mRNA (Scandalios et al. 1980; Kopczynski and Scandalios 1986). Another regulatory gene, *Car2*, influences the developmental program of catalase expression in the scutellum by decreasing the levels of CAT-1 protein synthesis. *Car2* exerts its effect by regulating the levels of translatable *Cat1* mRNA associated with scutellar polysomes (Chandlee and Scandalios 1984). *Car1* and *Car2* are genetically independent of each other. However, the two are similar in that each acts in a tissue-, time-, and structural gene-specific manner; both exhibit additive inheritance. Since *Car1* has been shown to be distally located to the *Cat2* structural locus and to be *trans*-acting, it may function by transmitting a diffusible molecular signal to the *Cat2* structural locus (or its transcript) to influence its expression.

Effect of Elevated Temperature on the Catalase Developmental Program

Chronic exposure of seedlings to elevated temperatures of 40°C during the first 10 days of development did not alter the catalase developmental program significantly in the "typical" lines (e.g., W64A). However, the high catalase activity variant lines D10 and R6-67 showed a significant decrease in catalase activity when grown at 35°C or 40°C to levels closely resembling the catalase activity profile in W64A (Fig. 1). It was further demonstrated that the observed decrease in activity is due to reduced amounts of the CAT-2 protein resulting from a decreased rate of CAT-2 synthesis in line R6-67 growing at

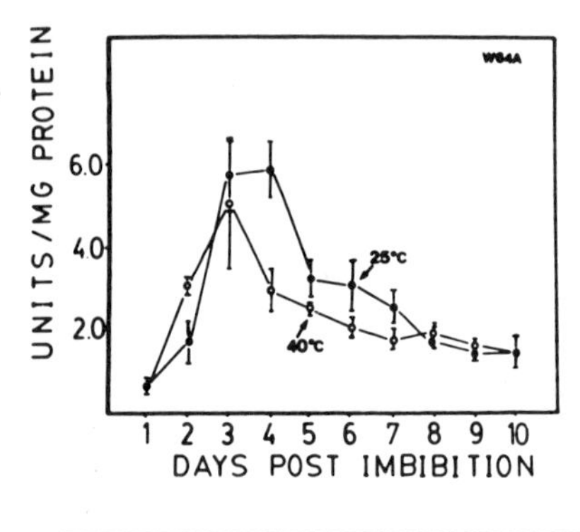

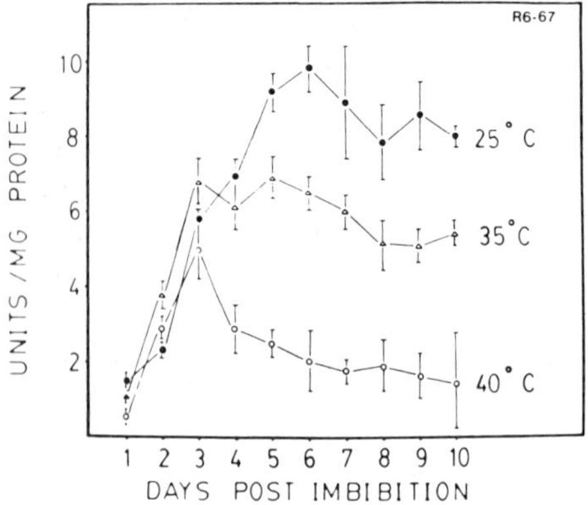

Figure 1

Total catalase activity in scutella of developing maize seedlings from days 1 to 10 post-imbibition. The inbred maize lines W64A (normal catalase activity profile) and R6-67 (high catalase activity profile variant) were germinated and grown at 25°C (●), 35°C (△), and 40°C (○). Bars represent the S.E.M. of five independent experiments. Note that under normal growing conditions (25°C), R6-67 exhibits scutellar catalase activity levels three- to fourfold higher than W64A, due to extended synthesis of CAT-2 in R6-67. However, when grown at elevated temperatures (35°C and 40°C), the high catalase activity is not observed in R6-67, its activity more closely resembling that observed for W64A. There is no apparent difference in catalase activity for W64A whether it is grown at 25°C or 40°C.

the higher temperatures (Matters and Scandalios 1986a). Since the levels of CAT-2 in R6-67 are controlled by the *Car1* regulatory gene, it is possible that the product of *Car1* is inhibited at high temperatures.

Spatial Pattern of *Cat2* Activation in the Scutellum

Even though the expression of *Cat2* and the accumulation of CAT-2 protein in the scutellum are substantiated with considerable experimental evidence, it was not known whether all scutellar cells are genetically programmed to express this gene synchronously or whether there is some specific pattern of *Cat2* activation within these cells. By using immunofluorescence microscopy and anti-CAT-2 IgG, we found that a gradient of *Cat2* activation occurs within the scutellar cell mass during postgerminative development (Tsaftaris and Scandalios 1986). The gradient *Cat2* gene activation occurs from the outer perimeter of the tissue inward toward the embryonic axis (Fig. 2).

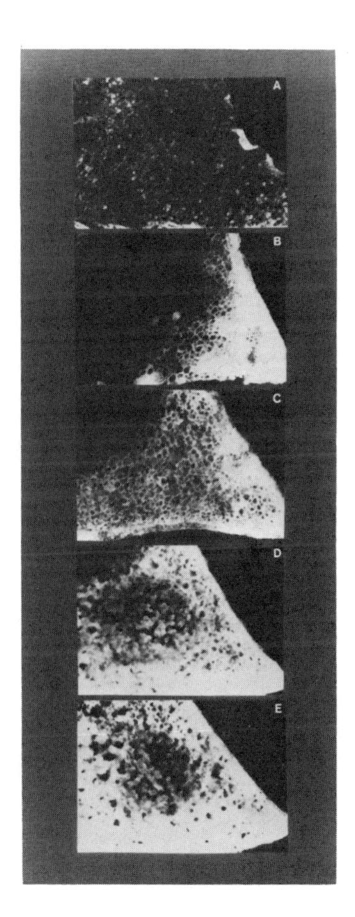

Figure 2

Immunofluorescence localization of CAT-2 in W64A scutellar cells at different days of postgerminative development. Thin sections of W64A scutella were incubated first with anti-CAT-2 IgG and then with fluorescein isothiocyanate (FITC)-labeled goat anti-rabbit IgG. (*A–E*) 0-, 2-, 4-, 6-, and 8-day-old scutella. Note the progressive expression of *Cat2* in scutellar cells from the outer to the inner perimeter with time.

Precocious Activation of *Cat2* in Aleurone

The possibility that a trigger for *Cat2* activation in the scutellum may derive from the single layer of aleurone cells prompted us to investigate whether or not the *Cat2* gene may be expressed in the aleurone at a much earlier period during kernel development following pollination, a time when *Cat1* was known to be predominantly expressed in most tissues (Scandalios 1979). Treatment of thin-sectioned 20-day-old W64A endosperm and aleurone specimens with anti-CAT-2 IgG resulted in prominent fluorescence labeling of the single layer of the peripheral aleurone cells, whereas similarly treated 20-day-old thin-sectioned kernels of A16 (*Cat2* null) with anti-CAT-2 IgG did not result in fluorescence. To further support the observation that the *Cat2* gene is specifically expressed in the aleurone at this early stage of kernel development, tissues from a single kernel were separated, and their homogenates were analyzed electrophoretically. Zymograms of catalase isozyme patterns in each of the three kernel tissues (endosperm, scutellum, aleurone) of W64A after pollination were examined. *Cat2* expression is evident only in the aleurone extract but not in the scutellum or the endosperm of the same kernel where *Cat1* is singularly expressed (Tsaftaris and Scandalios 1986).

To verify that the CAT-2 protein found specifically in the aleurone cells is synthesized in these cells from an active *Cat2* gene and not transported from the surrounding maternal pericarp tissue (which does not express CAT-2 itself), the mRNA present in the aleurone at this stage was isolated and subjected to RNA blot analysis (Pave et al. 1979) using a specific ^{32}P-labeled *Cat2* cDNA probe. Our results show that *Cat2* mRNA sequences are present in W64A aleurone cells at that early stage of kernel development, indicating active CAT-2 protein synthesis within these cells. When mRNA from aleurone cells of A16 (*Cat2* null) was similarly probed as a control, it contained no *Cat2* mRNA sequences (Fig. 3).

The aleurone cells are triploid, like the endosperm cells from which they are derived. As kernel development progresses (~15–20 days after fertilization), the outermost layer of the endosperm cells begins to differentiate from the rest of the endosperm cells, which play a role as a storage tissue,

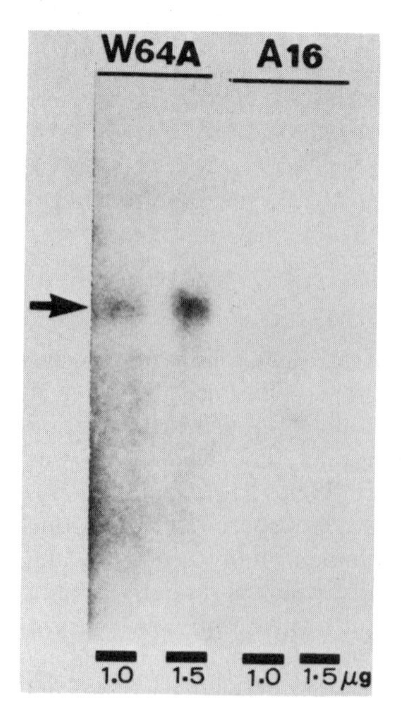

Figure 3

RNA blot analysis of poly(A)$^+$ mRNA extracted from 2 g of W64A or A16 aleurone cell tissue. After electrophoresis in 1.2% agarose/formamide/formaldehyde gels and transfer to nitrocellulose membranes, mRNAs were hybridized to a [^{32}P]dCTP-labeled *Cat2*-specific cDNA probe. Note the presence (→) and absence of *Cat2*-specific mRNA in the W64A and A16 aleurone poly(A)$^+$ mRNA pools, respectively. Amounts of poly(A)$^+$ RNA added per lane are indicated below each gel.

primarily for starch. The aleurone cells contain aleurone grains and oil but no starch. Following germination, the aleurone cells may furnish enzymes for digestion of the endosperm reserves (Kiesselbach 1949). However, we know that CAT-2 is synthesized de novo in the scutellum (Quail and Scandalios 1971). The aleurone itself is not utilized or degraded, and its cells appear intact after the remainder of the endosperm has been completely exhausted during seedling growth. The reason for the early and specific expression of the *Cat2* gene in aleurone cells is not clear, since these as well as the rest of the embryonic cells of the kernel are actively producing the *Cat1*-gene product (Scandalios et al. 1984). The aleurone is also the tissue in which other maize genes are specifically expressed during kernel development and includes the genes responsible for kernel color. Genetic analysis of the latter genes and their expression in aleurone cells led to the early discovery of transposable elements in maize (McClintock 1948).

It is conceivable that the aleurone cell layer, being the outermost layer of the embryonic cells of the kernel, is the first to sense environmental signals (e.g., water, temperature, etc.) and responds by furnishing a variety of secretory products to the rest of the kernel tissues to initiate germination and seedling growth. If there is such a response, perhaps a signal for *Cat2* expression is initiated in the aleurone and subsequently diffuses inward to activate *Cat2* in a sequential manner. This notion is supported by the fact that the first group of scutellar cells to express the *Cat2* gene are those nearest to the aleurone cells (Tsaftaris and Scandalios 1986). Alternatively, the possibility for the existence of an inhibitory substance(s) in scutellar cells, which must be inactivated for CAT-2 synthesis to occur, has not been eliminated. Such a mechanism could also explain the pattern of gradual *Cat2*-gene expression that we observe in the scutellum. Irrespective of the mechanism, it is evident that the expression of the *Cat2* structural gene in the scutellum during postgerminative development is not uniform in all scutellar cells but is spatially regulated, perhaps by a diffusible molecular signal originating in the aleurone cells.

Effect of Exogenous Factors on Catalase

In addition to the genetic factors directly affecting the temporal and spatial expression of the catalase structural genes, there is a significant response to various exogenously applied environmental signals, both chemical and physical. A variety of environmental factors that are known to create conditions of oxidative stress, including redox-active compounds and herbicides such as paraquat, hyperoxia, and pollutants (O_3, SO_2), have a significant effect on free-radical scavenging enzymes, including catalase (Halliwell and Gutteridge 1985; Matters and Scandalios 1986c).

Maize leaves treated with 10^{-5} M paraquat showed a significant increase in both total catalase (~25%) and superoxide dismutase (~40%) due to increases in the amount of polysome-bound mRNA encoding specific isozymes (Matters and Scandalios 1986b).

Photo-induced Expression of the *Cat2* Gene

The CAT-2 isozyme is absent in dark-grown leaves, but upon exposure to white light, CAT-2 protein levels increase, due to de novo synthesis (Fig. 4). When total poly(A)$^+$ RNA, polysomes, or isolated polysomal mRNA from light- and dark-grown leaves was translated in vitro, CAT-2 was detected only among the translation products from *Cat2* mRNA derived from the light-grown leaf extracts (Skadsen and Scandalios 1987). The absence of *Cat2* mRNA translation from polysomes isolated from dark-grown leaves could be explained if the mRNA was bound to ribonucleoprotein particles (RNPs) that were so large as to sediment with the polysomes. This possibility was checked by using a *Cat2* cDNA clone to probe slot blots of total mRNA and polysomal mRNA from light- and dark-grown leaves. The *Cat2* mRNA was found to be distributed in identical molecular weight fractions and in approximately equal quantities in polysomes from dark- and light-grown leaves (Skadsen and Scandalios 1987). Thus, it appears that the mRNA is bound to polysomes and not RNPs.

Our results show that *Cat2* mRNA is present in both light- and dark-grown leaves, but it is rendered translatable only when leaves are exposed to light. Although the underlying mechanisms for this phenomenon are not clear at

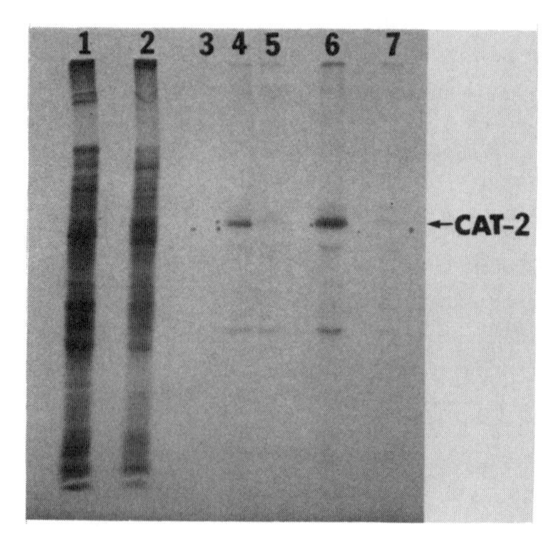

Figure 4

Fluorogram showing synthesis of CAT-2 in primary leaves of W64A seedlings grown in darkness (9 days), or in light (6 days dark plus 3 days light). Protein was labeled in vivo, in intact primary leaves, by topical application of [^{35}S]methionine. Labeled CAT-2 was isolated after reaction of the leaf homogenate with CAT-2-specific antibodies. (*1*) Total protein from light-grown tissue; (*2*) total protein from dark-grown tissue; (*3*) position, dots, of unlabeled pure CAT-2 protein, as marker; (*4,5*) equivalent counts per minute of immunoprecipitate of CAT-2 in light and dark leaves, respectively; (*6,7*) total immunoprecipitate per leaf of light and dark CAT-2, respectively.

this point, it is clear that the control of *Cat2* expression in maize leaves involves a unique form of translational regulation that prevents the formation of the CAT-2 polypeptides in etiolated leaves and that the essential positive "trigger" is light.

DISCUSSION

The maize catalase gene–enzyme system is a good model to study the differential expression of genes in higher eukaryotes. The structural genes (*Cat1*, *Cat2*, and *Cat3*) comprising this gene family are modulated in their temporal and spatial expression by both genetic (the *Car1* and *Car2* temporal regulatory genes) and environmental (light, chemicals, hyperoxia, etc.) signals.

The observed developmental program of catalase gene expression in the scutellum of maize is a consequence of extensive genetic and epigenetic regulation of *Cat1* and *Cat2*, as is the tissue-specific expression of all the maize *Cat* genes. The high degree of regulation of these genes and the fact that no totally acatalasemic organisms (barring strict anaerobes) have been found are suggestive of the biological importance of catalases. The *Cat* genes, along with other antioxidant enzyme genes, appear to serve a crucial role in protecting aerobic organisms against the toxic or lethal effects of activated oxygen species.

The fact that we have isolated and cloned the three maize *Cat* genes successfully (L. Bethards et al., in prep.; M. Redinbaugh et al., unpubl.) has permitted us to begin exploring the underlying molecular mechanisms for the temporal and spatial regulation of these genes in more detail. It is becoming clear from such studies that the expression of the *Cat* structural genes is a consequence of complex and intricate regulatory circuits at the transcriptional, translational, and posttranslational levels. We are currently attempting to identify and characterize promoters and *cis*-acting elements that may be involved in the regulation of the structural genes and to determine the level of control of these genes in different tissues of the maize plant grown in normal and stressed environments.

ACKNOWLEDGMENTS

This work was supported by research grants to J.G.S. from the National Institutes of Health (GM33817) and the U.S. Environmental Protection Agency (R-812404). I am indebted to my colleagues R.W. Skadsen, G.L. Matters, and A.S. Tsaftaris for their contributions to this project.

REFERENCES

Chandlee, J.M. and J.G. Scandalios. 1984. Regulation of *Cat1* gene expression in the scutellum of maize during early sporophytic development. *Proc. Natl. Acad. Sci.* **81:** 4903.

Chandlee, J.M., A.S. Tsaftaris, and J.G. Scandalios. 1983. Purification and partial characterization of three genetically defined catalases of maize. *Plant Sci. Lett.* **29:** 117.

Halliwell, B. and J.M.C. Gutteridge. 1985. *Free radicals in biology and medicine.* Clarendon Press, Cambridge, England.

Kiesselbach, T.A. 1949. The structure and reproduction of corn. *Research Bulletin* no. 161. University of Nebraska Agricultural Experimental Station, Lincoln, Nebraska.

Kopczynski, C.C. and J.G. Scandalios. 1986. *Cat2* gene expression: Developmental control of translatable CAT-2 mRNA levels in maize scutellum. *Mol. Gen. Genet.* **203:** 185.

Matters, G.L. and J.G. Scandalios. 1986a. Effect of elevated temperature on catalase and superoxide dismutase during maize development. *Differentiation* **30:** 190.

———. 1986b. Effect of the free radical-generating herbicide paraquat on the expression of the superoxide dismutase (*Sod*) genes in maize. *Biochim. Biophys. Acta* **882:** 29.

———. 1986c. Changes in plant gene expression during stress. *Dev. Genet.* **7:** 167.

McClintock, B. 1948. Mutable loci in maize. *Carnegie Inst. Wash. Year Book* **47:** 155.

Pave, N., R. Crkvenjakov, and H. Boetker. 1979. Identification of procollagen mRNAs transferred to diazobenzyloxymethyl paper from formaldehyde agarose gels. *Nucleic Acids Res.* **6:** 3559.

Quail, P.H. and J.G. Scandalios. 1971. Turnover of genetically defined catalase isozymes in maize. *Proc. Natl. Acad. Sci.* **68:** 1402.

Scandalios, J.G. 1979. Control of gene expression and enzyme differentiation. In *Physiological genetics* (ed. J.G. Scandalios), p. 63. Academic Press, New York.

Scandalios, J.G., A.S. Tsaftaris, J.M. Chandlee, and R.W. Skadsen. 1984. Expression of the developmentally regulated catalase (*Cat*) genes in maize. *Dev. Genet.* **4:** 281.

Scandalios, J.G., D.Y. Chang, D.E. McMillin, A.S. Tsaftaris, and R.H. Moll. 1980. Genetic regulation of the catalase developmental program in maize scutellum: Identification of a temporal regulatory gene. *Proc. Natl. Acad. Sci.* **77:** 5360.

Skadsen, R.W. and J.G. Scandalios. 1987. Translational control of photo-induced expression of the *Cat2* catalase gene during leaf development in maize. *Proc. Natl. Acad. Sci.* **84:** 2785.

Tsaftaris, A.S. and J.G. Scandalios. 1986. Spatial pattern of catalase (*Cat2*) gene activation in scutella during postgerminative development in maize. *Proc. Natl. Acad. Sci.* **83:** 5549.

COMMENTS

Markert: Dr. Scandalios, are these diffusible regulators able to go from cell to cell?

Scandalios: The data we have (i.e., the gradient type of activation of the

Cat2 gene in the scutellum) suggest that such diffusible regulators may go from cell to cell of a homogeneous nature. However, this is merely speculation at present. There was a similar gradient effect reported in hydra several years ago; but to my knowledge, ours is the first such gradient effect in a higher plant regarding a specific gene product during development.

Pedersen: This is a specific question but may have some general implications. Can you define the portion of the genomic clone that is responsible for the binding of the *trans*-acting factor?

Scandalios: Not yet, but this is a major reason for sequencing the gene. So far, we have only sequenced the 5′ end of the *Cat2* gene.

Pedersen: Right. To analyze the role of such protein DNA interactions, one would have to be able to put the genes back in the plant again. Is that possible? And how good a system is the plant for doing that kind of work?

Scandalios: In maize you cannot do that yet. We hope to use the Ti plasmid to put the gene into tobacco and express it. Another way is to possibly force the gene in by either microinjection or electroporation. But, at present, we can only transform petunia or tobacco plants.

McLachlan: What are the broader implications you were alluding to?

Pedersen: We have a plant session here so we can compare the benefits of doing genetic analysis in plants versus, perhaps, mammals. My initial impression was that the mouse has a distinct advantage in this regard because transgenic mice can be made readily, whereas transgenic corn cannot be made.

Scandalios: Transgenic plants can be made.

Pedersen: Is that possible in the other species of plants where these plasmids can be used to carry out the kind of genetic analysis that tells you, say, what portions of the *cis* sequences are actually binding the *trans*-acting factors?

Scandalios: Yes, such experiments are being attempted. Another aspect of catalase is that it is highly conserved. Thus, it may be possible to take a low-catalase mutant of a mouse, put the corn *Cat* gene into acatalasemic mouse cells, and see if it is expressed.

Iannaccona: Can microinjection be used in the plant zygotes?

Scandalios: I read a thesis by someone at Yale who had used microinjection successfully in plants. The main problem is to remove the cell walls. Once the walls are digested away, one has intact protoplasts with membranes. One can do this readily. The problem with corn is that whole plants cannot be regenerated from protoplasts.

Iannoccone: Must the Ti plasmid be used as a vector, or can other plasmids be used?

Scandalios: At present, Ti is really the only effective vector for transforming plants.

Iannoccone: But is there a constraint that requires that it be used for microinjection?

Scandalios: No, especially not into animals.

Kochhar: You mentioned light-sensitive forms. How did you do that?

Scandalios: You are referring to the induction of *Cat2* protein in the light. Dark-grown leaves do not have any *Cat2* protein, but, when exposed to white light, *Cat2* is synthesized de novo within a short period (< 1 hr). The interesting aspect of the light effect is that it is regulated translationally. The *Cat2* message is made, it associates with the ribosomes, but it is not translated in the dark. When polysomes from dark-grown leaves are probed with a *Cat2* clone, the *Cat2* mRNA is present, but it is not translated until the leaves are exposed to light. We are trying to explain that. It could have something to do with the capping of the mRNA or with the elongation of the polypeptides. Those are the two likely possibilities for translational control. Translational regulation may be a very important mechanism, especially where environmental signals are involved.

Kochhar: It's not a heat response?

Scandalios: You mean heat shock?

Kochhar: Not heat shock, but when you bring it into light.

Scandalios: No. We keep the temperature the same in both dark and light regimens. The heat response I discussed earlier is due to chronic exposure at 40°C. We do get a heat shock response with some other proteins but not with catalase.

Markert: But the heat shock is a kind of translational regulation too. It dissociates polysomes.

Scandalios: Yes, it could be.

Kochhar: This is not a heat shock, of course, but I don't see anything wrong with using the words "light shock" for this very abrupt change.

Scandalios: Some people believe that oxidative shock and heat shock proteins are the same.

Bournias-Vardiabasis: What is the molecular weight of the catalase form?

Scandalios: All of the catalases have a subunit of about 56,000 daltons. They vary in organisms in a range within 56–59 kD, so they are the same size.

Dencker: Why are two forms of catalase necessary?

Scandalios: There are three forms (possibly four) in maize. That is the million-dollar question. Not only are there differences in the tissue-specific expression of these forms of catalase but in the biochemical properties of these isozymes. Some are more heat stable than others.

If one looks at the physiological regimen during development and differentiation of corn, when these isozymes are differentially expressed, it makes some sense when one compares their properties. *Cat2,* for example, is expressed primarily during sporophytic development and under a number of shocks, including exposure to herbicides. When the plant is shocked with these chemical factors, one or the other—not all three—is elevated.

Markert: Is there anything that can be identified in maize, particularly, or in other plants that would be really parallel to teratological development in man?

Scandalios: There are quite a lot of mutants and phenocopies in corn that, in fact, parallel teratological development in man and other animals. An example is anther ears, where the male influorescence develops on the ears rather than as a separate organ. There are many such bizarre mutants in maize. We do not understand their molecular bases. In fact, maize is just waiting to be dealt with at the molecular level and by toxicologists at the toxicological level.

Evaluation of Abnormal Developmental Mechanisms in Hydra

E. MARSHALL JOHNSON, YONG H. CHUN, AND LINDA A. DANSKY
Daniel Baugh Institute
Jefferson Medical College
Philadelphia, Pennsylvania 19107

OVERVIEW

The differentiated and pluripotent undifferentiated cells of adult *Hydra attenuata* can be dissociated from one another and then randomly reaggregated into a pellet that will form new hydra in over 90 hours by means of an ordered ontogenic sequence of cellular actions, interactions, proliferation, and differentiation. Because development of this artificial "embryo" is a series of developmental events that follow one another with very little overlap, it is possible to identify, as well as compare and contrast, specific developmental events susceptible to perturbation by similar or dissimilar chemical agents. Also, whether or not the possible developmental site of action of a chemical actually is relatable to the effect on specific developmental events can be tested at least partially by use of transient exposure to identify the most vulnerable or critical stage of development affected. To the extent that the fundamentals of development are phylogenetically conserved, the hydra artificial embryo may be useful as a general indicator of developmental phenomena affected by individual chemicals acting on mammalian embryos.

INTRODUCTION

Some basic questions relevant to studies of perturbed developmental mechanisms are (1) what event does the putative developmental toxicant disrupt and (2) how does it achieve its effect? The first is at the macroscopic level of developmental processes or target, and the second is the biochemical means by which one or more of these are perturbed. This is not to say that cause and effect need to be associated so clearly because it could be that the environment within which a cell finds itself could change, for example, its membrane characteristics which, in turn, would have an effect on its chemistry and thereby on development. An important additional consideration is the developmental stage when the perturbation occurs. This is because, in addition to differences in vulnerability between cell types, there are differences in vulnerability evident at different developmental stages of the same cell group.

Throughout all evaluations of developmental mechanisms, there is the all-important concern of whether or not the measured occurrence is causally related to the perturbed developmental outcome observed. The endpoint assays evaluated could be concomitant events with little, if any, relationship to the developmental abnormality. This is an extremely difficult distinction to make, but it merits careful consideration. One possible way to obviate this kind of problem, at least partially, is to establish tests that could disprove a relationship if results inconsistent with the hypothesis are obtained. For example, if it were considered that a particular biochemical finding were related to a specific developmental event, perhaps that same biochemical occurrence could then be achieved by other mechanisms. If the chemical changes actually were related causally to the malformation, the malformation would then appear. Such hypothesis testing of developmental effects can be achieved by several means. Those available need to be used to the fullest extent possible.

In some regards, developmental perturbations in lower forms of life can be evaluated more readily than those in less accessible and more complex mammalian embryos. The relevance of these primitive systems to development of higher forms is a topic worthy of examination. We have been working with hydra for some years and have attempted to take advantage of its history of over 200 years as a model system for studies of developmental biology. It is easily manipulated. By means of, for instance, decapitation experiments, it has provided much valuable information regarding basic developmental phenomena. We have extended the system somewhat further to take advantage of its developmental biology and attempted to study abnormal development by manipulating its cells into a configuration from which they had to self-regulate or die.

An important aspect of the artificial preparation is the variety of developmental events that occur during its ontogeny. The breadth and diversity of the developmental biology evident in this system (Table 1) has the advantage of allowing assay of effects on a diverse spectrum of developmental events, any of which, if perturbed, potentially could lead to abnormal development.

Table 2 is a listing of the kinds of developmental events occurring at the specific developmental stages (times) at which we examine the artificial embryos of hydra. It is interesting to observe that, in general, this artificial preparation achieves its development in an ordered, serial progression of timed events, with very little overlapping of one type of developmental event with another, for example, cell migration with cell differentiation.

Table 1
Biological Events Achieved by Hydra

Event	Reference
Changes in cell size and shape[a]	(Webster and Hamilton 1972; Gierer 1977)
Selective cell death	(Gierer et al. 1972)
Cells must become spatially oriented,[a] recognize neighbors,[a] and form specialized junctions	(Filskie and Flower 1977; Wakeford 1979)
Form selective adhesive associations,[a] migrate	(Webster and Hamilton 1972; Wood 1980)
Induce differentiation of other cells less differentiated than themselves[a]	(Browne 1909; Lee and Campbell 1979) (Sugiyama and Fujisawa 1979)
Form intercellular matrix[a]	(Epp et al. 1979)
Be responsive to inductive stimuli and differentiate[a]	(Berking and Gierer 1977; Berking 1979) (Browne 1909; Bode et al. 1973; Schaller 1976)
Form cell-specific organelles and products	(Lentz 1965)
Undergo mitosis and then differentiate[a]	(David and Campbell 1972; Burnett et al. 1973)
Form organ fields	(Browne 1909; Gierer 1977; Otto and Campbell 1977) (Nishimiya et al. 1986)
Regulate organ field size[a]	(Webster and Hamilton 1972; Bode et al. 1976)
Become associated into tissues[a]	(Davis 1975)
Grow in size in a regulated manner	(Loomis and Lenhoff 1956) (Heimfeld and Bode 1985, 1986)

[a]Also observed in artificial embryos.

RESULTS

The technique for dissociating adult *H. attenuata* into its component cells and preparation of pellets (artificial embryos), consisting of random aggregations of cells, has been described previously (Johnson et al. 1982). Here, suffice it to indicate that these animals consist of at least six differentiated cell types and one undifferentiated pluripotent interstitial cell type.

The effects of vinblastine (Chun et al. 1983a,c), excess vitamin A (Y.H. Chun and E.M. Johnson, in prep.) and β-aminoproprionitrile (BAPN) (L.A. Dansky and E.M. Johnson, in prep.) on this embryo have been examined in some detail by a series of experiments. Vinblastine is a mitotic inhibitor that prevents microtubular development essential for cell division and changes in cell shape (Dustin 1978). Though it might do other things as well, at least it does this. Excess vitamin A affects differentiation adversely. Though it too may have other effects, perhaps at higher doses, at least it affects differentiation adversely (Sporn and Roberts 1983). BAPN is a lathyrogen that binds to the enzyme lysyl oxidase and prevents the cross-linking of collagen into fibers (Tang et al. 1983).

Table 2
Some Key Events in the Development of the Hydra Artificial Embryo

Time (hr) of development	Events	Designation[a]
0	Loosely arranged, randomly distributed cells; not all cell types evident	P
1	Cell sorting in progress; first indication of future ectoderm and endoderm; mesoglealike interface between ectoderm and endoderm.	P
2	Interstitial cells in ectoderm	P
3	Mesoglea present	P
6	First observation of musculoepithelial cells of basal disk	P
8	Beginning of central hollowing with thickening of ectoderm and endoderm	L
12	Attenuated ectoderm and endoderm	L
24	All cell types evident; tentacle buds indicated by aggregation of undifferentiated interstitial cells	B
48	Tentacles developed; elongated body axis present and regional distribution of cell types evident	E
70	Whorl of six tentacles around hypostome; extensive cell division evident	H

[a]Abbreviations: P, pellet; L, laminar; B, buds; E, elongation; H, hypostomes.

Table 3 is the dose-response relationship of this artificial embryo when it is exposed chronically to various vinblastine sulfate concentrations. The lowest, or minimal, affective concentration (MAC) was 2 mg of vinblastine sulfate per liter of hydra media. At this MAC, vinblastine interfered with developmental events that normally occur during the first hours of development. The embryo became laminar, but the first indication of tentacle bud development did not occur before the preparations disintegrated and died. This was taken as an indication that the aspect of development most vulnerable to vinblastine occurred early in development and before mitotic disruption would be involved. From this experiment alone, one cannot conclude, however, that mitosis would not have been perturbed at the MAC. The preparations were dead by the time of the 66-hr observation, and cell division does not become a factor until this time in development. Whether or not the mitotically active stages of development would have been more, less, or equally vulnerable to vinblastine would be better evaluated by a different type of experimental protocol.

Table 4 contains summary data from several experiments where exposure to vinblastine was transient and restricted to increasingly short time intervals.

Table 3
Developmental Stage Attained by Hydra Embryos Chronically Exposed to Graded Concentrations of Vinblastine Sulfate

Concentrations of vinblastine (mg/liter)	Stages[a]					
100	P	L	L	D		
10	P	L	L	D		
4	P	L	L	L	D	
3	P	L	L	L	D	
2	P	L	L	L	D	
1	P	L	B	B	E	H
0.9	P	L	B	B	E	H
0.1	P	L	B	E	H	H
0.0	P	L	B	E	H	H
Hours of development	4	18	26	42	66	90

The hours of development allowed to pass between evaluations are not constant among these studies made over a period of years. As our technique has improved for keeping the animals freer of bacteria, the number of hours needed to combine various developmental stages has shortened and also, continuing histologic studies have modified the time at which grossly observed events can be related to some developmental occurrences. Interestingly, rather large variations in temperature and illumination did not affect these parameters.

[a]Abbreviations: P, pellet smoothed on surface; L, lamination evident; B, buds of tentacles formed; E, elongation of tentacles evident; H, hypostome and body axis present.

Table 4
Effects of Transient Exposure to Vinblastine Sulfate

Developmental ages exposed (hr)	Exposure
	24 hr
0–24	Development stopped during early laminar stage
24–48	No effect
48–72	No effect
72–96	No effect
	8 hr
0–8	Development stopped during early laminar stage
8–16	Development stopped during early laminar stage
16–24	No effect or arrested during early laminar stage
	2 hr
−2–0	Development stopped during early laminar stage
0–2	Development stopped during early laminar stage
2–4	No effect

Exposure was the lowest concentration of vinblastine that produced developmental effects in chronically exposed embryos (developmental MAC).

Exposure for the first 24 hr of development, that is, from the time of pellet (embryo) formation through the next 24 hr, produced the same effect at the same rate as did the chronic exposure used in Table 3—the developmental MAC response. It should be noted in passing that it is necessary to take into account the phenomenon so well documented in pregnant mammals and represented by Karnovsky's law, that is, there usually exists some concentration that will produce an adverse effect. To avoid this confounder, the MAC is used with the goal of selectively detecting effects on the developmental event most susceptible to the agent and not concomitant events achieved at higher exposure levels.

When exposure to vinblastine was reduced to only 8 hr, exposure for the first 8 hr of development gave the MAC response. Exposure during the second 8 hr of development also stopped development at the laminar stage, but the preparations disintegrated and died somewhat more slowly than did those treated for the first 8 hr of development. An 8-hr treatment period extending from 16 to 24 hr of development gave variable results. The preparations either developed normally or, if arrested at the laminar stage, disintegrated between 90 and 120 hr after the commencement of development.

A 2-hr exposure starting during the preparation (dissociation) process and ceasing at the time of pellet formation stopped development at the laminar stage, and disintegration occurred by 42 hr. When the 2-hr exposure period began at the time of pellet formation and ended 2 hr later, the outcome was the same as that of the earlier 2-hr exposure, except that disintegration did not occur until 90 hr after pellet formation. Development progressed no further, just more time passed before the embryo disintegrated. From these results, one concludes that disruption of cell division and other later-occurring developmental events were less vulnerable to perturbation by vinblastine than were very early events, such as directional cell migrations.

Table 5 illustrates the relationship between embryo development and graded concentrations of all-*trans*-retinoic acid. The MAC with chronic exposure was 3 mg/liter, and development stopped at the time of initial tentacle bud formation. At this MAC, the first aggregates of undifferentiated interstitial cells that normally occur at the sites of tentacle formation took place in an apparently normal manner. The actual differentiation of tentacle buds and their elongation, however, did not occur. A slightly different effect was seen with a natural form of vitamin A, retinyl acetate, as is depicted in Table 6. The MAC of this test compound was 20 mg/liter, but the developmental stage affected, though similar, was not exactly the same as with all-*trans*-retinoic acid. That is, the initial and grossly obvious formation of tentacle buds did not take place and there was no further differentiation. (Actually there are only slight differences in the effects on differentiation of tentacles from these two dissimilar but related compounds. On histologic examination, some aggre-

Table 5
Developmental Stage Attained by Hydra Embryos Chronically Exposed to Graded Concentrations of All-*trans*-retinoic Acid

Concentrations of vitamin (mg/liter)	Stages[a]					
100	P	P	D			
10	P	L	L	B	D	
5	P	L	B	B	D	
4	P	L	B	B	D	
3	P	L	B	B	B	D
2	P	L	B	B	E	H
1	P	L	B	B	E	H
0.1	P	L	B	E	H	H
0	P	L	B	E	H	H
Hours of development	4	20	28	44	68	92

[a]See footnote to Table 3 for abbreviations.

gation of interstitial cells also occurred in the presence of the acetate, but a full discussion of this would require an exposition of the relevant histology [Chun et al. 1983b] that is beyond the scope of this report.)

As in the follow-up experiments with vinblastine, the effects of synthetic and natural vitamin A were examined by means of transient exposures. From the data in Table 7, it is clear that no 24-hr period of retinoic acid tested at the

Table 6
Developmental Stages Attained by Hydra Embryos Chronically Exposed to Graded Concentrations of Retinyl Acetate

Concentrations of retinyl acetate (mg/liter)	Stages[a]					
100	P	P	D			
40	P	L	L	D		
30	P	L	L	D		
20	P	L	L	L	D	
10	P	L	B	E	E	H
9	P	L	B	E	E	H
1	P	L	B	E	H	H
0	P	L	B	E	H	H
Hours of development	4	20	28	44	68	92

[a]See footnote to Table 3 for abbreviations.

Table 7
Effects of Transient Exposure to a Synthetic and a Natural Vitamin A Congener

Developmental ages exposed (hr)	Vitamin A
	Retinoic acid
0–24	No effect
24–48	No effect
48–72	No effect
72–96	No effect
	Retinyl acetate
0–24	Development stopped at end of laminar stage
24–48	Developmental delay w/normal ontogeny later
48–72	No effect
72–96	No effect

Exposure was at the lowest concentration of each agent that produced developmental effects in chronically exposed embryos (developmental MAC).

MAC concentration had any discernible effect on development. In marked contrast, exposure to retinyl acetate during the first 24 hr stopped development and did so at the end of the laminar stage. Exposure during the second 24 hr delayed development, but the embryos proceeded to develop normally and formed apparently normal adult hydra. Exposure during the third or fourth days had no effect on development.

The third chemical type we tested was undertaken to test the idea that there should be a marked difference in dependence of a developing system (vis-à-vis an adult system) on some biochemical events, that is, events that subserve essential developmental phenomena. Since cell-to-cell movement is an essential component of ontogenetic systems and since cell migration is influenced by the substratum on which cells move, alteration of the extracellular matrix should have clear effects on development. Accordingly, we treated both adult hydra and hydra artificial embryos with an inhibitor of collagen cross-linking, BAPN. The idea was that the embryo would have a narrow window during which cell migration would take place. If the collagen substratum were not as needed during that period, the cells would miss their opportunity to migrate and would be, therefore, irrevocably diverted into a pattern of abnormal or arrested development. If this type of unique or heightened vulnerability of a developing system actually existed, it might be related to the developmental hazard potential of the agent that would be indicated by an A/D ratio (adult MAC/developmental MAC) generally larger than 5 (Johnson 1984). That is, it could be held that before exposure began, adults would have sufficient cross-linked collagen to serve the needs of adults and, therefore, would be able to survive a period where there is little or no additional cross-linking.

Responses of artificial embryos to chronically exposed graded concentrations of BAPN are shown in Table 8. At treatment levels greater than or equal to the MAC (200 mg BAPN/liter of hydra media), embryos did not develop past the laminar stage and soon disintegrated. Table 9 shows the effects of transient exposures of BAPN on embryos. A constant exposure from 0 to 70 hr did not permit development of tentacle buds, whereas an equally long exposure from 22 to 92 hr had no effect. Since the exposure time was the same, both the length of exposure to the developmental MAC, as well as the developmental stage at which the chemical was exposed, were important factors. This was demonstrated again when the chemical exposure was shortened to 48 hr. Development stopped at the laminar stage when exposure was from 0 to 48 hr, had delayed but otherwise apparently normal ontogeny with exposure from 22 to 70 hr, and appeared unaltered with a 46- to 94-hour treatment period. There was no effect on development at exposures of 24 hr, regardless of when initiated.

From these results, it is concluded that the critical exposure period to BAPN for the developing embryo was the first 48 hr. Organ development (formation of tentacles) did not occur at this exposure. Contrastingly, adult hydra were not affected during this period except at a concentration ten times larger than the developmental MAC. This perhaps reflects a different need for collagen in the adult and the embryo. Whether or not blockage of collagen cross-linking is the mechanism of action involved, BAPN clearly is more hazardous to embryos than it is to adults and has an A/D ratio of 10.

Table 8
Developmental Stage Attained by Hydra Embryos Chronically Exposed to Graded Concentrations of BAPN

Concentration of BAPN (mg/liter)	Stages[a]					
2000	P	D				
400	P	L	L	D		
200	P	L	L	D		
100	P	L	B	E	E	H
90	P	L	B	E	H	H
0.0	P	L	B	E	H	H
Hours of development	4	18	26	42	66	90

[a]See footnote to Table 3 for abbreviations.

Table 9
Effects of Transient Exposure to BAPN

Developmental ages exposed (hr)	Exposure
	70 hr
0–70	Development stopped at laminar stage
22–92	No effect
	48 hr
0–48	Development stopped at laminar stage
22–70	Developmental delay but with normal ontogeny
46–94	No effect
	24 hr
0–24	No effect
4–28	No effect
22–46	No effect
46–70	No effect
70–94	No effect

Exposure was at the lowest concentration of BAPN that produced developmental effects in chronically exposed embryos (developmental MAC).

DISCUSSION

When the differentiated and undifferentiated cells normally present in hydra are manipulated into a configuration where they must reestablish the basic body pattern and also develop new organs, a large variety of developmental processes occur. Ontogeny under this circumstance is beyond simple homeoplasia. Some developmental events appear to be achieved by the terminally differentiated cells, for example, formation of laminar body plan; others involve in situ differentiation, for example, tentacle formation; and yet others apparently involve a combination of both classes of cells, for example, organ field formation represented by induction of the appropriate number of tentacles in proper relation to the hypostome. This combination of diverse developmental events achieved largely in serial manner, with minimal overlap between one type of event and another, provides a system that is potentially useful, at least in the earlier phases of mechanistic studies of abnormal development.

This artificial preparation appears to have a high degree of developmental relevance, and perhaps some of the developmental phenomena used by this preparation are common to all ontogenic systems. They may be conserved phylogenetically, at least with regard to the types of developmental events achieved, although they would be anticipated to be dissimilar from mammalian embryos in some of the means by which the effects are achieved.

The data of the experiments reported here are consistent with the thesis that hydra artificial embryos may be useful for studies of developmental mechanisms. At this very preliminary stage of examination, it is evident already that they share some generalized similarities to effects produced in mammalian embryos exposed to toxicants at suprathreshold levels. The artificial embryo responds to toxicants in a stereotypic manner, it demonstrates a threshold of effect or exposure levels below which developmental perturbation is not evident, and it has remarkable stage specificity for effects that are relatable both to the developmental event occurring and the duration of insult.

One is not confident of the relevance of this preparation for ultimate identification and characterization of abnormal developmental mechanisms, but it may point the way to identification of a vulnerable developmental target(s). If this were true, it would at least answer the question of whether similar or different developmental phenomena are the most vulnerable target of different developmental toxicants.

ACKNOWLEDGMENTS

This work was supported, in part, by Reproductive Hazards in the Workplace (grant 15-70) from the March of Dimes Birth Defects Foundation.

REFERENCES

Berking, S. 1979. Control of nerve cell formation from multipotent stem cells in hydra. *J. Cell Sci.* **40:** 192.

Berking, S. and A. Gierer. 1977. Analysis of early stages of budding in hydra by means of an endogenous inhibitor. *Wilhelm Roux's Arch. Dev. Biol.* **182:** 117.

Bode, H.R., K. Flick, and G. Smith. 1976. Regulation of interstitial cell differentiation in *Hydra attenuata. J. Cell Sci.* **20:** 29.

Bode, H.R., S. Berking, C.N. David, A. Gierer, H. Schaller, and E. Trenkner. 1973. Quantitative analysis of cell types during growth and morphogenesis in hydra. *Wilhelm Roux's Arch. Dev. Biol.* **171:** 269.

Browne, E.M. 1909. The production of new hydranths in *Hydra* by the insertion of small grafts. *J. Exp. Zool.* **7:** 1.

Burnett, A.L., R. Lowell, and M.N. Cyslin. 1973. Regeneration of a complete *Hydra* from a single, differentiated somatic cell type. In *Biology of hydra* (ed. A.L. Burnett), p. 255. Academic Press, New York.

Chun, Y.H., E.M. Johnson, and B.E.G. Gabel. 1983a. Relationship of developmental stage to effects of vinblastine on the artificial "embryo" of hydra. *Teratology* **27:** 95.

Chun, Y.H., E.M. Johnson, B.E.G. Gabel, and A.S.A. Cadogan. 1983b. Effect of vinblastine sulfate on the growth and histologic development of reaggregated hydra. *Teratology* **27:** 89.

———. 1983c. Regeneration by dissociated adult hydra cells: A histologic study. *Teratology* **27:** 81.

David, C.N. and R.D. Campbell. 1972. Cell cycle kinetics and development of *Hydra attenuata*. *J. Cell Sci.* **11:** 557.

Davis, L.E. 1975. Histological and ultrastructural studies of the basal disk of *Hydra*. III. The gastrodermis and the mesoglea. *Cell Tissue Res.* **162:** 107.

Dustin, P. 1978. *Microtubules*. Springer-Verlag, New York.

Epp, L.G., P. Tardent, and R. Banninger. 1979. Isolation and observation of tissue layers in *Hydra attenuata* pall (cnidaria, hydrozoa). *Trans. Am. Microsc. Soc.* **98:** 392.

Filskie, B.K. and N.E. Flower. 1977. Junctional structures in *Hydra*. *J. Cell Sci.* **23:** 151.

Gierer, A. 1977. Physical aspects of tissue evagination and biological form. *Q. Rev. Biophys.* **10:** 529.

Gierer, A., S. Berking, H. Bode, C.N. David, K. Flick, B. Hansmann, H. Schaller, and E. Trenkner. 1972. Regeneration of hydra from reaggregated cells. *Nat. New Biol.* **239:** 98.

Hemifeld, S. and H.R. Bode. 1985. Growth regulation of the interstitial cell population in hydra. II. A new mechanism for the homeostatic recovery of reduced interstitial cell populations. *Dev. Biol.* **111:** 499.

———. 1986. Growth regulation of the interstitial cell population in hydra. III. Interstitial cell density does not control stem cell proliferation. *Dev. Biol.* **116:** 51.

Johnson, E.M. 1984. Mechanisms of teratogenesis: The extrapolation of results of animal studies to man. In *Pregnant women at work* (ed. G. Chamberlain), p. 135. Royal Society of Medicine and Macmillan Press, London.

Johnson, E.M., R.M. Gorman, B.E.G. Gabel, and M.E. George. 1982. The *Hydra attenuata* system for detection of teratogenic hazards. *Teratog. Carcinog. Mutagen* **2:** 263.

Lee, H. and R. Campbell. 1979. Development and behavior of an intergeneric chimera of *Hydra* (*Pelmathohydra oligacis* interstitial cells: *Hydra attenuata* epithelial cells). *Biol. Bull.* **157:** 288.

Lentz, T.L. 1965. Fine structure in the nervous system of the regenerating *Hydra*. *J. Exp. Zool.* **159:** 181.

Loomis, W.F. and H.M. Lenhoff. 1956. Growth and asexual differentiation of *Hydra* in culture. *J. Exp. Zool.* **132:** 555.

Nishimiya, C., N. Wanek, and T. Sugiyamo. 1986. Genetic analysis of developmental mechanisms in hydra. XIV. Identification of the cell lineages responsible for the altered developmental gradients in a mutant strain (reg 16). *Dev. Biol.* **115:** 486.

Otto, J. and R. Campbell. 1977. Budding in *Hydra attenuata*: Bud stages and fate map. *J. Exp. Zool.* **200:** 417.

Schaller, H.C. 1976. Action of the head activator on the determination of interstitial cells in *Hydra*. *Cell Differ.* **5:** 13.

Sporn, M.B. and A.B. Roberts. 1983. Role of retinoids in differentiation and carcinogenesis. *Cancer Res.* **43:** 3034.

Sugiyama, T. and T. Fujisawa. 1979. Genetic analysis and developmental mechanisms in *Hydra.* VII. Statistical analysis of developmental-morphological characters and cellular compositions. *Dev. Growth Differ.* **21:** 361.

Tang, S., P.C. Trackman, and H.M. Kagen. 1983. Reaction of aortic lysyl oxidase with beta-aminoproprionitrile. *J. Biol. Chem.* **258:** 4331.

Wakeford, R.J. 1979. Cell contact and positional communication in hydra. *J. Embryol. Exp. Morphol.* **54:** 171.

Webster, G.W. and S. Hamilton. 1972. Budding in hydra: The role of cell multiplication and cell movement in bud initiation. *J. Embryol. Exp. Morphol.* **27:** 301.

Wood, R.L. 1980. Freeze fracture studies on cell junction formation in regenerating hydra. In *Developmental and cellular biology of coelenterates* (ed. P. Tardent and R. Tardent), p. 447. Elsevier/North-Holland, New York.

COMMENTS

Scandalios: What happens if you treat the animal with the drug vinblastine after the tentacles have formed? Is there any effect at all?

Johnson: No, not at the minimally effective concentration. At a higher concentration of vinblastine, the tentacles will form, they will be induced, but they do not elongate, because increased cell division begins just after that time, and it is stopped in this preparation at higher concentrations.

We tried to use minimal effective concentrations to produce effects in that developmental phenomenon that was the most sensitive to perturbation by the agent, and we tried to avoid concomitant events that might also occur at higher concentrations.

Scandalios: Is it known how vinblastine affects mitotic division, and what the mechanism is?

Johnson: No, I have no idea other than effects on microtubules.

Brent: Actually, these embryos are really disaggregated adults, aren't they?

Johnson: They are disaggregated, reaggregated cells of adults.

Brent: Have you ever tried other things? For instance, what happens in the same system if you just cut off the tentacles and look for regrowth?

Johnson: Yes, we tried that. Of course, there is a long, noble history of such experiments in hydra. The ability to say that gradients exist in this preparation came from such experiments.

What is created in those kinds of situations is known to developmental biologists as homeoplasia, development on a preexisting pattern. It would be a reasonable way to test agents effective at preventing cellular induction. They would be detected because induction does take place in decapitated hydra. However, if the agent had its primary effect on pattern formation, cell migration, intercellular recognition phenomena, or de novo formation of basement membrane, such a transection study would not be an adequate test.

Brent: In your introduction, you indicated that you had gone through an evolution with regard to the utilization of this technique. Do you think it is more useful, in a sense, as a simple tool to study developmental biology principles than it is as a method of assaying risk?

Johnson: I don't think it ever had any relevance to risk because the concentrations used in hydra don't predict effective doses in mammals. Risk estimation entails all interspecies differences, toxicokinetics, and so forth, and such are not what we see here. The assay is a very effective means for detecting hazards—things that are more harmful to developing systems than to mothers—or showing us that if one takes the concentration or exposure to the adult toxic dose in a segment II study, one will have developmental effects. I think that the assay can do human embryos a lot of good as a quick and dirty prescreening type of assay for setting priorities.

Kochhar: In discussing dosages, you mentioned that they are irrelevant to mammalian teratology. Take the example of retinoic acid. The dosage you used in hydra was 3 mg/liter which is 10 μM. In other tissue culture experiments, that would be ten times the effective dose. Such a dose would produce a lot of cytotoxicity in other systems. Did you find any evidence of that in hydra?

Johnson: No, we did not. To cause cell death, the compound would probably have to get inside of the cell or profoundly affect the cell membrane and its function. There must be very different degrees to which test agents gain access to the inside of cells. At the doses we used, we did not see cell death.

Kochhar: Just one general comment. I think your dosages could be compared with those used in tissue cultures by other investigators.

Johnson: Perhaps, for tissue culture, but I doubt it with hydra. The first several compounds that we tested in this assay some years ago were effective at concentrations very close to effective doses in mammals. That was a bit terrifying to me because it could not be true. We tested more chemicals and found that only the first ones used held up. Therefore, we don't see much relevance in hydra for predicting dosage levels in intact mammals.

Pratt: Since you presented evidence for specific effects of retinoic acid on the differentiation of cells, do you or other investigators have evidence for retinoic acid-binding proteins (receptors) that might mediate these effects in hydra?

Johnson: I don't know of such evidence. No one has looked closely for the mechanism, as we just found this 3 months ago.

Pratt: As a follow-up on that, have you had any success with your activating systems to detect certain teratogens, such as cyclophosphamide, which is one of the common ones used?

Johnson: Not that I would be willing to talk about.

Kimmel: In the hydra system, the exposure that you are giving to the so-called embryo is a direct exposure. It seems to me that if you show developmental effects with such a direct exposure and you also show adult effects with the direct exposure, you have to interpret the results as being direct effects on both systems. They may be different types of effects, but why do you try to differentiate between specifically targeted developmental effects versus effects that affect both "embryo" and adult?

Johnson: I don't think that a mitochondrion is really different in the artificial preparation than it is in an adult. If we had an uncoupler, I think it would affect both at about the same concentration. I think that is what is seen. If migrating cells had to have a matrix through which to migrate, or on which to migrate, they would be very dependent upon having, say, cross-linked collagen at the right time, at the right place; and if it is not there when they need it, but shows up later on, it's too late for them. Whereas, in an adult, if it cross-links no new collagen for some period of time, there might not be any adverse effect. I don't know whether that is actually the mechanism that has been achieved here, but the results of the experiment are consistent with that idea.

Kimmel: In the case of BAPN, are you aware of any data in vivo on BAPN treatment?

Johnson: No.

Kimmel: The problem with BAPN is that it probably never reaches the embryo. In unpublished studies that I did in the rabbit, it was so reactive that it caused tremendous toxicity in the mother and never reached the embryo. But when I delivered it directly to the embryo, it was teratogenic.

Kochhar: There is a lot of evidence that it is teratogenic. Bob Pratt has shown that.

Johnson: True.

Pratt: Yes, it directly affects the mammalian embryo and causes problems in cross-linking of collagen in a number of sites, including the palate.

Kimmel: That's in vivo?

Pratt: In vivo.

Kimmel: With what kind of exposure?

Pratt: Injection at day 13 or 14 without any toxicity to the mother.

Markert: You talked about the effect of these various chemicals on processes like induction and aggregation, and so forth; but, really, the effect has to be on some cell. You said there were six different kinds of cells in these preparations. Did they all react similarly?

Johnson: No, and I hope I did not say that they are affecting the processes. They are having effects at the time that these processes are taking place. I do not know specifically that induction is blocked.

Markert: But the effect on induction might be either on the inducing cell or the responding cell.

Johnson: Absolutely.

Markert: And as far as you could see, all the cells behaved alike?

Johnson: There are no obvious differences. With this preparation, I don't think one would be able to see it. You could do an explant kind of experiment and take an aggregate composed of inducing interstitial cells and move it to an epithelium that has not been insulted. We have not done that.

McLachlan: I have a question that is similar to that. You said that in the beginning when you disaggregate, you have a ratio of 6:1 of differentiated to undifferentiated cells?

Johnson: No. There are at least six types of differentiated cells in adult hydra and one class of undifferentiated cells.

McLachlan: When the reaggregate starts to differentiate and become another adult, does that ratio of six differentiated cells to one stay the same, or are the differentiated cells terminally differentiated and, therefore, don't become part of the growing pool of cells; and your differentiated cells take over. Is that right?

Johnson: Great question. The answer is both yes and no. The undifferentiated cell stays there; you can see it immediately. Some of the terminally differentiated cells stay there because they have cytoplasmic specific organelles by which they can be recognized. Some of the terminally differentiated cells are not there, and those cells are formed by the ectoderm. When they come into juxtaposition with the undifferentiated cells at the proper time, those kinds of cells will develop again. If that does not happen, those kinds of cells will not develop again.

Welsch: I have a comment regarding the question about cyclophosphamide activation. It is my understanding that Dr. Richard Wiger (Norwegian Institute of Public Health) has done this experiment in *Hydra attenuata*. Presumably the developmental toxicity of cyclophosphamide was lower in the presence of a 9000*g* mammalian liver-derived supernatant fraction.

Johnson: I have reviewed those unpublished data and I have written to the investigator suggesting how he might change that experiment. We have tested cyclophosphamide in a preliminary and, as yet, unpublished form and we got very different results. The S-9 preparations were highly toxic, but microsomes in a dialysis bag were tolerated. With them the A/D ratio did not change; it was an order of magnitude lower dose at which the effect was achieved. This is a very different result than was mentioned by the Norwegian laboratory.

Use of Tissue Culture in the Analysis of Limb Chondrogenesis

MICHAEL SOLURSH
Department of Biology
University of Iowa
Iowa City, Iowa 52242

OVERVIEW

The embryonic chick limb is used as a model for studying mechanisms involved in normal chondrogenesis. High density cultures of limb mesenchymal cells have the capacity to form several types of differentiated cells, including cartilage tissue. This process resembles that observed in situ during normal development in many respects, including the formation of precartilage cell aggregates. Required homotypic cell interactions can be bypassed by maintaining cells briefly in a rounded configuration by culture in gels or treatment with cytochalasin D. A diffusible factor derived from the limb ectoderm inhibits chondrogenesis in gel cultures by promoting cell spreading. These in vitro results are consistent with a role for diffusible factors in the control of patterning of chondrogenesis through alterations in cytoskeletal-extracellular matrix interactions.

INTRODUCTION

The developing limb is a useful model system for mechanistic approaches to abnormal human development (Krey et al. 1984). Limb malformations in human populations are common birth defects, and the limb is susceptible to a number of known teratogens. The limb exhibits many of the basic processes involved in the developing human, as well. These include the mechanisms involved in tissue differentiation and patterning.

The studies described here are concerned with the normal mechanisms involved in cartilage differentiation in the chick embryonic limb bud. Cartilage elements form in a distinct pattern and in a harmonious relationship to other tissue types during normal development. To approach the underlying mechanisms that are involved in this process, several types of tissue culture methods have been developed in my laboratory. These include high density cultures of dissociated cells in which cartilage tissues form. These can be utilized for the analysis of the social behavior of the cells. Low density cultures in or on gels can be used for the analysis of cell-extracellular matrix interactions, environmental requirements for cytodifferentiation, and lineage relationships among different cell types. Heterologous tissues, such as specific

Banbury Report 26: Developmental Toxicology: Mechanisms and Risk

epithelia, can also be added to these cultures for studies on tissue interactions, which might be involved in the development of tissue patterning. The in vitro studies lead to a model for the patterning of developing cartilage in the limb bud, which requires testing in situ during normal development. The model can provide insights into the possible mechanisms involved in abnormal development in the human.

RESULTS

High Density Cultures

When mesenchyme cells from several sources, including mouse (Umansky 1966) and chick limb buds (Caplan 1970), are plated at confluent density, a variety of cell types differentiate in culture. These cell types include cartilage, skeletal muscle, and fibroblastic cells (Fig. 1).

To obtain numerous replicate cultures from small amounts of embryonic tissue, we developed a micromass culture system (Ahrens et al. 1977). Several

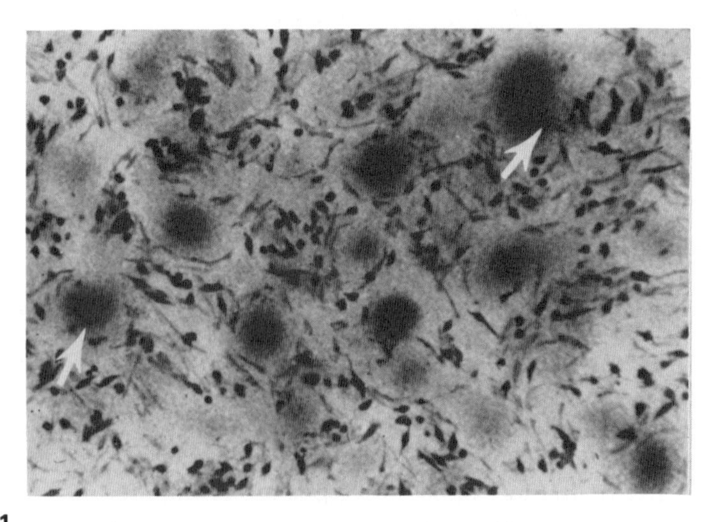

Figure 1

Day-3 high density culture, prepared from stage-23 (Hamburger and Hamilton 1951) chick embryo wing buds and doubly stained with Alcian blue at pH 1 to localize sulfated proteoglycans associated with cartilage and a monoclonal antibody directed against sarcomere myosin (MF-20) to localize skeletal muscle cells. Note that cartilage tissue is present in large multicellular groups (nodules). A large number of MF-20-positive cells also differentiate during culture and are located primarily around the nodules, although myoblasts in or on top of nodules are not uncommon (→). Magnification, 65 × . (Reprinted, with permission, from Swalla and Solursh 1986.)

cultures can be prepared from even a single embryo, since microliter volumes of cell suspension are inoculated per culture. This micro method has been used for cells from a variety of sources besides chick limb buds, which include mammalian limb buds (Lewis et al. 1978; Owens and Solursh 1981) and chick cranial neural crest cells (Wedden et al. 1986).

This culture system has been adapted for use as a screening method for teratogens. It is particularly useful for establishing such cultures in multiwell culture plates for rapid screening procedures (Hassell and Horigan 1982). As shown in Figure 2, we have developed this method for rapid teratogen testing by growing cells in 96-well culture plates in a defined culture medium. After elution of specifically stained components, such as Alcian-blue-stained proteoglycans, results can be quantitated in the same plate in an ELISA reader (Paulsen and Solursh 1987).

In our studies on the mechanisms involved in initiating chondrogenesis, we have found that chondrogenesis in high density cultures closely parallels the process in situ. As shown in Figure 1, cartilage forms as a tissue consisting of groups of chondrocytes, rather than single chondrocytes. The cell interactions involved can be studied more readily in mechanistc terms in culture than in situ. In high density cultures, areas in which cells appear to be more extensively multilayered than in surrounding regions are apparent during the first day of culture. These aggregates have been shown to represent precursors to the cartilage nodules (Ahrens et al. 1977). It is noteworthy that the precartil-

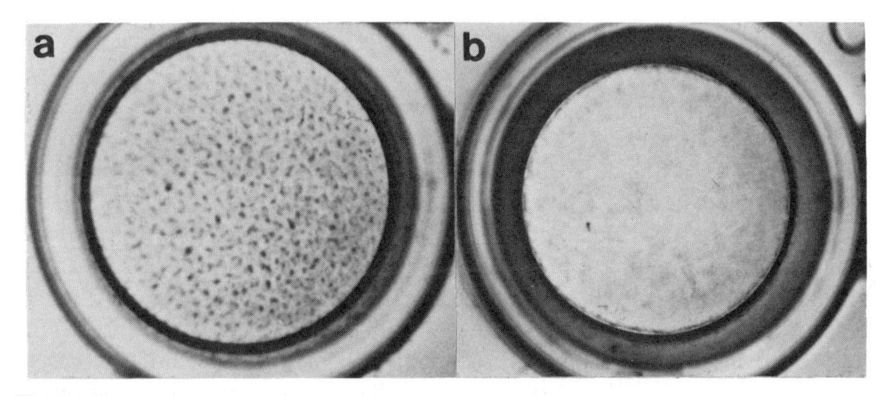

Figure 2

Stage-24 chick wing bud mesoderm cells plated at 3.5×10^6 cells per well in a 96-well tissue culture microtiter test plate. The cultures were grown for 3 days in a defined medium of a 60 : 40 mixture of F12 and Dulbecco's modified Eagle's medium, containing only insulin, transferrin, dihydrocortisone, and antibiotics as additives, and fixed and stained with Alcian blue at pH 1. (*a*) Control culture. Note the small size and relatively even distribution of cartilage nodules. (*b*) Similar 3-day culture in defined medium containing 250 μg/ml of retinoic acid. Note the complete absence of cartilage nodules. Magnification, $6.3 \times$.

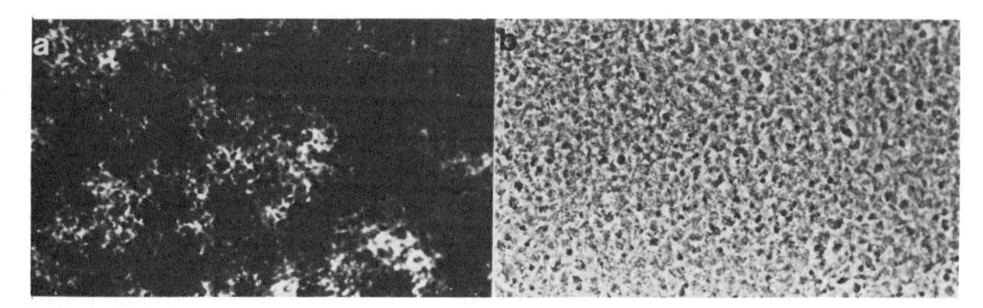

Figure 3

A 16-hour high density cell culture prepared from stage-24 chick wing buds, as in Fig. 1. (*a*) Precartilage cellular aggregates are clearly visible after incubation with PNA-rhodamine. (*b*) A phase-contrast photograph of the culture shown in *a*. Magnification, 89×. (Reprinted, with permission, from Aulthouse and Solursh 1987.)

age aggregates in culture are associated with an extracellular matrix component that binds the galactose-specific lectin, peanut agglutinin (PNA) (Fig. 3) (Aulthouse and Solursh 1987). PNA-binding material is detected before 16 hr, whereas cartilage-specific components, such as type-II collagen, are not detected before 24 hr of culture. Similarly, skeletal rudiments in situ bind PNA prior to overt chrondrogenesis (Zimmermann and Thies 1984; Aulthouse and Solursh 1987). The role of this extracellular matrix component in mediating cell-cell interactions involved in chondrogenesis is currently under investigation.

Low Density Cultures

There is considerable evidence supporting the idea that homotypic cell interactions mediate the onset of chondrogenesis. As mentioned above, one of the first indications of the cartilaginous skeleton in the embryo is the appearance of precartilage cell condensations or blastemata (Fell and Canti 1934). More direct evidence is provided by the observation that cartilage fails to form at all if high density cultures are prepared from a cell mixture in which 25% of the cells or less have the ability to form cartilage in culture and the other 75% are some other cell type (e.g., adrenal cell line) (Solursh and Reiter 1980).

As a means of understanding the mechansism by which homotypic cell interactions promote chondrogenesis, we have found culture situations in which single-limb mesenchyme cells can differentiate into chondrocytes. One such condition is low density cultures of limb mesenchyme in or on hydrated type-I collagen gels or in agarose (Solursh et al. 1982). In these cultures, a number of cell types can be identified by use of immunohistochemistry with antibodies directed against cell-type-specific molecules. In Figure 4, type-II collagen is used to recognize cartilage cells, sarcomere myosin for skeletal

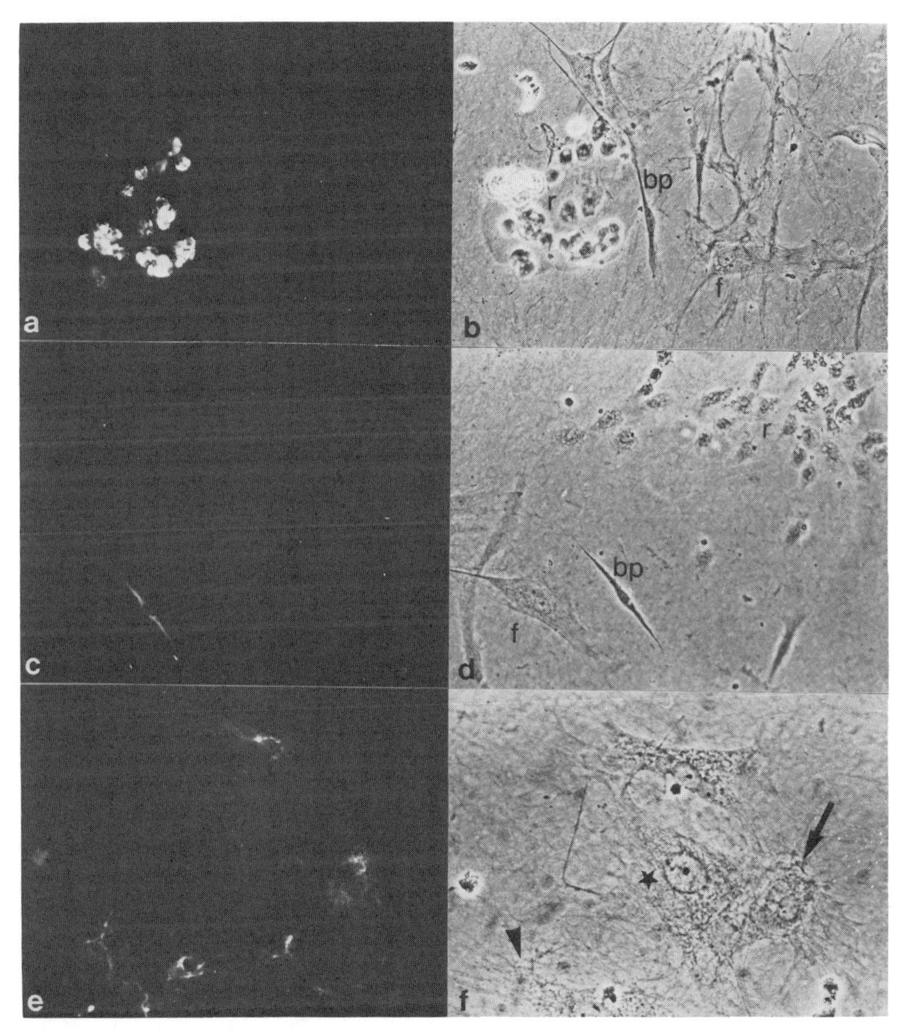

Figure 4

Indirect immunofluorescence (*a,c,e*) and corresponding phase contrast (*b,d,f*) micrographs of three morphologically distinct cell types observed in low density cultures of limb mesenchyme on hydrated collagen gels. Cultures were fixed 5–6 days after plating and stained with monoclonal antibodies directed against type-II collagen (*a*), sarcomere myosine (MF-20) (*c*), or myotendinous antigen (MI) (*e*). Rounded cells (r), which occur singly or as cartilage colonies, stain with anti-type-II collagen but not with MF-20. Other cells are bipolar (bp) and stain with MF-20 but not with anti-type-II collagen or M1. A third type of cell has a flattened morphology (f) and neither stains with MF-20 nor anti-type-II collagen. M1 antibody stains some of the flattened cells and is distributed intracellularly (arrow) or as a fibrillar matrix (arrowhead). Some flattened cells (*) do not stain with this antibody. Magnifications: (*a–d*) 90 ×; (*e,f*) 227 ×. (Reprinted, with permission, from Zanetti and Solursh 1986a.)

muscle, and myotendinous antigen for a subclass of connective tissue fibroblasts.

It is widely believed that the skeletal muscle precursors are derived from somites, and limb connective tissue cells are derived from the lateral plate mesoderm (Chevallier et al. 1977; Christ et al. 1977). The factors that determine the type of connective tissue formed are not known. It is noteworthy that the proportions of cartilage and fibroblastic cells can be manipulated readily in collagen gel cultures. The addition of fibronectin, for example, decreases the number of cartilage cells and increases the number of fibroblastic cells (Swalla and Solursh 1984).

There is a clear correlation between a rounded cell shape and chondrogenic differentiation (see Solursh 1983). In gel cultures, rounded cells can be shown to begin to produce type-II collagen (Solursh et al. 1982). Furthermore, brief treatment of low density cultures of limb mesenchyme with cytochalasin D causes disruption of the actin cytoskeleton, cell rounding, and the subsequent appearance of immunologically detectable type-II collagen (Zanetti and Solursh 1984). Other cytoskeletal-disrupting agents, such as colchicine, affect neither cell shape nor chondrogenesis. Cytochalasin also alleviates the inhibition of chondrogenesis by fibronectin. These studies suggest that the actin cytoskeleton is intimately involved in regulating the onset of chrondrogenesis and is one component involved in mediating the effects of cell-cell interaction on gene expression.

Tissue Interactions

We have utilized these culture systems to begin to explore possible mechanisms involved in establishing the spatial patterns of tissue differentiation in the developing limb bud. In the case of chondrogenesis, it is clear that potentially chondrogenic cells are widely distributed in the limb bud, since cartilage can form in explant (Zwilling 1966) or micromass (Ahrens et al. 1979) cultures of the normally nonchondrogenic, peripheral limb mesenchyme. The limb bud ectoderm appears to inhibit chondrogenesis in these regions. Chondrogenesis is more extensive in limb bud explants after removal of the ectoderm (Kosher et al. 1979) and is inhibited in micromass cultures by the addition of limb ectoderm (Solursh et al. 1981). It is noteworthy that cartilage always forms in situ at some distance from the ectoderm. One mechanism for the restriction of cartilage to the core region of the limb bud might be that the ectoderm establishes a peripheral zone that is inhibitory to chondrogenesis (Solursh 1984).

We have utilized the collagen gel culture system to study the mechanism of action of the ectodermal inhibition of chondrogenesis. The inhibitory activity appears to be a diffusible factor that acts by flattening the mesenchyme cells. As seen in Figure 5, the activity can diffuse from ectoderms lying on Millipore

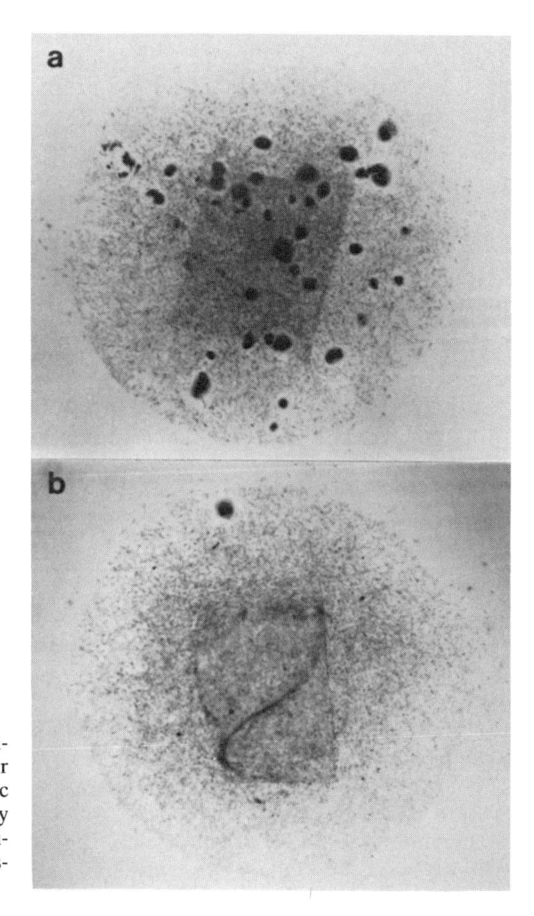

Figure 5
Hematoxylin-stained collagen gel cultures with Millipore filter only (*a*) or filter plus limb ectoderm (*b*). Note the drastic reduction of cartilage foci formation by the presence of the ectoderm. Magnification, 9×. (Reprinted, with permission, from Solursh et al. 1984.)

filters placed on a collagen gel. Since the inhibition is observed when mesenchyme cells are in collagen gels but not in agarose gels, the effect must be mediated through cell-extracellular matrix interactions (Solursh et al. 1984). The inhibition is alleviated by cytochalasin D as well (Zanetti and Solursh 1986a). More recently, we have found that ectoderm-conditioned medium causes the early flattening of mesenchymal cells on collagen gels (Zanetti and Solursh 1986b). These results are consistent with a role for diffusible factors in controlling the pattern of chondrogenic differentiation through alterations of cytoskeletal-extracellular matrix interactions.

DISCUSSION

The tissue culture approaches described here have permitted mechanistic approaches to chondrogenesis at several levels of biological organization. These include studies of the effects of one tissue on another, such as epithelial-mesenchymal interactions, the social behavior of limb mesenchymal cells, and the environmental requirements for chondrogenesis at the level of a single cell. By use of this model system, it might now be possible to isolate molecules that play regulatory roles in chondrogenesis. These include, for example, the PNA-binding, precartilage matrix protein and the ectodermally derived chondrogenic inhibitor. In addition, the mechanisms by which the cytoskeleton mediates effects of the extracellular matrix on gene activity are of central importance to understanding cytodifferentiation. All of these processes are potentially involved in abnormal development.

Tissue culture approaches clearly can be useful in the development of initial models and hypotheses for further testing. However, it is important to remember that it is essential to return ultimately to the intact embryo in order to assess critically the relevance of the hypotheses to normal and abnormal development.

ACKNOWLEDGMENTS

The studies described were largely supported by National Institutes of Health grant HD05505. I thank D.F. Paulsen for making this unpublished data available.

REFERENCES

Ahrens, P.B., M. Solursh, and R.S. Reiter. 1977. Stage-related capacity for limb chondrogenesis in cell culture. *Dev. Biol.* **60:** 69.

Ahrens, P.B., M. Solursh, R.S. Reiter, and C.T. Singley. 1979. Position related capacity for differentiation of limb mesenchyme in cell culture. *Dev. Biol.* **69:** 436.

Aulthouse, A.L. and M. Solursh. 1987. The detection of a precartilage, blastema-specific marker. *Dev. Biol.* **120:** 377.

Caplan, A.I. 1970. Effects of nicotinamide-sensitive teratogen 3-acetylpyridine on chick limb cells in culture. *Exp. Cell Res.* **62:** 341.

Chevallier, A., M. Kieny, and A. Mauger. 1977. Limb-somite relationship: Origin of the limb musculature. *J. Embryol. Exp. Morphol.* **41:** 245.

Christ, B., H.J. Jacob, and M. Jacob. 1977. Experimental analysis of the origin of the wing musculature in avian embryos. *Anat. Embryol.* **150:** 171.

Fell, H.B. and R.G. Canti. 1934. Experiments on the development in vitro of the avian knee joint. *Proc. R. Soc. Lond. B* **116:** 316.

Hamburger, V. and H.L. Hamilton. 1951. A series of normal stages in the development of the chick embryo. *J. Morphol.* **88:** 49.

Hassell, J.R. and E.A Horigan. 1982. Chondrogenesis: A model developmental system for measuring teratogenic potential of compounds. *Teratog. Carcinog. Mutagen.* **2:** 325.

Kosher, R.A., M.P. Savage, and S.-C. Chan. 1979. *In vitro* studies on the morphogenesis and differentiation of the mesoderm subjacent to the apical ectodermal ridge of the embryonic chick limb-bud. *J. Embryol. Exp. Morphol.* **50:** 75.

Krey, A.K., D.H. Dayton, and P.F. Goetinck. 1984. NICHD research workshop: Normal and abnormal development of the limb. *Teratology* **29:** 315.

Lewis, C.A., R.M. Pratt, J.P. Pennypacker, and J.R. Hassell. 1978. Inhibition of limb chondrogenesis *in vitro* by vitamin A: Alterations in cell surface characteristics. *Dev. Biol.* **64:** 31.

Owens, E.M. and M. Solursh. 1981. *In vitro* histogenic capacities of limb mesenchyme from various stage mouse embryos. *Dev. Biol.* **88:** 297.

Paulsen, D.F. and M. Solursh. 1987. Microtiter micromass cultures of limb-bud mesenchymal cells. *In Vitro* (in press).

Solursh, M. 1983. Cell-cell interactions and chondrogenesis. In *Cartilage* (ed. B.K. Hall), vol. 2, p. 121. Academic Press, New York.

———. 1984. Ectoderm as a determinant of early tissue pattern in the limb bud. *Cell Differ.* **15:** 17.

Solursh,, M. and R.S. Reiter. 1980. Evidence for histogenic interactions during *in vitro* limb chondrogenesis. *Dev. Biol.* **78:** 141.

Solursh, M., T.F. Linsenmayer, and K.L. Jensen. 1982. Chondrogenesis from single limb mesenchyme cells. *Dev. Biol.* **94:** 259.

Solursh, M., C.T. Singley, and R.S. Reiter. 1981. The influence of epithelia on cartilage and loose connective tissue formation by limb mesenchyme cultures. *Dev. Biol.* **86:** 471.

Solursh, M., K.L. Jensen, N.C. Zanetti, T.F. Linsenmayer, and R.S. Reiter. 1984. Extracellular matrix mediates epithelial effects on chondrogenesis *in vitro*. *Dev. Biol.* **105:** 451.

Swalla, B.J. and M. Solursh. 1984. Inhibition of limb chondrogenesis by fibronectin. *Differentiation* **26:** 42.

———. 1986. The independence of myogenesis and chondrogenesis in micromass cultures of chick wing buds. *Dev. Biol.* **116:** 31.

Umansky, R. 1966. The effect of cell population density on the developmental fate of reaggregating mouse limb bud mesenchyme. *Dev. Biol.* **13:** 31.

Wedden, S.E., M.R. Lewin-Smith, and C. Tickle. 1986. The patterns of chondrogenesis of cells from facial primordia of chick embryos in micromass culture. *Dev. Biol.* **117:** 71.

Zanetti, N.C. and M. Solursh. 1984. Induction of chondrogenesis in limb mesenchymal cultures by disruption of the actin cytoskeleton. *J. Cell Biol.* **99:** 115.

———. 1986a. Epithelial effects on limb chondrogenesis involve extracellular matrix and cell shape. *Dev. Biol.* **113:** 110.

———. 1986b. Diffusible factor produced by limb ectoderm inhibits chondrogenesis by promoting cell spreading. *J. Cell Biol.* **103:** 97a.

Zimmermann, B. and M. Thies. 1984. Alterations of lectin binding during chondrogenesis of mouse limb buds. *Histochemistry* **81:** 353.

Zwilling, E. 1966. Cartilage formation from so-called myogenic tissue of chick embryo limb buds. *Ann. Med. Exp. Fenn.* **44:** 134.

COMMENTS

Saxén: Could we get back to the first of your results, Dr. Solursh? I think it is remarkable that you get single cartilage cells in collagen. As you said, people have failed to show that in nonconfluent cultures. So, the only "goody" you added into your cultures was collagen type I. Are you willing to speculate how this would affect single cells? Is there collagen type I present prior to chondrogenesis in vivo in the chondrogenic mesenchyme?

Solursh: From studies of Dessau et al., it has been shown that type-I collagen is present throughout the early mesenchyme prior to chondrogenesis. Right before the onset of chondrogenesis, as judged by immunofluorescence, which is nonquantitative, there appears to be an increased concentration of type-I collagen within the precartilage aggregates. The type-I collagen is clearly there at the right time and in the right places. Whether or not it plays any special role is not clear. In these cultures, type-I collagen appears to provide a permissive environment for chondrogenesis by single cells. As I mentioned, agarose will work just as well and, in fact, even a little better in a quantitative sense. Apparently, there are many ways of doing the same thing. Presumably, there is some underlying change in the cell or some environmental feature that the cells recognize and are programmed to respond to in a certain way.

But it is interesting that the early mesenchymal cells respond to type-I collagen by remaining spherical. Some other cells, like dermal fibroblasts, for instance, will flatten very quickly. So their spreading response to this particular matrix is apparently programmed.

Pratt: As you are well aware, retinoic acid can replace the polarizing region in the chick limb, which influences the differentiation and development of the digits. You mentioned a diffusible morphogen, and since retinoids inhibit chondrogenesis, is it possible that this may be a natural morphogen that would regulate the areas that would undergo chondrogenesis in the chick limb?

Solursh: That is a very prevalent idea. But retinoids may, in fact, play a normal role in patterning. How they might work, or the extent, just is not clear.

In the case of the effects of retinoids on amniote limb development, it is very exciting that exposure of early limbs to retinoids can cause effects on patterning, something that is not often produced by teratogens or environmental agents. One can get limb duplications, for instance. Just how the retinoids work in that situation is not known. One prevalent idea is that retinoids may act directly on the ectodermal ridge to cause a change in its configuration.

Work of Eichele et al. suggests that beads impregnated with retinoids have the ability to provide directional release. The direction of release needs to be toward the ectodermal ridge to get these limb duplications.

But, on the other hand, Summerbell and co-workers have suggested that a retinoid-like molecule may play a role in establishing the anterior/posterior axis of the limb because mirror image duplication is produced by retinoids at the anterior end of a limb; in this way, two posterior organizing zones are produced. What is actually happening remains to be worked out.

Pratt: What about your system? Do you have any idea as to what this morphogen may be, that is, the substance transmitted from the ectoderm to the mesoderm?

Solursh: In the case of retinoids, or the influence that establishes the anterior/posterior axis of the limb, the activity appears to be localized in the mesoderm rather than in the ectoderm, although that is also clearly a diffusible type of influence. The ectodermal influence that we have been studying is very widespread. It does not have some of the tissue specificity that one often sees in embryonic inducers. Epithelia from a variety of sources produce this same type of inhibitory effect, at least embryonic epithelia.

We have not isolated the component, and we really know very little about it except that it is heat and protease sensitive. The activity appears to be blocked by agents that prevent attachment or spreading. It interacts with fibronectin. We are in the process of trying to find out more about the component(s).

Kochhar: I was just going to reiterate what Dr. Solursh said, that the effects of retinoids on mammalian systems are not very different from those on chick embryos. There is no doubt that retinoic acid-like teratogens inhibit chondrogenesis, but retinoids also inhibit chondrogenesis in avian systems too. All-*trans* retinoic acid and most of its analogs inhibit chondrogenesis. So the question is how do they produce duplication of the limb parts in certain situations. The answer has something to do with an effect on the ectodermal ridge.

In fact, Tickle and co-workers have found that the first effect that they see after local application of retinoic acid is an extensive lengthening of the ectodermal ridge and, secondarily, on limb duplication. So I think the ectodermal ridge or epithelium is providing a lot of guiding factors. The question is, what is that factor that the epithelium is providing?

All retinoids have long been known to intervene and actually maintain differentiation of epithelia, but all of those effects were presumed to be mediated by an effect on cell multiplication and cell division. Now, an effect on some aspects of genomic regulation has been proposed. We do not know whether any of these gene-modifying effects are operating in the embryo, but apparently an explanation should be sought at that level.

Saxén: Did you ever try to culture your chondroblasts on or in the presence of the basement membrane components like laminin, fibronectin, or collagen type IV or V?

Solursh: We know, for instance, that fibronectin either added to high-density cultures or added to mesenchyme cells suspended in collagen gel can inhibit chondrogenesis and that inhibition can be alleviated by cytochalasin D. However, laminin has no detectable effect one way or the other on chondrogenesis. We have not tried type-IV collagen or other components. Chondrogenesis normally occurs at some distance from basement membranes.

Welsch: Several recent studies, primarily in the field of carcinogenesis and some observations made in my laboratory in limb mesenchymal cell micromass cultures indicate that retinoic acid derivatives are fast cell–cell channel blockers. We use fluorescent dye injections and dye coupling during chondrogenic differentiation of these cells to assess the functional status of cell–cell channels. All-*trans* and 13-*cis*-retinoic acid are effective within minutes, depending on the concentration, in blocking cell–cell communication, which is presumably mediated by gap junctions. The electromicroscopical observations reported by Merker, Zimmermann, and Barrach in Berlin (*Acta Biol. Acad. Sci. Hung.* **35:** 195 [1984]), have associated the appearance of gap junctions close in time with the onset of chondrogenic differentiation, suggesting that there may be a relationship.

The factor that you described is apparently a macromolecule that is heat labile and protease sensitive. A molecule of this size would not be compatible with gap junction-mediated effects, because that would require a small compound with a molecular weight of 800 or 1000 at most. The factor described by you would not fit through cell–cell channels, but it is possible that junctional communication is involved in chondrogenic differentiation.

Solursh: It would be interesting to know the possible relationship between cytoskeletal organization and the extent of cell communication; somehow the influences outside the cell and on the cell surface must be mediated and must extend to within the cell, thinking of it as a continuum.

Kochhar: Could the PNA-binding protein be one of the cytoskeletal proteins? Is that what you are suggesting?

Solursh: No. It is clearly extracellular—on the surface of the cells.

Kochhar: You mentioned that single cells can embark on chondrogenesis, but because there is type-II collagen, is that enough of a criterion to call a cell a chondrocyte?

Solursh: With very few exceptions, type-II collagen is diagnostic of chondrocytes. We have also used other markers at the same time, however, because we find type-II collagen message very early in the embryo and very widely distributed. Using other markers, like antibodies for cartilage proteoglycan, however, we essentially find the same results. But I agree, it is very difficult to find cell type-specific molecules, and, to be careful, one must use multiple endpoints.

Dencker: Did you ever study the effect of retinoic acid on single cells during differentiation and was it effective?

Solursh: Yes, too effective. The work of Hassell and others shows that in high-density cultures, chondrogenesis is inhibited by retinoids, as is the continued expression of chondrogenesis by already differentiated cartilage cells. We found that if we treat the early embryonic mesenchymal cells with retinoids, either in suspension or collagen gels, they die, even at very low concentrations. They seem to be extremely susceptible, whereas chondrocytes become flattened, start to make type-III collagen, and stop making type-II collagen. There is something peculiarly different about these early mesenchymal cells that makes them very susceptible to very low concentrations of retinoids. We have been unable to explore the effects of retinoids at the onset of chondrogenic differentiation because the cells die.

It is quite different if they are treated when they are attached. If they are put on plastic, they do not die; rather, they flatten, but they do not become chondrocytes.

Comparative Evaluation of a Short-term Test for Developmental Effects Using Frog Embryos

THOMAS D. SABOURIN AND ROBIN T. FAULK
Environmental Sciences Department
Battelle Columbus Division
Columbus, Ohio 43201-2693

OVERVIEW

FETAX (Frog Embryo Teratogenesis Assay: *Xenopus*) was evaluated as a candidate in vitro teratogenesis assay by testing 35 chemicals listed in a consensus National Toxicology Program (NTP) teratogenesis chemical repository. The most promising endpoints were embryo malformation and growth during the 4-day test. Seventeen of 20 in vivo mammalian teratogens tested positive, and 12 of 15 nonvariable mammalian teratogens tested negative in FETAX, for an overall predictive accuracy of 83%. Growth of the embryos is a sensitive index and generally tracks results observed in malformation/lethality comparisons. Nineteen of the 20 in vivo mammalian teratogens tested produced malformations in frog embryos. For the 19 teratogens, 67% of the primary malformations reported in mammalian embryos were detected in FETAX. The fact that *Xenopus* is a vertebrate, with many developmental features in common with mammals, coupled with preliminary data on observed embryo abnormalities, makes FETAX a strong candidate for further consideration as a teratogen screen.

INTRODUCTION

Perceptions among scientific, regulatory, and industrial communities have changed during this decade on the issue of alternative methods for toxicity testing (National Research Council 1985; U.S. Environmental Protection Agency 1985; Office of Technology Assessment 1986). There is a growing interest to incorporate novel in vivo and in vitro methods that screen toxic substances. This change can be attributed, in part, to the overwhelming number of chemicals in the marketplace that require safety evaluations and to the resultant research focus on alternative test systems. The need for routine teratogenicity testing has led to the development of a number of in vitro teratogenesis assays that may prove useful in prioritizing compounds for further testing (New 1978; Johnson 1980; Bournias-Vardiabasis and Teplitz 1982; Greenberg 1982; Schuler et al. 1982; Goss and Sabourin 1985). One

Banbury Report 26: Developmental Toxicology: Mechanisms and Risk

such system is FETAX, which was developed by Dumont and co-workers and applied to screening complex environmental mixtures, as well as pure compounds (Dumont et al. 1983b).

FETAX meets most of the criteria set forth by Kimmel et al. (1982) for the validation of in vitro teratogenesis assays (Dumont et al. 1983b; Courchesne and Bantle 1985; Sabourin et al. 1985). Endpoints such as mortality, malformation, growth, development, and motor impairment can be quantified rapidly and exhibit a dose–response relationship with the establishment of narrow confidence limits. Since many of the stages of amphibian development are similar to mammalian development, the "developmental relevance" (Smith et al. 1983) of FETAX is higher than that of many of the other in vitro teratogenesis assays.

The objective of the research presented in this paper was to evaluate FETAX using a number of chemicals (Table 1), the majority of which are NTP repository chemicals designated for use in validating teratogenesis screening assays (Smith et al. 1983). The results are compared with those obtained in published research on in vivo mammalian systems.

The following procedures are used to obtain data in the FETAX assay system. Twelve hours prior to mating, a female *Xenopus* is given 1000 I.U., and a male, 400 I.U. of human chorionic gonadotropin injected into the dorsal lymph sac. Amplexus ensues in 4–6 hours, and fertilized egg deposition in 9–12 hours from the time of injection. Normally developing mid-cleavage blastulae are selected for a test. For each toxicant concentration, two groups of 20 blastulae each are placed in either covered glass or plastic petri dishes. Developing embryos are continuously exposed to the test solution for 96 hours, with daily test solution renewals. Mortality and stage of development are checked at 24, 48, 72, and 96 hours, and remaining endpoints are recorded only at 96 hours. The endpoints include LC_{50} (mortality), EC_{50} (malformation–teratogenesis), no observable effects concentration (NOEC), growth (both length and developmental stage obtained in a given time period), motor behavior (ability to swim across a specified field), pigmentation (relative change in experimental embryos), and gross anatomy.

Several stages of normal development, from blastula to 96-hour tadpole, are shown in Figure 1. The 96-hour embryo is approximately 10 mm in total length. At this stage, the animal is capable of swimming and feeding. Metamorphosis occurs in *Xenopus* approximately 2 months postfertilization.

The primary endpoints in FETAX are mortality and malformation. Embryos from each test chemical concentration are (1) compared with controls/solvent controls and (2) assessed for teratogenicity. Teratogenicity is suggested in dose–response data when the concentration effective in producing embryo malformations is lower than that producing embryolethality. Com-

Table 1
Test Chemicals Assayed in the FETAX Test System

Chemical	CAS[a] no.	Concentration range (mg/liter, except where noted*)
Acetazolamide	59-66-5	0–1,000
Amaranth	915-67-3	0–10,000
9-Aminoacridine[b]	90-45-9	0–10
6-Aminonicotinamide	329-89-5	0–5,600
Aspirin	50-78-2	0–1,000
Cadmium chloride[b]	10108-64-2	0–10
Caffeine	58-08-2	0–1,000
Carbon tetrachloride	56-23-5	0–0.056%*
Chlorambucil	305-03-3	0–100
Coumarin	91-64-5	0–560
Cyclophosphamide	50-18-0	0–10,000
Dexamethasone-21 acetate	50-02-2	0–100
Diethylstilbestrol	56-53-1	0–3.2
Dilantin	57-41-0	0–80
Diphenhydramine HCl	147-24-0	0–10
Doxylamine succinate	562-10-7	0–1,800
Ethyl alcohol	64-17-5	0–10%*
Ethylenethiourea	96-45-7	0–1,000
N-ethyl-*N*-nitrosourea	759-73-9	0–100
5-Fluorouracil	51-21-8	0–1,000
Formaldehyde	50-00-0	0–0.01%*
Hydroxyurea[b]	127-07-1	0–1,800
Isoniazid	54-85-3	0–5,600
Methotrexate	59-05-2	0–560
Nitrilotriacetate	5064-31-3	0–560
Penicillin G	69-57-8	0–32,000
L-Phenylalanine	63-91-2	0–25,000
Phthalimide	85-41-6	0–560
t-Retinoic acid	302-79-4	0–0.320
Saccharin	81-07-2	0–24,000
Sodium arsenate	7631-89-2	0–7,500
Sodium cyclamate	139-05-9	0–32,000
Urethane	51-79-6	0–5,600
Vinblastine sulfate[b]	143-67-9	0–10
Vincristine sulfate	2068-78-2	0–18

[a]Chemical Abstract Service.
[b]Results for work completed on these chemicals have been published previously (Sabourin et al. 1985).

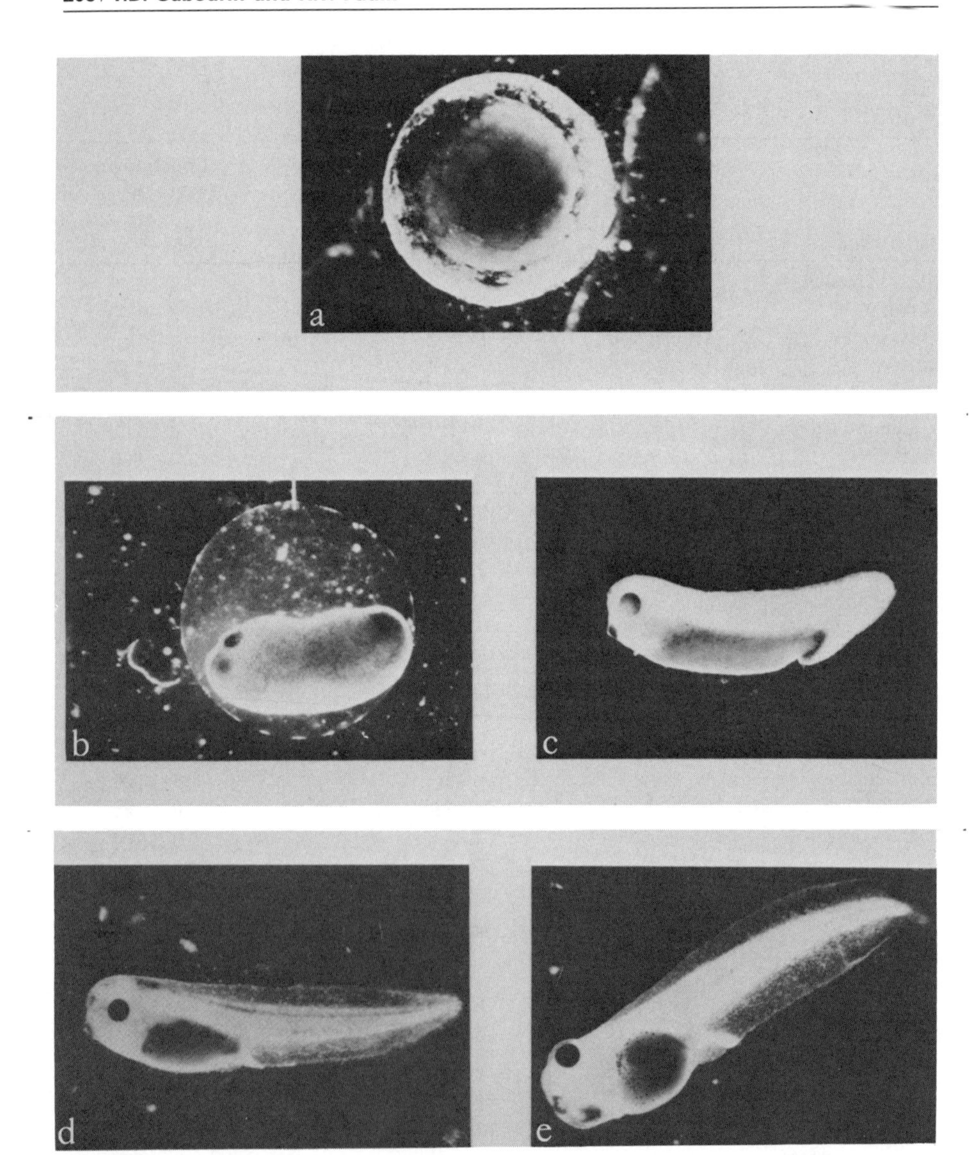

Figure 1
Normal development of *Xenopus*: Age: 5 hr (*a*); 20 hr (*b*); 48 hr (*c*); 72 hr (*d*); 96 hr (*e*).

monly occurring malformations observed in 96-hour embryos include edema, blistering, abnormalities of the axial skeleton, loss of pigmentation, and optic defects (Fig. 2).

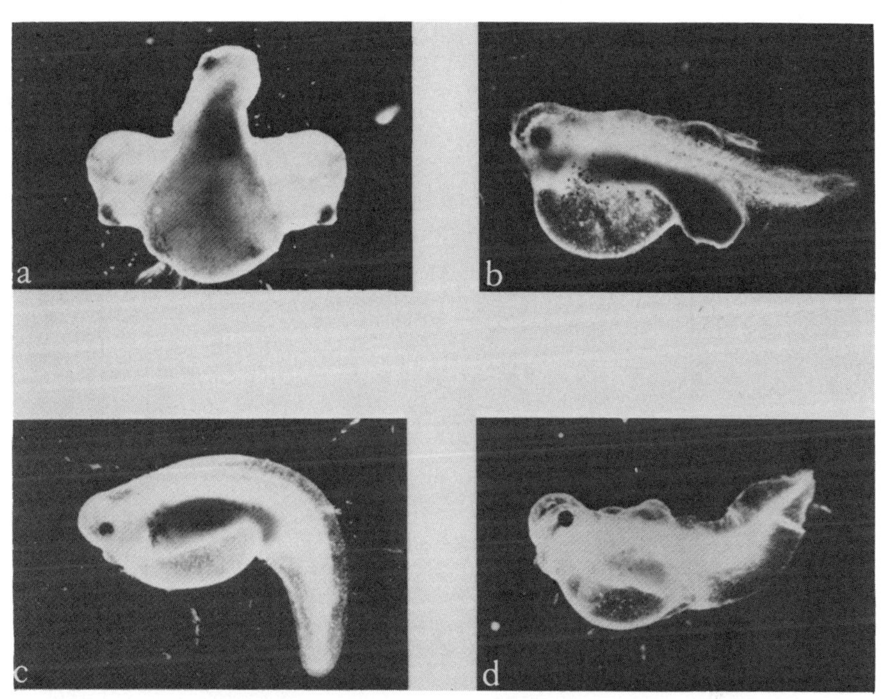

Figure 2

Examples of malformations observed in 96-hr embryos: (*a*) Caffeine (180 mg/liter); (*b*) 5-fluorouracil (320 mg/liter); (*c*) hydroxyurea (100 mg/liter); (*d*) *t*-retinoic acid (0.018 mg/liter).

RESULTS

Results obtained for one compound that tested nonteratogenic (penicillin G) and another that tested teratogenic (vinblastine sulfate) in FETAX are illustrated in Figure 3. Percent embryolethality and embryo malformation dose–response curves are essentially identical for penicillin G, indicating that malformed embryos are only observed at lethal concentrations. In contrast, results for vinblastine showed that 100% (40/40) of the frog embryos were malformed at a concentration where only 5% (2/40) mortality occurred. The likelihood of teratogenicity of the test chemical in FETAX increases in direct proportion to the divergence of the embryo malformation and embryolethality curves.

The comparison of embryolethality and embryo malformation provides an index of teratogenicity (TI_{50}). Various teratogenic indexes have been used to categorize chemicals on the basis of measured responses. The TI_{50} in FETAX is the test concentration effective in producing embryolethality in 50% of the

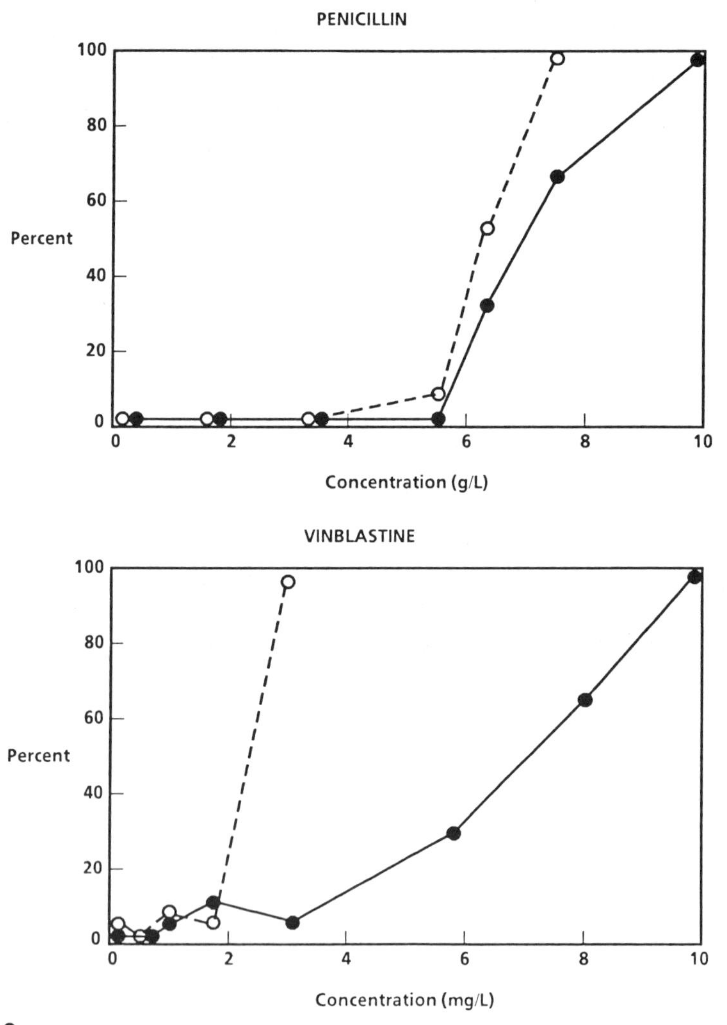

Figure 3
Embryolethality (●) and embryo malformation (○) in *Xenopus* for a representative non-teratogen, penicillin (*top*), and a teratogen, vinblastine sulfate (*bottom*). Values are given as means.

frog embryos divided by the test concentration that results in malformation in 50% of the embryos. A TI_{50} above 1.5 indicates potential teratogenicity in the FETAX screen. A test chemical in this category should warrant further assessment in another in vitro teratology system or in an in vivo system. A

TI_{50} range of 1.6–2.5 is considered slightly teratogenic, and a range above 2.5 indicates high potential for teratogenicity.

The 35 chemicals tested in FETAX elicited the results given in Table 2. Chemicals listed in Table 2 are divided into teratogens and nonvariable teratogens on the basis of their consensus actions in in vivo mammalian test systems (Smith et al. 1983). The upper half of the table lists in vivo mammalian teratogens. Using the TI_{50}, 17 of the 20 chemicals tested positive for teratogenesis in FETAX; 12 of the 15 nonvariable teratogens tested negative. On the basis of this list of 15 chemicals and the consensus in vivo actions, FETAX has a sensitivity of 85% (15% false negatives) and a specificity of 80% (20% false positives) for an overall predictive accuracy of 83%.

Table 2
FETAX Responses

Category 1 (2.6 and up)	Category 2 (1.6–2.5)	Category 3 (<1.6)
	In vivo mammalian teratogens	
6-Aminonicotinamide	Chlorambucil	Aspirin
Caffeine	Cyclophosphamide	Ethanol
Dexamethasone	Diethylstilbestrol	L-Phenylalanine
Ethylnitrosourea	Dilantin	
5-Fluorouracil	Ethylenethiourea	
Hydroxyurea	Methotrexate	
t-Retinoic Acid	Sodium Arsenate	
Vinblastine Sulfate	Urethane	
Vincristine Sulfate		
	In vivo mammalian nonvariable teratogens	
Diphenhydramine HCl	Phthalimide	Acetazolamide
Isoniazid		Amaranth
		9-Aminoacridine HCl
		Cadmium chloride
		Carbon tetrachloride
		Coumarin
		Doxylamine succinate
		Formaldehyde
		Nitrilotriacetate
		Penicillin G
		Saccharin
		Sodium cyclamate

A similar comparison of the 35 test chemicals, using data found in the NIOSH Registry of Toxic Effects of Chemical Substances (RTECS) (Lewis and Sweet 1984), shows that the published in vivo data base includes some conflicting results (Table 3). Table 3 was compiled by searching RTECS for appropriate toxicity and reproductive effects data, where available, for the same species and by the same route of administration. For a given test chemical, data for more than one species and/or more than one route of administration were used if these data were available. A number of chemicals (e.g., dilantin and ethanol) appear both in a teratogenic category and in the nonteratogenic, category 3. In addition, 5 of the 15 nonvariable teratogenic chemicals appear in category 1. These data underscore the difficulties en-

Table 3
In Vivo Mammalian Responses

Category 1 (2.6 and up)	Category 2 (1.6–2.5)	Category 3 (<1.6)	Appropriate data not found
	In vivo mammalian teratogens		
6-Aminonicotinamide	Ethanol	Caffeine	Dexamethasone
Aspirin	Ethylnitrosourea	Dilantin	L-Phenylalanine
Chlorambucil	Urethane	Ethanol	Sodium arsenate
Cyclophosphamide		Ethylenethiourea	
Diethylstilbestrol		5-Fluorouracil	
Dilantin		Urethane	
Ethylenethiourea			
Ethylnitrosourea			
5-Fluorouracil			
Hydroxyurea			
Methotrexate			
t-Retinoic acid			
Vinblastine sulfate			
Vincristine sulfate			
	In vivo mammalian nonvariable teratogens		
Amaranth	Acetazolamide	Carbon tetrachloride	9-Aminoacridine
Cadmium chloride		Coumarin	Doxylamine succinate
Diphenhydramine HCl		Saccharin	Formaldehyde
Isoniazid		Sodium cyclamate	Nitrilotriacetate
Sodium cyclamate			Penicillin G
			Phthalimide

From NIOSH (Lewis and Sweet 1984).

countered in attempts to evaluate a screen using the in vivo data base as a comparison.

Nineteen of the 20 in vivo teratogens produced malformations in *Xenopus* embryos. L-Phenylalanine was the only chemical tested from this group that did not exhibit toxicity. Concentrations of L-phenylalanine up to 25 g/liter produced no signs of lethality or teratogenicity. Of the remaining 19 test chemicals, aspirin and ethanol did not test positive for teratogenesis in FETAX using a TI_{50}, but both caused embryo malformations. The primary malformations observed were edema and microcephaly for aspirin and ethanol, respectively. A comparison of FETAX results with the published in vivo data for the 19 in vivo mammalian teratogens indicates that approximately 2 out of every 3 malformations observed in mammals can be detected in FETAX (Table 4). Although this table is preliminary, it provides evidence that FETAX test results may reveal more about the potential action of a test chemical than a simple plus or minus. The value of FETAX in predicting abnormalities can be reassessed when an adequate data base is obtained.

An endpoint of FETAX that is not used in standard malformation/embryolethality comparisons is growth. Growth during the experimental period is determined by measuring length achieved in 96 hours. Chemicals that test teratogenic in FETAX usually inhibit growth in the affected individuals. Conversely, nonteratogenic compounds generally do not affect growth until test concentrations are in the lethal range. Results for a teratogen (hydroxyurea) and a nonteratogen (cadmium chloride) are illustrated in Figure 4. Hydroxyurea inhibited growth in embryos at test concentrations below 5% of the LC_{50}, whereas cadmium chloride did not inhibit growth until test concentrations approached the LC_{50}. Excluding L-phenylalanine, for which no response was obtained, 16 of the 19 in vivo

Table 4
Comparative Qualitative Responses for 19 Mammalian Teratogens

Malformation	Mammalian frequency	FETAX frequency	No. of matches	FETAX performance (%)
Skeletal	18	16	15	83
Visceral	3	10	2	67
Nervous	10	8	6	60
Optic	5	4	1	20
Osmoregulatory	3	4	2	67

$$\frac{\text{No. of matches}}{\text{Total mammalian malformations}} = \frac{26}{39} = 67\%$$

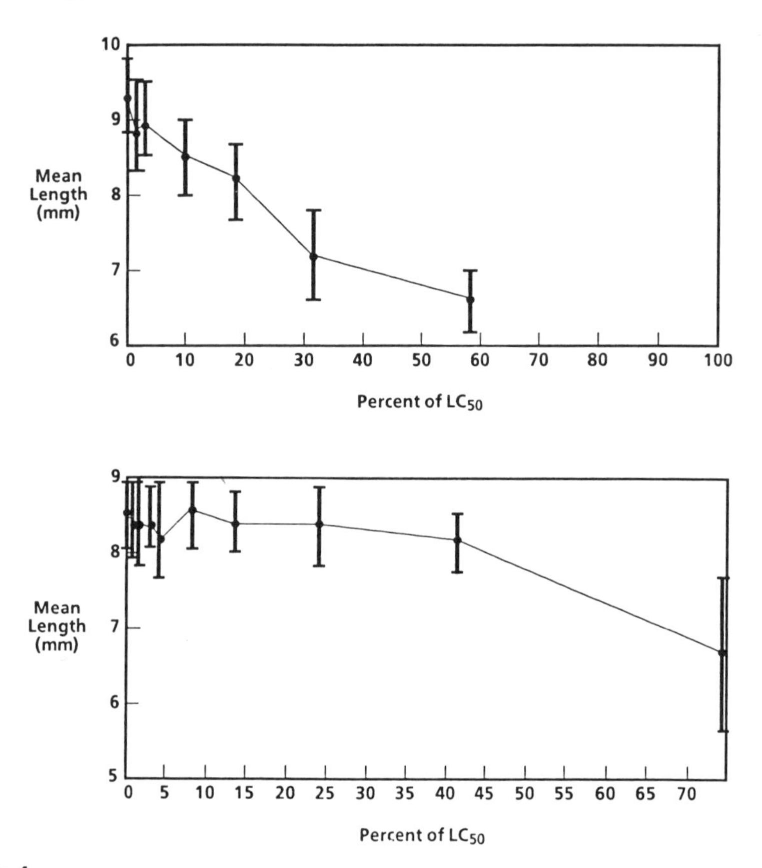

Figure 4
Length in *Xenopus* embryos after a 96-hr exposure to a teratogen, hydroxyurea (*top*), and a nonteratogen, cadmium chloride (*bottom*). Values are given as means and 1 S.D.

teratogens showed the trend of hydroxyurea, and 10 of the 15 nonteratogens tracked the trend of cadmium chloride (Table 5). Growth is a simple and sensitive indicator of embryotoxicity in *Xenopus*.

Table 5
Growth Inhibition

	Inhibition well below LC_{50}	Inhibition at or near LC_{50}
Mammalian teratogens	16	3
Mammalian nonteratogens	5	10

DISCUSSION

The predictive accuracy of 80% + obtained for the list of 35 chemicals has encouraged further work with FETAX in our laboratories. J.N. Dumont (pers. comm.) has obtained similar correspondence between FETAX and published mammalian results. Correlation between laboratories has also been encouraging. For example, Courchesne and Bantle (1985) and Sabourin et al. (1985) obtained TI_{50}s for hydroxyurea of 4.3 and 4.5, respectively.

One advantage of using FETAX is the apparent developmental relevance of the test system. Using the same stock of dilantin, we determined that the primary endpoint was pericardial edema in both cultured whole rat embryos (Clapper et al. 1986) and frog embryos (present study). Dumont et al. (1983a) similarly found that meclizine induced hydrocephaly in both frogs and mammals. Courchesne and Bantle (1985) reported that several genotoxic chemicals caused the same general types of malformations in both *Xenopus* and rodent embryos. *Xenopus* has also been shown to exhibit certain organogenic teratogenic effects demonstrated previously in mammalian embryos (Schultz et al. 1985). The frequency of these frog–mammal correlations for 19 teratogens reported here (67%) is adequate to warrant further investigation. Confirmation of these inferences would mean the assay has use beyond that of a simple screen. The 35 compounds tested in this study cover a range of genotoxic, cytotoxic, and organotoxic effectors. The ability to detect these abnormalities will be a distinct advantage when working with unknowns.

The strength of FETAX is the number of endpoints that can be measured. For example, data generated on embryo malformation can be supported by data on reduced growth. Growth data, in turn, correlates with data on developmental stage achieved by the embryo at a specified time. The FETAX endpoints exhibit good dose–response curves to allow comparison between test chemicals for ranking purposes.

Broad-scale use of FETAX, as well as other in vitro teratogenesis assays, will depend on the establishment of a bioactivation system. Proteratogenic substances, such as cyclophosphamide, may test as false negatives in a test system that lacks bioactivation. Preliminary work with *Xenopus* indicates that a limited degree of basal P-450 activity and subsequent induced P-450 activity is present in the 96-hour embryo (T.D. Sabourin, unpubl.). The addition of an external metabolic activation system, such as isolated *Xenopus* hepatocytes or uninduced rat liver microsomes, will likely be required. Bantle and Dawson (1986) have recently reported successes with uninduced rat liver microsomes, as demonstrated by increased cyclophosphamide teratogenesis in the activated versus nonactivated systems.

We have presented evaluation data for 35 chemicals, most of which are listed in an established NTP teratogenesis chemical repository. FETAX results were compared with in vivo mammalian results found in several research and review publications (Harbison 1978; Shepard 1980; Smith et al. 1983; Beckman and Brent 1984). The quantitative use of these results for the sake of comparison is unconvincing, as exemplified by Table 3, which is based on mammalian data found in RTECS. Validation for FETAX and other in vitro teratogenesis assays should consist of a list of representative consensus chemicals tested concurrently in the frog embryo system and in an appropriate mammalian model system. Results from such a controlled validation program would facilitate qualitative and quantitative comparisons. Our evaluation indicates that FETAX satisfies the criteria for an in vitro teratogenesis screening assay and should be placed into a rigorous validation program.

ACKNOWLEDGMENTS

Technical assistance in conducting FETAX assays was provided by Ms. Michele Moore Gatz. The study was supported by funding from Battelle Memorial Institute.

REFERENCES

Bantle, J.A. and D.A. Dawson. 1986. The use of *Xenopus laevis* embryos for the detection of agents teratogenic in humans. In *Proceedings of the Tenth Symposium on Aquatic Toxicology and Hazard Assessment*, p. 33. American Society for Test and Materials, New Orleans, Louisiana.

Beckman, D.A. and R.L. Brent. 1984. Mechanisms of teratogenesis. *Annu. Rev. Pharmacol. Toxicol.* **24:** 483.

Bournias-Vardiabasis, N. and R.L. Teplitz. 1982. Use of *Drosophila* embryo cell cultures as an in vitro teratogen assay. *Teratog. Carcinog. Mutagen.* **2:** 333.

Clapper, M.L., M.E. Clark, N.W. Klein, P.J. Kurtz, B.D. Carlton, and R.S. Chhabra. 1986. Cardiovascular defects in rat embryos cultured on serum from rats chronically exposed to phenytoin. *Teratog. Carcinog. Mutagen.* **6:** 151.

Courchesne, C.L. and J.A. Bantle. 1985. Analysis of the activity of DNA, RNA, and protein synthesis inhibitors on *Xenopus* embryo development. *Teratog. Carcinog. Mutagen.* **5:** 177.

Dumont, J.N., T.W. Schultz, and R.G. Epler. 1983a. The response of the FETAX model to mammalian teratogens. *Teratology* **27:** 39A.

Dumont, J.N., T.W. Schultz, M.V. Buchanan, and G.L. Kao. 1983b. Frog embryo teratogenesis assay: *Xenopus* (FETAX)—A short-term assay applicable to complex environmental mixtures. In *Short-term bioassays in the analysis of complex environmental mixtures* (ed. M. Waters et al.), p. 393. Plenum Press, New York.

Goss, L.B. and T.D. Sabourin. 1985. Utilization of alternative species for toxicity testing: An overview. *J. Appl. Toxicol.* **5:** 193.
Greenberg, J. 1982. Detection of teratogens by differentiating embryonic neural crest cells in culture: Evaluation as a screening system. *Teratog. Carcinog. Mutagen.* **2:** 319.
Harbison, R.D. 1978. Chemical-biological reactions common to teratogenesis and mutagenesis. *Environ. Health Perspect.* **24:** 87.
Johnson, E.M. 1980. A subvertebrate system for rapid determination of potential teratogenic hazards. *J. Environ. Pathol. Toxicol.* **4:** 153.
Kimmel, G.L., K. Smith, D.M. Kochhar, and R.M. Pratt. 1982. Overview of *in vitro* teratogenicity testing: Aspects of validation and application to screening. *Teratog. Carcinog. Mutagen.* **2:** 211.
Lewis, R.J. and D.V. Sweet. 1984. *Registry of toxic effects of chemical substances.* U.S. Department of Health and Human Service National Institute of Occupational Safety and Health, Washington, D.C.
National Research Council. 1985. *Models for biomedical research: A new perspective.* National Academy Press, Washington, D.C.
New, D.A.T. 1978. Whole-embryo culture and the study of mammalian embryos during organogenesis. *Biol. Rev.* **53:** 81.
Office of Technology Assessment. 1986. *Alternatives to animal use in research, testing, and education;* U.S. Congress. U.S. Government Printing Office, Washington, D.C.
Sabourin, T.D., R.T. Faulk, and L.B. Goss. 1985. The efficacy of three non-mammalian test systems in the identification of chemical teratogens. *J. Appl. Toxicol.* **5:** 227.
Schuler, R., B.D. Hardin, and R. Niemeier. 1982. *Drosophila* as a tool for the rapid assessment of chemicals for teratogenicity. *Teratog. Carcinog. Mutagen.* **2:** 293.
Schultz, T.W., J.N. Dumont, and R.G. Epler. 1985. The embryotoxic and osteolathyrogenic effects of semicarbizide. *Toxicology* **36:** 183.
Shepard, T.H. 1980. *Catalog of teratogenic agents.* Johns Hopkins University Press, Baltimore.
Smith, M.K., G.L. Kimmel, D.M. Kochhar, T.H. Shepard, S.P. Spielberg, and J.G. Wilson. 1983. A selection of candidate compounds for in vitro teratogenesis test validation. *Teratog. Carcinog. Mutagen.* **3:** 461.
U.S. Environmental Protection Agency. 1985. *Alternative methods for toxicity testing: Regulatory policy issues.* U.S. EPA Office of Policy, Planning and Evaluation, Washington, D.C.

COMMENTS

Iannaccone: Is there environmental evidence for embryonic teratogenicity?

Sabourin: Yes, there is. The *Xenopus* assay has been used in the field within the last couple of years in some environmental hot spots. From environmental standpoint, adult fish and fish embryos have been observed more often than frogs, simply because they are more prevalent.

Iannaccone: What kind of water do you use? How do you make up the solutions that the frogs swim around in?

Sabourin: We are using our regulatory laboratory water, which is a product of a reverse-osmosis system. Actually, the *Xenopus* will grow well in tap water.

Iannaccone: Have you worked with high-performance liquid chromatography (HPLC) waters?

Sabourin: We have what is called a FETAX saline that we use.

Iannaccone: But one must start with water to make that up.

Sabourin: Yes. One can use water from a still, Barnstead water, or Millipore water.

Scandalios: From your data showing these gross defects in the frog, there was one example in each case. I presume you exposed a population of frogs. Were there any frogs that did not show any defects?

Sabourin: What I showed were the gross defects; but, generally, we see a nice dose–response. If we use ten different concentrations and we have one effective concentration, out of 40 embryos, we may see 3 or 4 that are affected at low concentration. That would be right around our no-effect level; then at a higher concentration, we might see 20 or 30, and so forth, until all 40 are affected.

Scandalios: No. What I am asking is when you expose a population of frogs to an effective concentration, are any frogs normal at the end of the treatment?

Sabourin: I would have to answer yes.

Scandalios: Okay. To me, as a developmental biologist, it is important to find out why these frogs are resistant and, perhaps, get at the mechanism from the positive side rather than from the negative side. Are you or is someone else doing that?

Sabourin: Terry Schultz at the University of Tennessee has been working with collagen formation and has done some work along those lines. That is as close as anybody is to working on something like that right now. If you noticed my example of vinblastine, the embryo malformation curve had a few points of no malformation, and all of a sudden, it shot right up to 100% malformation. This happened in roughly one quarter of the chemicals that we ran through the system; it was all or none. My answer, then, would be no to your question; but, in some cases, it does spread out slightly. The type of malformations observed are usually the same for a given chemical.

Shepard: I would like to comment on that last discussion very briefly. There are pretty good examples in the whole embryo culture test, where a one-somite difference will make the difference in a teratogenic response. So, there is a very, very critical time period, more than we have discussed previously.

Pedersen: Could you remind me when you start your treatment? I think this is critical in *Xenopus*.

Sabourin: It is at mid-cleavage blastula, the 64-cell stage.

Pedersen: I asked about timing because John Gerhart and his associates have demonstrated a number of treatments that will cause malformations, including no heads and two heads. The treatments are all administered within the first hour following fertilization; the abnormalitites seem to be the result of disturbances in setting up the axis.

Sabourin: Again, we are going to be dependent upon the data base that is built for it. There is that window on the front end, that 5-hour period before we cull out our blastulas and set those up for hour zero of the test. We have no knowledge of what happens before that. In addition to that, the process of limb formation happens within the first 30 days or so of development.

There is a publication by our sister laboratory in Frankfurt on *Xenopus,* carrying an assay through the time of metamorphosis. EM-12 and thalidomide were effective in the system but showed their teratogenic effects down the line, around day 25 or so. So, a short-term prescreen like FETAX will miss some effectors.

Pedersen: To elaborate on the results of the Gerhart lab, the consequences of their treatments seem to be particularly important for a group of cells in the endoderm, which are destined to become the Spemann organizer. So, the kind of double-headed monsters that you see may be a result of affecting a very small group of cells.

Saxen: Would you repeat exactly when you monitored the effect, at what stage?

Sabourin: At 96 hours; that would be 101 hours postfertilization.

Saxen: I am sure that you have considered following them for a longer time because they are still extremely immature at that stage. You could even consider letting the larvae go through the metamorphosis. That would mean examining them after 2 months.

Sabourin: That is correct. In *Xenopus*, it is 60 days.

Hanson: I am invariably struck by the differences in the way one chooses to analyze these types of test systems statistically. There is no real generation of any power curves to detect them. And, beyond that, I am always somewhat distressed at the choice of a way of characterizing anomalies, because they are usually categorized by organ system. This does not really make much sense, at least in any mechanistic approach. I wonder why you do not choose to look for specific types of anomalies that may represent certain kinds of pathogenic mechanisms in your system and group them in that fashion. Such data might be generalized to other nonfrog systems.

Sabourin: What I have shown you is very preliminary. We can do a lot more work in terms of the statistics. I did not show you the curves that we have generated to analyze these data, but we do use such methods.

Hanson: In applying those to field situations then, it seems to me that one would have serious problems in generalizing. With the kinds of stratifications that would be necessary, as well as a variety of other issues, how do you get adequate numbers? In your presentation, you commented that there were large numbers of animals studied; I think you mentioned the number 40.

Sabourin: Yes, for setting up the test.

Hanson: I guess I don't view that as a very large number or one that would have very much statistical power.

Sabourin: What I am saying is there are 40 animals per test concentration. And there are 400 for the test.

Hanson: Even so, when you start to stratify by individual types of anomalies or processes, you are really not talking about a great deal of power unless you do it with an extremely potent agent or a very high concentration of it.

Oakley: I wanted to raise another statistical point. It is not really just related to your presentation. It seems to me that in almost all of the screening test systems that have been discussed, there have been words like "highly specific" and "reliable tests," and we see numbers like 80%. I believe that we could be misled by thinking that the results from screening a large number of unknown chemicals for teratogenesis would be falsely positive only 20% of the time. It is quite likely to be much higher than 20% and is a function of what proportion of screened unknowns are really human teratogens. If only 1 out of 100 unknown chemicals screened is a human teratogen, and 20 out of 100 of nonhuman teratogens test positive, the false-positive rate is 20/21 or 95%. If, on average, the rate of human teratogens is only 1/1000 screened unknowns, the rate of false positives will be 200/201, or 99.5%. Thus, for screening tests to be reliable, developing tests with specificities approaching 99% will be required.

I think we need to put that concept on the table and think about it. I know how difficult it is to develop a test, and I don't mean to sound negative about it, but I don't think we ought to assume that because we have 80% specificity, we have a great test. We really have a terrible test at that level.

Sabourin: The other problem is that the data base we are working with is difficult. I think the only way that this system or any other is going to be validated realistically will be in a side-by-side comparison with the traditional test. In other words, take 100 blind compounds, run those through five assays, and then do the comparison. From the literature, the results could be +15% or −15%; thus, on the low side, the results could be near 99%.

Kimmel: I think that is an important issue—very important in terms of any of the test systems that we might use as prescreens—but, in fact, we do not know what the specificity of the whole animal test systems is either, mainly because we do not have the human data to go along with it. There are two steps in the problem that we have of the validity of the test systems that we use. All we can do is make attempts at it, about what the comparison between the human and animal situation is.

Brent: Yes, but as is seen in clinical screening situations, if someone has a screening test that is 80% specific, they tend to think it is a wonderful test, and it turns out to be a terrible test clinically.

Kimmel: I agree.

Brent: This is only discovered when you give a substance to patients with conditions that have not been diagnosed; then, what does the test result mean? I think this is even more difficult to deal with.

Morrissey: Dr. Sabourin, you have come up with a ratio of your malformations to the embryo lethality. Have you looked at any other way of analyzing those data? You have three categories.

Sabourin: Do you mean with the adult frogs?

Morrissey: Well, you are not really working with an adult. You do not have an adult/embryo ratio.

Sabourin: No. Are you asking whether we are considering taking a number of adult frogs and repeating essentially some of the tests to see what sort of a ratio we can establish?

Morrissey: Is there a better way? Is there some way that you can look at the LC_{50} values and make some comparisons to what we know the embryo sees in animal experiments, and how closely do those numbers correlate? Have you made any attempt at such an analysis?

Sabourin: No. We have thought of a number of things. I guess, in putting it on paper, this is the best ratio we have come up with for what we are doing.

Bournias-Vardiabasis: I have two questions. First, is the 96-hour exposure the minimum effective exposure time? Can you expose them, say, for 48 hours and get similar results, or do you have to expose the embryos for 96 hours?

Sabourin: The 96-hour exposure allows us to cover a significant portion of development.

Bournias-Vardiabasis: So, this is the least amount of time that would give you the results. Do you ever follow them to adulthood? Do you let these embryos develop?

Sabourin: Not in the regimen of a test, but we have in several cases let them develop through metamorphosis.

Bournias-Vardiabasis: Finally, I would like to add to the previous discussion. I also find it very frustrating that we have nothing to go back to in order to make any kinds of comparisons. We come back and say 85% specificity, reliability, or whatever, and the data base is not there for us to make any kinds of conclusions.

Sabourin: There are probably two things. First, the data base may be there, but somebody has to take the time to sort it out. Second, back to the comment that I made, I really think that in order to do an adequate validation program, we need to stack it against the existing system.

Miller: Should you really separate the animal teratogens from human teratogens and actually make those separate comparisons with a smaller subpopulation? For instance, I noticed coumarin was in your negative, and you said everything fit with coumarin in your assay system. As one particular example, then, should you perhaps separate them as known or possible human teratogens and suspected nonhuman teratogens?

Sabourin: I guess we have not worked with enough teratogens to do that. I thought of breaking them up, for this presentation, but we did not really have enough of a data base.

Welsch: My comment relates closely to what Dr. Morrissey said. Among the endpoints assessed under current in vivo teratogenicity testing protocols is embryolethality. This is one of the three major manifestations of developmental toxicity. In the frog embryo assay, the embryo is also the target used to determine the concentration lethal to 50% (LC_{50}). However, embryolethality could be one of the manifestations of specific developmental toxicity such as lethal mutations. One would be misled in the interpretation of the assay.

Pedersen: You will miss those entirely.

Welsch: Yes.

Brent: I think, on the issue of a comparison of in vivo systems and animal studies, the first issue is that even the best animal studies are not entirely predictive. I would put on the table a provocative statement: I cannot believe that an in vitro system will ever approach the animal studies.

Does anybody here really believe that an in vitro system is going to be better than the whole animal study which, itself, is not predictive? There is no in vitro system that is going to be predictive. I don't see how it can be, when even a whole animal study cannot predict human teratogenicity. We have seen historial evidence again and again showing the inability to predict human teratogenicity from animal studies.

I don't know much about studying ponds and soil and other areas. It sounds like your system may be useful as a screen because you are looking for more than one compound at a time. In other words, if you were going to use a whole animal system, you would have to go into the soil and do gas chromatography and liquid chromatography and find out that there are 427 pollutants and test them all. I think maybe your system would be helpful in environmental toxicology, but I cannot imagine a pharmaceutical house, which may have spent 10 years developing a compound that has some type of great potential benefit, throwing out the compound on the basis of a screen before going to an animal study.

Sabourin: These tests are used on the front end. They are used well before the pesticide manufacturer files for an experimental use permit and goes through that process.

Brent: You are talking about relative costs that are quite minor compared to what these companies have to deal with. I just cannot see how an in vitro assay will be the sole test.

Sabourin: We are not talking about replacing, and it is not the only test. I think what we are doing is prioritizing these things. As the last slide indicated, we may be working with a battery of these tests. There are a number of drug companies that are looking into this area and definitely investing resources in it.

Brent: But if you end up with four or five batteries, it may be more expensive than a whole animal study.

One other point from the human standpoint, what you are interested in with regard to human malformations is not whether you are able to produce disorganization in an embryo, but whether it is potentially viable. In other words, that is the risk that you are worried about.

Does a human being who develops to term have a congenital malformation? Many agents will produce embryo lethality through disorganization of the embryo. One of the problems with the in vitro tests is that they do not tell you whether it is a total process; that is one of the reasons why you need a whole animal study.

Manson: I would like to make a quick comment on the previous conversation. The way I think these in vitro tests have been used successfully, in screening modes by companies and such, is when a compound has been tested in vivo and found to be positive, and there are 300 backups (O.P. Flint, ICI). The in vitro systems can be incredibly valuable in identifying structure activity relationships (SAR). Then you go back into the whole animal and confirm those relationships. For virtually every teratogen that has been identified in humans and animals that ultimately ends up marketed, the company takes a positive agent and they do SAR. They change the structure of the molecule—sometimes blindly, sometimes not. They do not go back to understand the mechanism of action. We still do not know what thalidomide does. They go back and change the structure of the molecule. So, I think that guidelines and SARs can be valuable.

But, other than that, there is a central problem, and I hope some of this will be addressed in more detail. We are talking about two very different things when we are talking about teratogenesis, which is studying malformation, versus developmental toxicology, where the most severe manifestation of prenatal insult is embryonic death. A lot of the confusion is based on the fact that in one case, we are using malformations as an outcome, and in another case, we are using all the various possible manifestations of prenatal insult that can have an impact.

Johnson: Assume that we are counting the number of premises that are made—somewhere around seven or eight primary premises were made in the past 20 minutes—and on those premises, a lot of others were brought forth. Dr. Brent certainly did a beautiful job of this; but I think that almost all of the primary premises are wrong, and we are very carefully setting up a straw man in each of the primary premises and shooting it down with secondary premises based on a wrong first premise.

Shepard: I am more optimistic. I only have one comment. I think that if these in vitro screens attract good developmental biologists into the area of teratology, that is enough.

Experimental Animal-Human Comparisons

Craniofacial Malformations Induced by Retinoids in Mouse Embryo Culture

ROBERT M. PRATT, BARBARA D. ABBOTT, TOSHIAKI WATANABE, AND EUGENIA H. GOULDING
Experimental Teratogenesis Section
Laboratory of Reproductive and Developmental Toxicology
National Institute of Environmental Health Sciences
National Institutes of Health
Research Triangle Park, North Carolina 27709

OVERVIEW

Recent clinical observations have shown that isotretinoin (13-*cis* retinoic acid [RA]) is a human teratogen that causes primarily heart and craniofacial malformations, including ear and palatal defects. Our studies were conducted in order to determine whether *cis* RA could induce similar craniofacial malformations in cultured mouse embryos. Day-8 CD-1 mouse embryos were cultured for 48 hr in 100% rat serum in the presence or absence of various concentrations of *cis* RA. At 2×10^{-6} M *cis* RA, growth retardation was minimal, and approximately one third of the embryos exhibited very specific defects, including a dramatic reduction in the size of the first and second visceral arches, which eventually give rise to the maxilla, mandible, and ear. Similar observations were also made with 4-oxo-13-*cis* RA (4-oxo RA), which is the major metabolite of *cis* RA in the human. Specific changes in protein synthesis were detected in embryos exposed to *cis* RA and labeled methionine for 6 hr of culture. Using day-10 mouse embryos cultured for 48 hr in Waymouth's medium containing 50% fetal calf serum, we observed that *cis* RA at 2×10^{-5} M produced a high percentage of embryos with median cleft lip. Our results suggest that *cis* RA directly interferes with cranial neural crest cells, resulting in craniofacial anomalies.

INTRODUCTION

Congenital malformations have recently been associated with the use of isotretinoin (13-*cis* RA, Accutane®) during early pregnancy (Lammer et al. 1985). Isotretinoin is a synthetic analog of vitamin A that is used primarily in the treatment of cystic acne (Peck et al. 1979). There have been a number of case reports of malformed children (Braun et al. 1984; Fernhoff and Lammer 1984). In addition, Lammer et al. (1985) have examined the malformed children born to women who took Accutane® and have found that all of them have defects of the ear.

The administration of large amounts of vitamin A to mammals during early pregnancy causes a variety of malformations in the offspring (Cohlan 1953; Giroud and Martinet 1959; Geelen 1979). The exact nature of the defect is dependent on the time of administration during gestation and the dose, and a great deal of interest has centered on limb and palatal defects (Kochhar et al. 1984). Malformations in other tissues have also been reported, including those in the nervous system, circulatory and respiratory systems, as well as defects of the urogenital and digestive systems and of the skeleton (Geelen 1979). Retinoic acid has been administered in whole-embryo culture and resulted in defects in various tissues, including the neural folds (Morriss and Steele 1974), heart (Davis and Sadler 1981), and limb (Kochhar 1975).

The aim of our recent studies was to examine the teratogenicity of isotretinoin in mouse whole-embryo culture in order to ascertain whether or not isotretinoin exerts a direct effect on the developing embryo and to shed some light on the mechanism of craniofacial anomalies induced by isotretinoin.

RESULTS

Day-8 Cultured Embryos

Day-8 CD-1 mouse embryos were at the three- to five-somite, prerotation stage at the initiation of the culture (Goulding and Pratt 1986). After the 48-hr culture period, growth and development were comparable to that of day-10 in vivo embryos. There was excellent development of the heart, limbs, and craniofacial region; Table 1 summarizes the data obtained from the day-8- to day-10 cultures (Tables 1–4 are taken from Goulding and Pratt 1986). Embryos were examined using several parameters, including crown-rump length, head length, total DNA, and total protein content. There were significant decreases in all parameters examined except for the head length at 2×10^{-6} M 13-*cis* RA. A 10–20% reduction in both crown-rump and head length and a 30% reduction in DNA and protein content were observed; these effects were dose-responsive. There were no malformations observed in the vehicle-treated controls, whereas there was a 26% and 56% malformation incidence in the 2×10^{-6} and 2×10^{-5} M *cis*-RA-treated embryos, respectively. All malformed embryos had small or absent visceral arches, and two embryos at 1×10^{-5} M *cis* RA had misshapen otic vesicles. No otic vesicle defects were observed at 2×10^{-6} M *cis* RA. The visceral arch defects (Fig. 1) were characterized by a short, flattened appearance of the first arch and apparent fusion to the second arch; the latter was usually smaller than controls, and the third arch was occasionally missing.

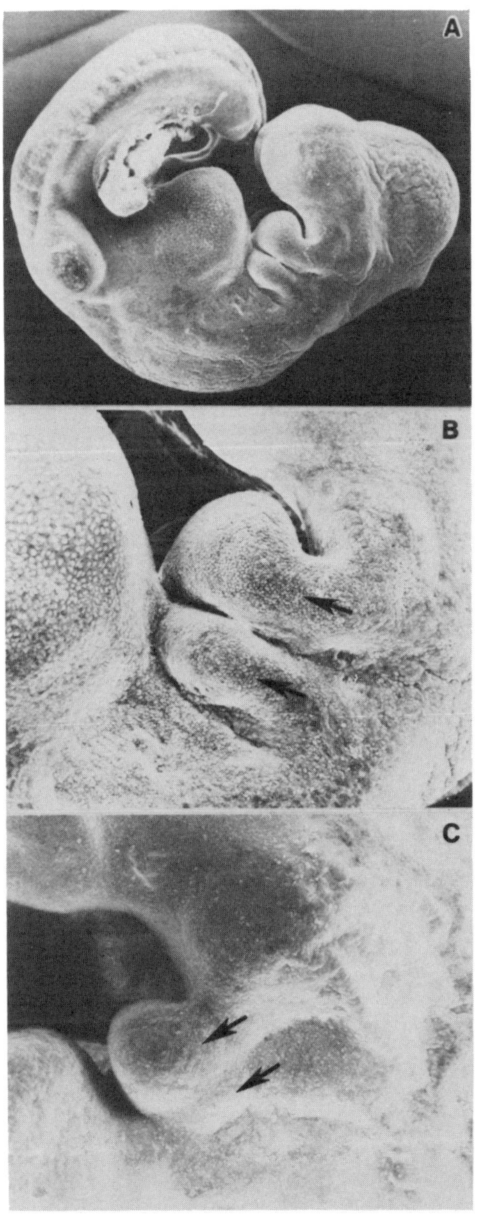

Figure 1

Scanning electron micrograph of representative CD-1 mouse embryos cultured for 48 hr, beginning on day 8 in the absence (*A*,*B*) or presence (*C*) of 2×10^{-6} M 13-*cis* RA. → Visceral arches. Note the short, flattened appearance of the first arch and apparent fusion to the second arch in embryo *C*, as compared to the visceral arches in embryos *A* and *B*. Magnifications: 42 × (*A*); 118 × (*B*,*C*).

Table 1
Effect of 13-*cis* RA on Day-8 to Day-10 CD-1 Mouse Embryos In Vitro

	DMSO controls ($n = 14$)	13-cis RA 2×10^{-6} M ($n = 19$)	13-cis RA 1×10^{-5} M ($n = 16$)
Crown-rump length (mm)	4.45 ± 0.06[a]	4.02 ± 0.07 (90)[b,c]	3.65 ± 0.08 (82)[c,d]
Head length (mm)	2.19 ± 0.03	2.05 ± 0.05 (91)	1.78 ± 0.05 (81)[c,d]
Total DNA (μg)	32.5 ± 1.1	23.0 ± 1.1 (71)[c]	22.0 ± 1.0 (68)[c,d]
Total protein (μg)	461 ± 24	328 ± 13 (71)[c]	330 ± 16 (72)[c,d]
No. abnormal/ total No. (%)	0/14 (0)	5/29 (26)[e]	9/16 (56)[d,e]

[a]Values are presented as mean ±S.E.M.
[b]Percentage of dimethylsulfoxide (DMSO)
[c]$P < 0.01$.
[d]$P < 0.01$, dose-response trend.
[e]All malformations were small or absent visceral arches, with or without malformed otic vesicles.

Table 2
Effect of 4-Oxo RA on Day-8 to Day-10 CD-1 Mouse Embryos In Vitro

	DMSO controls ($n = 14$)	4-Oxo RA 4×10^{-6} M ($n = 10$)	4-Oxo RA 1×10^{-5} M ($n = 22$)	4-Oxo RA 2×10^{-5} M ($n = 15$)
Crown-rump length (mm)	4.46 ± 0.06[a]	4.03 ± 0.07 (90)[b,c]	3.54 ± 0.15 (79)[c]	3.17 ± 0.13 (71)[c–e]
Head length (mm)	2.19 ± 0.03	2.15 ± 0.04 (98)	1.52 ± 0.12 (69)[c]	1.70 ± 0 (78)[c–e]
Total DNA (μg)	32.5 ± 1.12	ND	18.6 ± 2.70 (57)[c]	8.78 ± 0.73 (27)[c,e]
Total protein (μg)	461 ± 24	ND	283 ± 24 (61)[c]	290 ± 30 (63)[c,e]
No. abnormal/ Total No. (%)	0/14 (0)	0/10 (0)	15/22 (68)	15/15 (100)[e]

[a]Values are presented as mean ±S.E.M.
[b]Percentage of dimethylsulfoxide (DMSO) controls.
[c]$P < 0.01$.
[d]Represents only three embryos—abnormal flexion prevented measurements.
[e]$P < 0.01$, dose-response trend.

Table 3
Developmental Anomalies Induced by 4-Oxo RA on Day 8–Day 10 In Vitro

	1×10^{-5} M ($n = 22$)	2×10^{-5} M ($n = 15$)
No. malformed/ total no. (%)	15/22 (68)	15/15 (100)
Small or absent visceral arch(es)	14/15	13/15
Malformed otic vesicles	6/15	8/15
Open neural tube	2/15	1/15
Defective flexion	2/15	12/15

No malformations observed in DMSO controls or 4×10^{-6} M 4-oxo RA.

Similar results were obtained with the metabolite 4-oxo RA (Table 2), with 20–30% reductions in crown-rump and head length. Larger decreases were observed in the DNA and protein content, especially the DNA content of the embryos treated with 4-oxo RA at 2×10^{-5} M. The percentage of malformations was similar at 1×10^{-5} M between the two compounds. However, 4-oxo RA at 2×10^{-5} M produced 100% malformed embryos; this concentration of *cis* RA was embryolethal, and the results obtained with 4-oxo RA were also dose-responsive. The visceral arch defects observed in the *cis*-RA-treated embryos were also present in most of the malformed embryos treated with 4-oxo RA (Table 3); however, more 4-oxo-RA-treated embryos exhibited malformed otic vesicles (47%). Other defects that occurred after exposure to 4-oxo RA and were not observed in the *cis* RA embryos included open neural tube and defective flexion, giving the embryos a somewhat S-shaped appearance.

Day-10 Cultured Embryos

After 48 hr, overall growth of day-10 embryos was comparable to that of day-12 in vivo embryos; complete development of the primary palate was obtained under these culture conditions. Embryos exposed to 13-*cis* RA at 2×10^{-5} M for 48 hr exhibited little, if any, growth retardation (Table 4); the head length was the only parameter significantly reduced (87% of control). Although this concentration was ineffective as far as growth inhibition, it did result in 100% malformed embryos compared to 14% in the vehicle-treated controls. The defects in the control embryos were lateral cleft lips, and one embryo had short limbs. All the *cis*-RA-treated embryos had median cleft lips

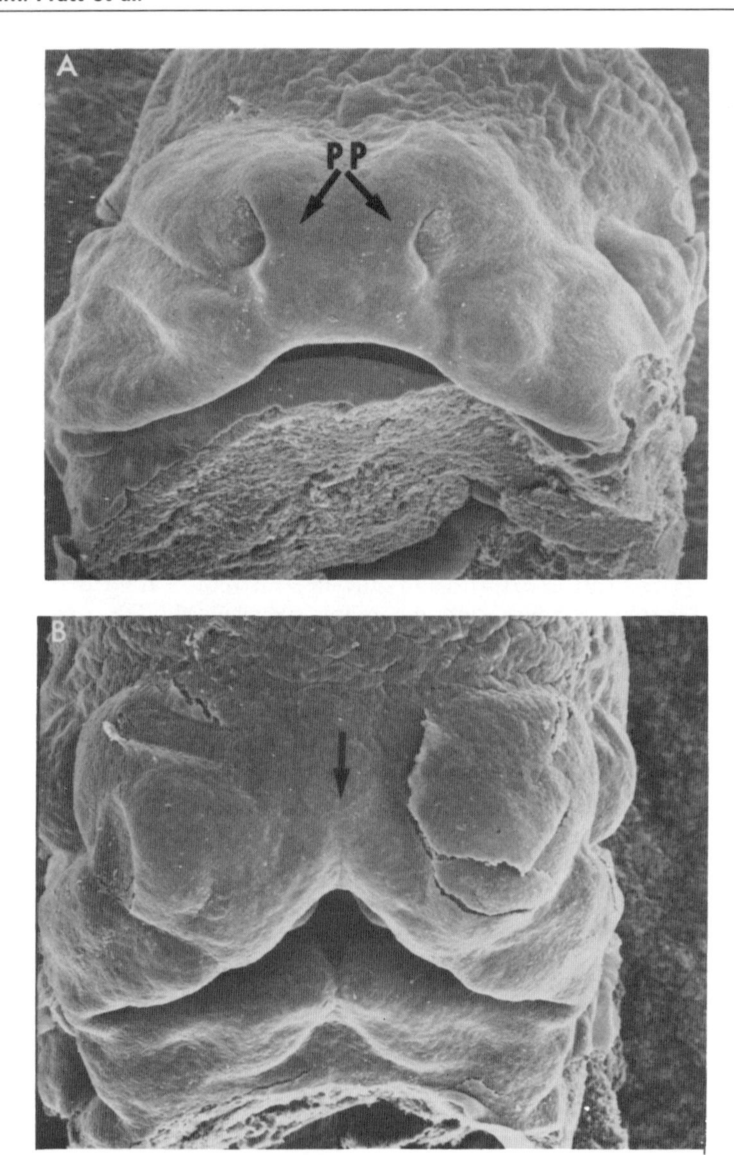

Figure 2

Scanning electron micrograph of CD-1 mouse embryos cultured for 48 hr, starting on day 10, in the absence (*A*) or presence (*B*) of 2×10^{-5} M 13-*cis* RA. → Normal development of primary palate (PP) (*A*) and the median cleft (*B*). Magnification, 50 ×. (Reprinted, with permission, from Goulding and Pratt 1986.)

(Fig. 2), with or without lateral clefts, and most (84%) also had severe limb reduction deformities (Fig. 3).

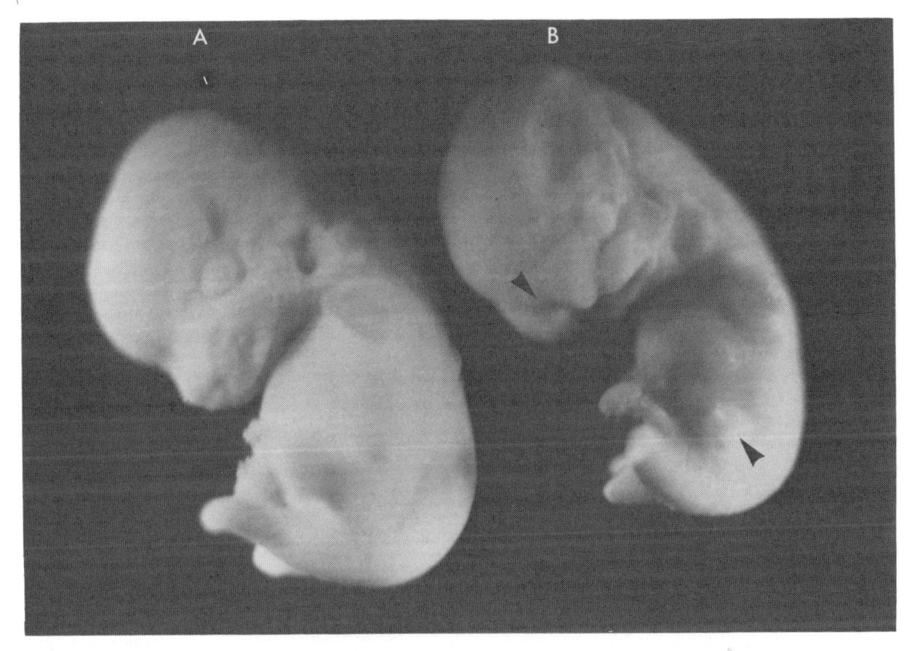

Figure 3
Bouin's fixed embryos cultured for 48 hr, beginning on day 10 in the absence (*A*) or presence (*B*) of 2×10^{-5} M 13-*cis* RA. Arrowheads indicate median cleft lip and short hindlimb in embryo *B*. Magnification, 29 ×. (Reprinted, with permission, from Goulding and Pratt 1986.)

Table 4
Effect of 13-*cis* RA on Day 10–Day 12 CD-1 Mouse Embryos In Vitro

	DMSO Controls ($n = 21$)	2×10^{-5} M 13-*cis* RA ($n = 19$)
Crown-rump length (mm)	8.5 ± 0.10	8.3 ± 0.18 (98)
Head length (mm)	4.7 ± 0.07	4.1 ± 0.10 (87)[a]
Total DNA (μg)	464 ± 38	375 ± 36 (81)
Total protein (μg)	3.0 ± 0.15	2.8 ± 0.17 (93)
No. abnormal/ total no. (%)	3/21 (14.3)	19/19 (100)

[a]$P < 0.01$.
[b]Malformed limbs and/or cleft lip.

Embryonic Protein Synthesis

Since retinoids are thought to alter protein synthesis in a number of responsive tissues and cells, we examined protein synthesis in embryo culture. Embryos were cultured on day 8 for 6 hr in the presence or absence of *cis* RA at 2×10^{-5} M; [^{35}S]methionine was added during the last 4 hr. The embryos were freed from their membranes and processed for two-dimensional gel electrophoresis. In the control embryos (Fig. 4A), we observed the absence of a 52-kD protein that was present in the *cis*-RA-treated embryos (Fig. 4B).

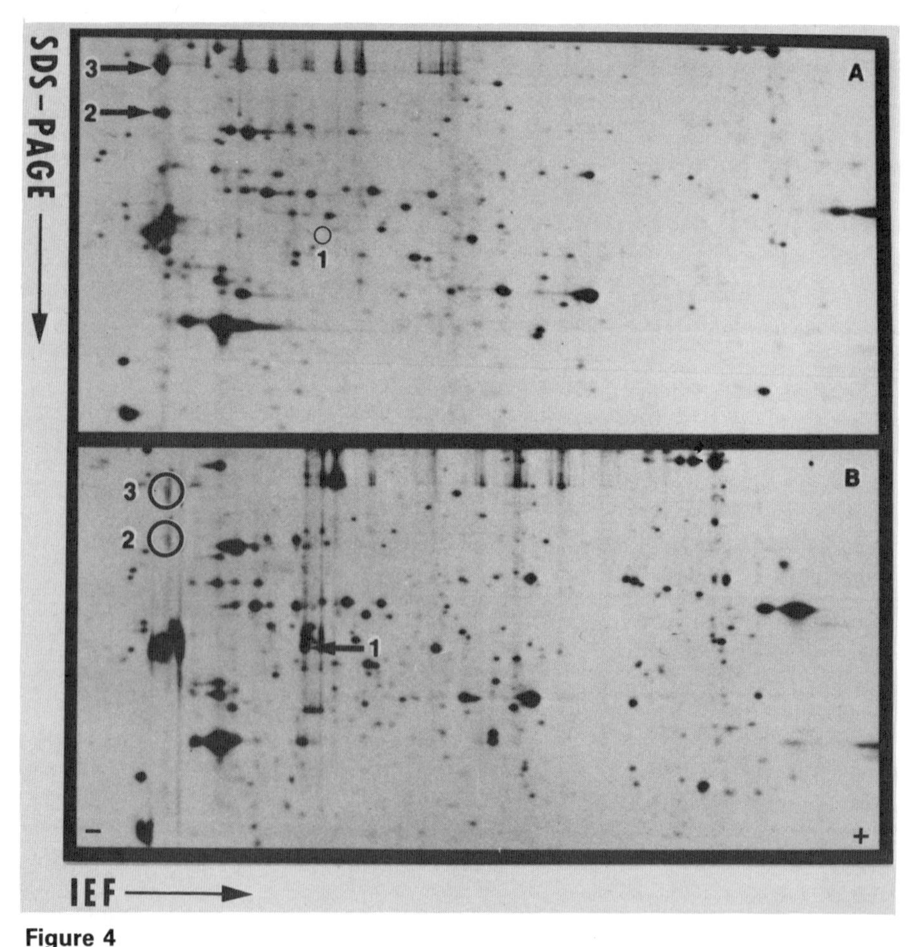

Figure 4
Two-dimensional SDS-gel electrophoretic analysis of [^{35}S]methionine-labeled embryos (day 8 for 6 hr) in the absence (*A*) or presence (*B*) of *cis* RA at 2×10^{-5} M. Separation of labeled proteins by molecular weight was from top to bottom and by isoelectric focusing from left to right. (*1*) A 52-kD protein; (*2*) 72-kD protein; and (*3*) 78-kD protein.

The *cis*-RA-treated embryos had very low amounts of 72-kD and 78-kD proteins present in the control embryos.

DISCUSSION

Congenital malformations have been associated recently with the use of 13-*cis* RA, Accutane®, during early pregnancy (Lammer et al. 1985). Most of these newborns have craniofacial abnormalities, with the ear being severely malformed. Other abnormal features include a flat nasal bridge, mandibular hypoplasia, isolated cleft palate, and cranial defects. Johnston et al. (1985) reported similar malformations in vivo in the C57BL/6J mouse when isotretinoin was administered at 200 mg/kg between days 7 and 9 of gestation. The predominant malformations were those of the external ear, absence or hypoplasia of the thymus, ventricular septal defects, and interrupted aortic arch.

It was not clear from these animal or human reports whether or not isotretinoin exerts a direct effect on the developing embryo to produce these malformations. Our results (Pratt et al. 1985; Goulding and Pratt 1986) indicate clearly that isotretinoin is teratogenic in CD-1 mouse whole-embryo culture, and the malformations we obtained are similar to those observed in vivo in the C57BL/6J mouse. All malformed embryos that had been exposed from days 8–10 to 2×10^{-6} M 13-*cis* RA had small or absent visceral arches, and a few had defective otic vesicles. The first arch gives rise to the mandible and maxilla, and the second gives rise to the ear. These malformations would be expected to result in defects similar to those observed in the human and the C57BL/6J mouse at birth. Day-10 embryos exposed to *cis* RA at 2×10^{-5} M for 48 hr exhibited a high percentage of embryos with limb defects and median cleft lip.

Our results demonstrate that labeled *cis* RA is readily available to the developing embryo both in vivo and in vitro (Goulding and Pratt 1986). A detailed study of the pharmacokinetics of labeled *cis* RA under these conditions has not been performed. A major metabolite of *cis* RA in the human is 4-oxo RA (Brazzell et al. 1983), and it accumulates to three to five times the concentration of isotretinoin. Following the administration of 13-*cis* RA to the NMRI mouse on day 9 of gestation, significant amounts of 13-*cis* RA, 4-oxo RA, and all-*trans* RA were present in the embryo after 2–4 hr (Creech-Kraft et al. 1987). The results of the present study demonstrate that 4-oxo RA exerts similar teratogenic effects in mouse whole-embryo culture as observed in *cis* RA. These results suggest that *cis* RA and/or 4-oxo RA are capable of exerting a direct effect on the human and mouse embryo to produce craniofacial malformations.

This direct effect of *cis* RA on the developing embryo may be confined to specific cell types such as the neural crest cells, which give rise to all of the facial mesenchyme (Johnston 1975). In the mammalian embryo, cranial neural crest cells are initially located in the margins of the neural folds. During the process of neural-fold elevation and approximation, the crest cells migrate to various sites. The forebrain and midbrain neural crest cells give rise to the frontonasal processes and first visceral arch, whereas the anterior hindbrain crest cells give rise primarily to second arch mesenchyme. Cranial neural crest migration is initiated in the mouse embryo at the three- to four-somite stage (Nichols 1981), which corresponds to the stage at which the day-8 CD-1 mouse embryos are explanted into culture. Johnston et al. (1985) found that with in vivo administration of *cis* RA into the C57BL/6J mouse, changes could be seen within 4 hr in the cranial neural crest cells; by 12 hr, clumps of neural crest cells were still attached to the neural folds in some regions such as the anterior hindbrain. After 24 hr, the cellular changes had disappeared, but severe underdevelopment of the visceral arches had occurred. Our studies using light and transmission electron microscopy indicate that similar changes occur in the cranial crest cells from CD-1 mouse embryos exposed to *cis* RA on day 8 in whole-embryo culture (Pratt et al. 1986). The exact nature by which *cis* RA interferes with crest migration is unknown but presumably may involve either a direct perturbation of the crest cell protein synthesis or an alteration in an extracellular matrix component, such as fibronectin, which is critical for cranial neural crest cell migration. It is of interest that labeled RA administered in vivo to the day-7 to day-8 pregnant mouse appears to localize primarily in the neuroepithelium and neural crest (Dencker et al. 1987).

The craniofacial defects and median and lateral cleft lip observed after *cis* RA administration to day-10 embryos cannot be explained by an alteration in cranial neural crest cell migration. By day 10 in the mouse, the cranial crest cells have already migrated from the forebrain to the frontonasal region. However, these cells are proliferating rapidly, both during and after migration, in order to provide the mesenchyme that will form structures such as the primary and secondary palate. Our studies (Watanabe et al. 1986) in whole-embryo culture, using labeled thymidine autoradiography, suggest that *cis* RA decreases the proliferation of these neural-crest-derived mesenchyme cells located in the frontonasal region, which subsequently results in clefts of the primary palate. It is also of interest that administration of all-*trans* RA to the chick embryo (stage 20–21) results in a dramatic reduction in the size of the frontonasal mass, thus resulting in clefts of the primary palate (Tamarin et al. 1984). Cell-culture studies using day-10 mouse frontonasal neural-crest-derived mesenchyme cells also support this hypothesis by demonstrating a *cis*-RA-induced dose-dependent inhibition of growth (Goulding and Pratt 1986).

Wiley et al. (1983) reported that the cranial neural crest cells of the hamster embryo were selectively vulnerable to *all*-trans RA administered in vivo at stages when the major migrations of the crest cells were complete. Using whole-body autoradiography, Dencker (1979) observed an accumulation of labeled all-*trans* RA in the mid-facial region of the day-10 mouse embryo, which suggests that cytoplasmic receptors for RA are present in selective regions of the embryo. These receptors presumably serve an important role in normal development and also may be responsible, in part, for causing certain embryonic cells and tissues to be sensitive to the teratogenic effects of high doses of retinoids. Preliminary biochemical observations (R.M. Pratt et al., unpubl.) demonstrate the presence of such receptors in the craniofacial region of day-10 to day-12 CD-1 mouse embryos. Kwarta et al. (1985) have reported the presence of RA receptors in day-19 mouse forelimb buds, and Maden and Summerbell (1986) have characterized the RA receptor in the stage 20–35 chick embryo limb bud.

The specific nature of the *cis* RA-induced defects may be due, in part, to the presence of retinoid receptors located in specific areas of the embryo, such as the cranial neural crest cells. These retinoid receptors in the cranial neural crest cells may serve a critical role in their normal differentiation and migration by regulating cell-surface proteins, such as neural cell-adhesion molecules (N-CAM). Thiery et al. (1982) (see also, Edelman et al., this volume) have reported that N-CAM, a cell-surface glycoprotein important for embryonic cell-to-cell interactions, is modulated during neural crest cell migration. N-CAM is present in crest cells prior to their migration from the neuroepithelium, disappears while these cells migrate to various sites in the embryo, and then reappears when the cells reach their final destination and undergo further differentiation. Our observations that retinoid-treated cranial crest cells are aggregated together near their site of origin at the neuroepithelium suggests that these cells may not have decreased their cell-surface N-CAM, which appears to be a prerequisite for cell migration. This certainly could explain the retinoid-induced inhibition of cranial neural crest cell migration, and studies are in progress in our lab to test this hypothesis.

REFERENCES

Braun, J.T., R.A. Franciosi, A.R. Mastri, R.M. Drake, and B.L. O'Neil. 1984. Isotretinoin dysmorphic syndrome. *Lancet* **I:** 506.

Brazzell, R.K., F.M. Vane, C.W. Ehmann, and W.A. Colburn. 1983. Pharmacokinetics of isotretinoin during repetitive dosing to patients. *Eur. J. Pharmacol.* **24:** 695.

Cohlan, S.Q. 1953. Excessive intake of vitamin A as a cause of congenital anomalies in the rat. *Science* **117:** 535.

Creech-Kraft, J., D.M. Kochhar, W.J. Scott, and H. Nau. 1987. Low teratogenicity of 13-cis retinoic acid (isotretinoin) corresponding to low embryo concentrations during organogenesis in the mouse: Comparison to the all-trans isomer. *Toxicol. Appl. Pharmacol.* **87:** 474.

Davis, L.A. and T.W. Sadler. 1981. Effects of vitamin A on endocardial cushion development in the mouse heart. *Teratology* **24:** 139.

Dencker, L. 1979. Embryonic-fetal localization of drugs and nutrients. *Adv. Study Birth Defects* **1:** 18.

Dencker, L., R. d'Argy, B.R.G. Danielson, H. Ghantous, and G. Sperber. 1987. Saturable accumulation of retinoic acid in neural and neural crest derived cells in embryonic early development. *Dev. Pharmacol. Ther.* **10:** 212.

Fernhoff, P.M. and E.J. Lammer. 1984. Craniofacial features of isotretinoin embryopathy. *J. Pediatr.* **105:** 595.

Geelen, J.A.G. 1979. Hypervitaminosis A induced teratogenesis. *CRC Crit. Rev. Toxicol.* **4:** 351.

Giroud, A. and M. Martinet. 1959. Tératogénèse par hypervitaminose A chez le rat, la souris, le cobaye et le lapin. *Arch. Fr. Pediatr.* **16:** 971.

Goulding, E.H. and R.M. Pratt. 1986. Isotretinoin teratogenicity in mouse whole embryo culture. *J. Craniofacial Genet. Dev. Biol.* **6:** 99.

Johnston, M.C. 1975. The neural crest in abnormalities of the face and brain. *Birth Defects Orig. Artic. Ser.* **7:** 1.

Johnston, M.C., K.K. Sulik, W.S. Webster, and B.L. Jarvis. 1985. Isotretinoin embryopathy in the mouse model: Cranial neural crest involvement. *Teratology* **31:** 26A.

Kochhar, D.M. 1975. The use of in vitro procedures in teratology. *Teratology* **11:** 273.

Kochhar, D.M., J.D. Penner, and C.I. Tellone. 1984. Comparative teratogenic activities of two retinoids: Effects on palate and limb development. *Teratog. Carcinog. Mutagen.* **4:** 377.

Kwarta, R.F., C.A. Kimmel, G.L. Kimmel, and W. Slikker. 1985. Identification of the cellular retinoic acid binding protein (cRABP) within the embryonic mouse (CD-1) limb bud. *Teratology* **32:** 103.

Lammer, E.J., D.T. Chen, R.M. Hoar, N.D. Agnish, P.J. Benke, J.T. Braun, C.J. Curry, P.M. Fernhoff, A.W. Grix, Jr., I.T. Lott, J.M. Richard, and S.C. Sun. 1985. Retinoic acid embryopathy: New human teratogen and a mechanistic hypothesis. *N. Engl. J. Med.* **313(14):** 837.

Maden, M. and D. Summerbell. 1986. Retinoic acid-binding protein in the chick limb bud: Identification at developmental stages and binding affinities of various retinoids. *J. Embryol. Exp. Morphol.* **97:** 239.

Morriss, G.M. and C.E. Steele. 1974. The effect of excess vitamin A on the development of rat embryos in culture. *J. Embryol. Exp. Morphol.* **32:** 505.

Nichols, D.H. 1981. Neural crest formation in the head of the mouse embryo as observed using a new histological technique. *J. Embryol. Exp. Morphol.* **64:**105.

Peck, G.L., T.G. Olsen, F.W. Yoder, J.S. Strauss, D.T. Downing, M. Pandya, D. Butkus, and J. Arnaud-Battandier. 1979. Prolonged remission of cystic and conglobate acne with 13-cis retinoic acid. *N. Engl. J. Med.* **300:** 329.

Pratt, R.M., B.D. Abbott, T. Watanabe, and E.H. Goulding. 1986. Isotretinoin-induced morphological and biochemical changes during mouse embryo culture. *Teratology* **33:** 49C.

Pratt, R., T. Watanabe, M. Russell, W. Willis, S. Perry, and E. Goulding. 1985. Isotretinoin teratogenicity in mouse whole embryo culture. *Teratology* **31:** 27A.
Tamarin, A., A. Crawley, J. Lee, and C. Tickle. 1984. Analysis of upper beak defects in chicken embryos following retinoic acid treatment. *J. Embryol. Exp. Morphol.* **84:** 105.
Thiery, J.-P., J.-L. Duband, U. Rutishauser, and G.M. Edelman. 1982. Cell adhesion molecules in early chicken embryogenesis. *Proc. Natl. Acad. Sci.* **79:** 6737.
Watanabe, T., W.D. Willis, E.H. Goulding, and R.M. Pratt. 1986. Effects of isotretinoin in vitro on mouse facial development. *Teratology* **33:** 73C.
Wiley, M.J., P. Cauwenbergs, and I.M. Taylor. 1983. Effect of retinoic acid on the development of the facial skeleton in hamsters: Early changes involving cranial neural crest cells. *Acta Anat.* **116:** 180.

COMMENTS

Oakley: I have two questions. How long does it take for the neural crest cells to migrate? The second question is, how does the response to retinoids vary by the strain of mouse?

Pratt: It takes about 6 hr from the start of their initial migration for the cranial neural crest cells to begin to populate the region of the first visceral arch. The neural crest cells continue to migrate into that region and, at the same time, are proliferating at a high rate; this builds up the mass of mesenchymal cells that will ultimately form the different visceral arches. This is a continuous process that takes place over a number of hours, and the population of cells is expanding at the same time that they are migrating. We have not examined different strains of mice, but I think most strains are susceptible to RA-induced malformations.

Oakley: I think there are three mutant strains that are predisposed to neural tube defects. We have found that treatment with RA on day 8 results in a higher percentage of abnormal neural tubes than treatment on day 9.

Pratt: Defects in the union of the neural folds do not necessarily involve a defect in the neural crest cells; other factors could be involved, perhaps related to cytoskeletal elements. The union of the neural folds occurs simultaneously, with the neural crest cells migrating from the neuroectoderm.

Miller: Is there a differential localization of RA binding in neural crest cells compared to other epithelial cells surrounding them?

Pratt: The only evidence we have related to the distribution of this binding protein, in the early embryo at least, comes from the studies that Dr. Dencker has just published. He found that labeled RA was localized to the neuroectoderm and the neural crest but not to other regions of the early mouse embryo. It is also interesting that RA is preferentially localized to the region of the primary palatal processes in more advanced embryos. Day-10 mouse embryos cultured for 48 hr with RA exhibit defects in the primary palate. The mesenchymal cells in the primary palate are derived from the cranial neural crest. We know, from studies performed with Dr. Anton Jetten at NIEHS, that day 10–12 embryos have cellular RA-binding proteins (receptors) that are comparable in levels to many of the other responsive tissues, including the developing limb bud. There have been some recent studies done with the chick limb bud showing that the receptor is present throughout a rather extensive period of embryonic development during which time it is susceptible to RA-induced malformations.

Miller: I thought that the distribution of the binding protein was more general.

Pratt: It may indeed have a general distribution. Dr. Dencker may have detected the most intense areas of localization. There may be a very low level of receptor that is necessary for a number of other embryonic tissues.

Dencker: At different stages of development, an accumulation of labeled RA can be seen in different areas along the neural tube. At one stage, accumulation might be found preferentially in the anterior neural folds. One day later in development, the accumulation might be seen further back along the neural tube. So, there is a temporal expression of this binding protein that is responsible for the observed accumulation. What I have described is one piece of evidence of a receptor being involved in retinoid-induced abnormalities.

Pratt: Is it true that the more advanced embryos and fetuses you examined displayed less of this localization? If you examine day-13, -14, and -15 embryos, do you see this intense localization in specific areas?

Dencker: There is a more general distribution at that time. After day 11, when the central nervous system is developed extensively, the concentration of retinoid decreases. Another piece of evidence to support the receptor hypothesis is that when higher doses of nonlabeled RA are given to an animal before a low dose of labeled retinoid, the uptake in the embryonic central nervous system is inhibited dramatically.

Scott: You mentioned that all-*trans* RA was more potent than 13-*cis* RA in your whole-embryo culture system. That's exactly the situation in vivo in mice, as reported by Dr. Kochhar. It takes a much higher dose, 200 mg/kg, before noticeable malformations appear with 13-*cis* RA in the mouse given on day 11 of pregnancy. The levels of those retinoids found in the 11-day embryo, measured by Dr. Heinz Nau in Berlin, give an explanation as to why these two were differentially teratogenic. All-*trans* RA reaches very high levels in the embryo, whereas 13-*cis* RA does not. The conclusion has been that the difference in vivo is a pharmacokinetic one.

Kochhar: In limb bud organ culture and in the high-density cell culture systems of the type Dr. Solursh reported, the two analogs are equally potent.

Pratt: The differential response of the cultured embryos to 13-*cis* and all-*trans* RA surprised us. The growth inhibitory effects of these two retinoids are very similar in our line of human embryonic palatal mesenchymal cells. But, it is important to remember that we are using 100% rat serum as the medium in our embryo culture system. I think there is a distinct possibility that there are binding proteins in the serum that may absorb 13-*cis* RA preferentially over all-*trans* RA. To answer that question, we should microinject specific amounts of 13-*cis* and all-*trans* RA into cultured embryos and compare the effects.

Manson: Do you feel you can make a link between the cellular RA-binding protein and the production of the N-CAM? How much is really known about those two processes?

Pratt: We noticed the specificity in the distribution of the labeled RA and have speculated that it is related to the receptors and to N-CAM; very little is known of these in the mouse embryo.

Kochhar: In your 2-dimensional gels, you found differences in protein synthesis. Can you detect low-molecular-weight proteins, such as the binding proteins, which are about 14–16 kD?

Pratt: The conditions under which these gels were run were not suitable for examining low-molecular-weight proteins. We are planning to examine specific cell types and, currently, we are using neural crest-derived mesenchymal cells from the mouse and human secondary palate to determine changes in protein synthesis. There are numerous cell types in these day-8 cultured embryos, and it is difficult to attribute these changes in protein synthesis to any one cell population such as the neural crest.

Saxen: Do you know the normal role of the binding proteins or where they are found? There are migratory cells all over the organism, including various kinds of neural crest cells. Are there any factors that appear to affect the neural crest specifically?

Pratt: Some unpublished information, just released from Dr. Carol Erickson's lab, strongly indicates that epidermal growth factor is one of the necessary components for the migration of neural crest cells. EGF stimulates extracellular matrix production, which is important for crest cell migration. There have been a number of studies from Dr. Jetten's lab at NIEHS supporting this, and we have obtained results using epithelial cells from the secondary palate indicating that retinoids have a distinct influence on the binding of EGF to its receptor.

Morrissey: Neural crest cells have many functions in development, and you have identified one possible defect here. Have you noticed any other organs or systems exhibiting a defect that might be attributable to a neural crest cell deficit?

Pratt: I think that question should be deferred to Dr. Lammer's talk because he will discuss some of the other embryonic structures that are affected, including the thymus, which appears to be related to altered neural crest migration.

Dencker: There is a study from Britain showing that if pregnant rats are given a vitamin A-deficient diet they will abort. These pregnancies can be maintained with RA supplementation throughout gestation. That is an indication that RA has a physiological role in early gestation.

Welsch: I have a comment regarding the potency differences between 13-*cis* and all-*trans* RA. Our studies on gap junctional communication suggest that the actions on dye coupling at low-exposure concentrations (0.1–1.0 mg/ml) seem to reflect the in vivo differences. However, at higher concentrations (10 mg/ml), there are no detectable differences. The onset of action is so fast that within a minute or two after exposure, the cell-cell channels apparently are closed by either compound.

Brent: Do you want to discuss the difference between a binding protein and a receptor?

Pratt: It's unfortunate that the people working in this area have termed this the "cellular RA-binding protein," because clearly it is very similar, if not identical, to a cytoplasmic receptor for retinoic acid.

Patterns of Malformation among Fetuses and Infants Exposed to Retinoic Acid (Isotretinoin)

EDWARD JAMES LAMMER
Embryology-Teratology Unit
Massachusetts General Hospital
Boston, Massachusetts 02114

OVERVIEW

A synthetic retinoid used for treating cystic acne, isotretinoin, became available in North America in 1982. An unusually high percentage of the offspring of mothers who used isotretinoin inadvertently during early pregnancy have major malformations. The pattern of malformation involves craniofacial, cardiac, thymic, and central nervous system (CNS) structures. Similar malformations were induced in experimental animal models using vitamin A and synthetic retinoids. The experimental models support the hypothesis that a major mechanism of retinoid teratogenesis is a deleterious effect on the activities of cranial neural crest cells during embryogenesis. Coordinated efforts to correlate the human outcomes and the experimental animal models may improve our understanding of the roles of cranial neural crest cells in human craniofacial, cardiac, and thymic morphogenesis and dysmorphogenesis.

INTRODUCTION

Hundreds of synthetic retinoids are under development for their potential preventive and therapeutic value in dermatologic diseases and cancer (Goodman 1984). In September 1982, the first synthetic retinoid, isotretinoin (Accutane, 13-*cis* retinoic acid), was licensed for the treatment of cystic acne in the United States. It was known that retinoic acids and vitamin A were teratogenic in experimental animals; thus, isotretinoin was not intended for use during pregnancy. Nonetheless, inadvertent use of isotretinoin has occurred during several hundred pregnancies. More than 50 malformed infants have been born following fetal exposure to isotretinoin.

We have followed two populations of infants who were exposed to isotretinoin in utero: one group that was prospectively ascertained and one that was retrospectively ascertained. Infants were considered prospectively ascertained if the exposed pregnancy was reported to a member of the research group in the absence of any knowledge of the fetal outcome (with respect to

Banbury Report 26: Developmental Toxicology: Mechanisms and Risk

structural or functional abnormalities). Almost all infants who were reported retrospectively had major malformations and were probably reported for that reason. We published the outcomes of the first 28 prospectively identified pregnancies that reached 20 weeks gestation or more; among these 28 infants, 5 (18%) had at least one major malformation (Lammer et al. 1985), an unusually high absolute risk for major malformation. Currently, we are studying the outcomes of a prospective cohort of more than 50 infants exposed to isotretinoin. We hope that the larger number of participants will help to narrow the confidence limits about this point measure of risk.

Among malformed infants exposed to isotretinoin, we found that their malformations were not a random selection of all possible developmental abnormalities. The infants had a specific pattern of malformation. I will describe the pattern of malformation that we found among the first 21 malformed infants who were exposed to isotretinoin during the first trimester of pregnancy and review the current evidence that one of the major mechanisms of retinoid teratogenesis is an adverse effect on the roles of cranial neural crest cells during embryogenesis (Lammer et al. 1985).

RESULTS

The study population was composed of 5 infants from the prospective cohort and 16 infants who were reported retrospectively. The mortality was high. Three infants were stillborn, and 11 others died from complications related to their birth defects. The pattern of malformation involved the morphogenesis of four structures: cranium-face, heart, thymus, and brain. The craniofacial malformations were the most common. External ear development appeared to be particularly susceptible to isotretinoin. Fifteen infants had small, malformed, low-set ears, usually associated with a stenotic or atretic external ear canal (Figs. 1 and 2). Characteristically, parts of the external ear derived from the hillocks of the second branchial arch were more commonly and more severely affected than the parts that were derived from the hillocks of the first branchial arch (Lammer and Sulik 1985). This pattern of ear malformation closely resembles that reported in the thalidomide embryopathy, suggesting that the mechanisms producing these malformations may be similar in the two embryopathies (Kleinsasser and Schlothane 1964; Lammer and Sulik 1985). Preauricular skin tags were not found. Both ears were almost always affected, although the severity was often asymmetric. Six infants had micrognathia (Fig. 2); three of these infants also had a cleft palate. Cleft lip was not seen. Facial asymmetry was a relatively common finding. Less commonly, maxillary hypoplasia occurred. Neither clefts of the mouth nor macrostomia were found.

Twelve infants had congenital heart defects. The defects were categorized

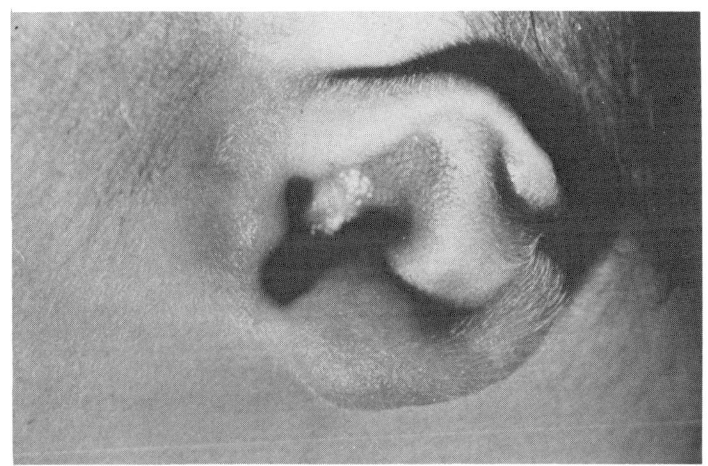

Figure 1

Malformed left external ear. The primary structure derived from hillocks of the first branchial arch, the tragus, is well formed. Structures from the second-arch hillocks are malformed. The helix and antihelix are crudely formed. Superiorly, the crus helicis meets the face in an abnormal manner that is characteristic of this embryopathy.

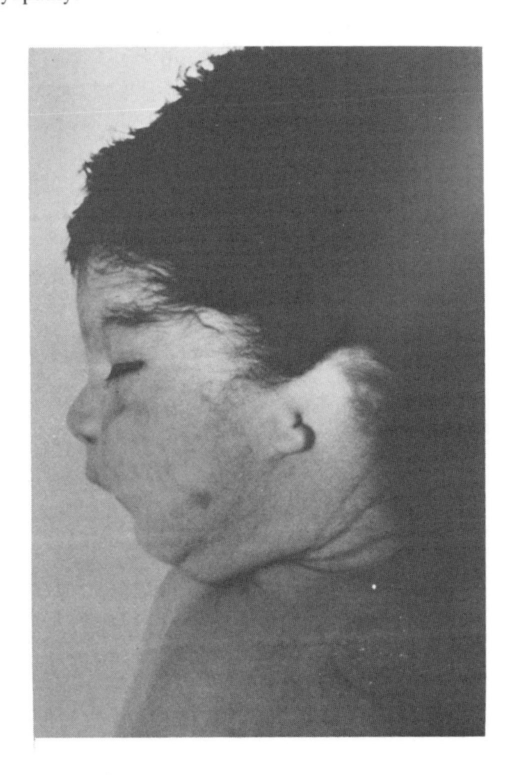

Figure 2

Severe external ear malformation. No canal was present. The embryonic origin of the low-set, poorly differentiated rudimentary ear structure that is present is unclear. Micrognathia is also evident.

into two main mechanistic categories: abnormalities of aorticopulmonary (conotruncal) septation and branchial arch mesenchymal tissue defects (Lammer and Opitz 1986). Five different types of conotruncal abnormalities were found. Branchial arch mesenchymal tissue defects are abnormalities of major arteries or abnormal patterns of regression and persistence of the original branchial arch arteries. Of these 12 infants, 6 also had abnormal thymic morphogenesis—ectopia, hypoplasia, or aplasia. We have also observed thymic dysmorphogenesis among infants who did not have congenital heart defects.

The CNS malformations included ventriculomegaly, hydrocephalus, microcephalus, cerebellar micro- and macrodysgenesis, and neuromigrational abnormalities. The types of hydrocephalus varied: obstruction secondary to a fourth ventricle cyst, hydranencephaly, aqueductal stenosis, and unknown. The cerebellar abnormalities included cerebellar hypoplasia, complete or partial agenesis of the vermis, nuclear abnormalities, and heterotopias. Several infants had peripheral facial nerve pareses.

The types of malformations found among the infants exposed to isotretinoin closely resembled those induced in experimental animal models by vitamin A and retinoic acid (Kochhar 1967; Shenefelt 1972; Geelen 1979; Newell-Morris et al. 1980). The major exceptions were the absence of neural tube and limb reduction defects among the human cases. The similarity between the human embryopathy and the experimental animal models suggested that the mechanisms of teratogenesis were probably similar. Poswillo (1975) reported initially that vitamin-A-induced craniofacial malformations might be attributed to deficiencies of cranial neural crest cells. Others showed that vitamin A and retinoic acid induced deficiencies of neural crest cells entering the branchial arches (Morriss 1975; Hassell et al. 1977; Wiley et al. 1983). We extended this mechanistic hypothesis by suggesting that the craniofacial malformations were not the only structures that were dysmorphic as a result of the neural crest deficiencies but that the cardiac and thymic abnormalities were also induced by the same mechanisms (Lammer et al. 1985). We were led to this conclusion by several studies that demonstrated the importance of cranial neural crest cells in normal aorticopulmonary septation and thymic morphogenesis (Kirby et al. 1983; Bockman and Kirby 1984). Kirby and co-workers found that they could induce those malformations in chick embryos by excising preotic neural tube and crest tissue prior to the time of onset of neural crest migration.

Recently, Webster et al. (1986) provided experimental evidence to support this hypothesis. Administration of isotretinoin to pregnant mice prior to or during cranial neural crest migration induced malformations of the ears, mandible, heart, and thymus that closely resembled those of the affected infants. These malformations were correlated with adverse effects on premigratory and migratory cranial neural crest cells using scanning and transmission electron microscopy. This work suggested that isotretinoin inhibited crest cell migration and may have affected differentiation of premigratory crest cells. Similar work by Goulding and Pratt (1986) using whole-embryo culture showed that the effects of isotretinoin were direct and probably not the result of growth retardation. They also found a reduced rate of proliferation among frontonasal mesenchymal cells, presumably of neural crest origin.

It is possible that cell populations other than cranial neural crest are susceptible to isotretinoin. However, nearly all of the organs whose development has been adversely affected are derived largely from neurectodermal cells. The factors determining susceptibility are unclear at this time. They may include the distribution and concentration of intracellular retinoic-acid-binding protein or the, as yet, undetected retinoid receptor(s).

DISCUSSION

It is apparent that fetal exposure to isotretinoin does not increase the risk for all types of birth defects. The birth defects have been limited to craniofacial, cardiac, thymic, and CNS structures. Whereas the mechanisms responsible for producing the CNS malformations are unclear, it appears that the other malformations, though they involve several organ systems, are related by common mechanisms acting on cranial neural crest cells. The findings from experimental animal models of retinoic acid embryopathy have been correlated with adverse effects on cranial neural crest populations both before, during, and after the onset of cranial neural crest cell migration. Effects on other cell populations are possible and likely, especially neurepithelium. The effects of isotretinoin are undoubtedly expressed through mechanisms other than inhibition of crest cell migration. Since retinoic acid can promote differentiation of embryonic and cancer cell lines in vitro, it is possible that another mechanism is an alteration of ectodermal cell differentiation during embryogenesis (Sporn and Roberts 1983; Lammer 1985). In addition, Goulding and Pratt (1986) found that retinoic acid can inhibit the rate of proliferation of some presumably crest-derived mesenchymal cells of the frontonasal region, suggesting a role for another mechanism.

The pattern of craniofacial, cardiac, and thymic malformation has been seen sporadically and as part of other human dysmorphogenic disorders; it is called the DiGeorge anomaly or sequence (Lammer and Opitz 1986). The correlation of the occurrence of this pattern of malformation among infants exposed to retinoic acid, with the results of experimental work that showed adverse effects on cranial neural crest cells, strongly suggests that this pattern of human malformation is a neurocristopathy (Lammer and Opitz 1986). Because the DiGeorge sequence is etiologically heterogeneous, it suggests that adverse effects on cranial neural crest cells may be a common mechanism of dysmorphogenesis and teratogenesis.

Critical analysis of the patterns of malformation among infants exposed to retinoic acid and correlation with the experimental models has been very productive and suggests several avenues of potentially fruitful exploration. First, careful correlation of the human embryopathy and the experimental models has already improved our understanding of the roles of cranial neural crest cells in normal and abnormal morphogenesis. Previously, several craniofacial disorders were attributed to abnormalities of cranial neural crest cells, for example, Treacher-Collins syndrome and oculoauriculovertebral dysplasia (Goldenhar syndrome). Similarities and differences between the findings of retinoic acid embryopathy and of those disorders will help sort out some of these controversies.

Second, the experimental models offer the opportunity to explore some of the regulation and control of embryonic cell differentiation. A large number of in vitro studies have shown the remarkable ability of retinoic acid to alter differentiation of malignant and embryonic cell lines. Webster et al. (1986) found that retinoic acid inhibited neural crest cell migration and prevented the crest cells from separating from the closing neural tube. This suggests that this model may be useful for studying the determinants that permit and inhibit the initiation of crest cell migration. This process may involve the turnover of cellular membrane constituents such as cell-adhesion molecules (CAMs), and so forth.

Third, the experimental models may be useful for studying induction. It is unclear what role the neural crest plays in the morphogenesis of the thymus gland, but in the absence of a crest cell component, the thymus may not develop at all. The crest cells may be inducing the formation of a specific stroma by interacting with endoderm of the more caudal branchial pouches and/or ectoderm of the caudal branchial arches and grooves. Without the crest cell contribution, it appears that leukocytes do not sequester and mature in the thymic promordia.

Finally, this focus of research may improve our understanding of the role of vitamin A and retinoids in normal morphogenesis. Dencker (1979) and Dencker et al. (1986) found that labeled retinoic acid accumulated in neurectodermal cells and presumptive cranial nerve ganglia cells of early mouse embryos. Presumably, the labeled retinoic acid was localized to sites of intracellular retinoic-acid-binding protein (or possibly to a retinoid receptor). How do the temporal and tissue-specific distributions of this binding protein influence morphogenesis? What is the relationship of the binding protein to the, as yet, unidentified "retinoid receptor"? Are there several different retinoid receptors, and does this heterogeneity explain the variability in teratogenic potency of the various synthetic retinoids? If there is only one retinoid receptor, why are some synthetic retinoids efficacious for certain dermatologic conditions and not for others? All of these questions pose exciting avenues for future exploration.

ACKNOWLEDGMENTS

This work could not have been done without the cooperation and support of Hoffman-LaRoche, Inc. and the Food and Drug Administration. Donna Bisazza ably provided the secretarial support.

This paper is dedicated to Dr. Jim Firestone, for the encouragement that he gave to me and for his unbridled zest for life.

REFERENCES

Bockman, D.E. and M.L. Kirby. 1984. Dependence of thymus development on derivatives of the neural crest. *Science* **223:** 498.

Dencker, L. 1979. Embryonic-fetal localization of drugs and nutrients. In *Advances in the study of birth defects* (ed. T.V.N. Persaud), vol. 1, p. 1. MTP Press, Lancaster, England.

Dencker, L., B.R.G. Danielsson, H. Ghantous, R. d'Argy, and G. Sperber. 1986. Saturable accumulation of retinoic acid in neural and neural crest derived cells in early embryonic development. *Teratology* **33:** 14A.

Geelen, J.A.G. 1979. Hypervitaminosis A induced teratogenesis. *CRC Crit. Rev. Toxicol.* **6:** 351.

Goodman, D.S. 1984. Vitamin A and retinoids in health and disease. *N. Engl. J. Med.* **310:** 1023.

Goulding, E.H. and R.M. Pratt. 1986. Isotretinoin teratogenicity in mouse whole embryo culture. *J. Craniofacial. Genet. Dev. Biol.* **6:** 99.

Hassell, J.R., J.H. Greenberg, and M.C. Johnston. 1977. Inhibition of cranial neural crest cell development by vitamin A in the cultured chick embryo. *J. Embryol. Exp. Morphol.* **39:** 267.

Kirby, M.L., T.F. Gale, and D.E. Stewart. 1983. Neural crest cells contribute to normal aorticopulmonary septation. *Science* **220:** 1059.

Kleinsasser, O. and R. Schlothane. 1964. Die Ohrmisbildungen im Rahmen der Thalidomid-Embryopathie. *Z. Laryngol. Rhinol. Otol.* **43:** 344.

Kochhar, D.M. 1967. Teratogenic activity of retinoic acid. *Acta Pathol. Microbiol. Scand.* **70:** 398.

Lammer, E.J. 1985. On the plausibility of retinoids adversely influencing neural crest cell activity. *Proc. Greenwood Genet. Cent.* **4:** 29.

Lammer, E.J. and J.M. Opitz. 1986. The DiGeorge anomaly as a developmental field defect. *Am. J. Med. Genet.* (Suppl.) **2:** 113.

Lammer, E.J. and K.K. Sulik. 1985. On a pattern of external ear malformation. *Proc. Greenwood Genet. Center* **5:** 107.

Lammer, E.J., D.T. Chen, R.M. Hoar, N.D. Agnish, P.J. Benke, J.T. Braun, C.J.

Curry, P.M. Fernhoff, A.W. Grix, I.T. Lott, J.M. Richard, and S.C. Sun. 1985. Retinoic acid embryopathy. *N. Engl. J. Med.* **313:** 837.

Morriss, G.M. 1975. Abnormal cell migration as a possible factor in the genesis of vitamin-A-induced craniofacial anomalies. In *New approaches to the evaluation of abnormal embryonic development* (ed. D. Neubert and H.J. Merker), p. 678. Thieme, Stuttgart.

Newell-Morris, L., J.E. Sirianni, T.H. Shepard, A.G. Fantel, and B.C. Moffett. 1980. Teratogenic effects of retinoic acid in pigtail monkeys (Macaca nemestrina). II. Craniofacial features. *Teratology* **22:** 87.

Poswillo, D. 1975. The pathogenesis of the Treacher Collins syndrome (mandibulofacial dysotosis). *Br. J. Oral. Surg.* **13:** 1.

Shenefelt, R.E. 1972. Morphogenesis of malformations in hamsters caused by retinoic acid: Relation to dose and stage at treatment. *Teratology* **5:** 103.

Sporn, M.B. and A.B. Roberts. 1983. The role of retinoids in differentiation and carcinogenesis. *Cancer Res.* **43:** 3034.

Webster, W.S., M.C. Johnston, E.J. Lammer, and K.K. Sulik. 1986. Isotretinoin embryopathy and the cranial neural crest: An in vivo and in vitro study. *J. Craniofacial. Genet. Dev. Biol.* **6:** 211.

Wiley, M.J., P. Cauwenbergs, and I.M. Taylor. 1983. Effects of retinoic acid on the development of the facial skeleton in hamsters: Early changes involving cranial neural crest cells. *Acta Anat.* **116:** 180.

COMMENTS

Johnson: After looking at those magnificent pictures of the migration waves of neural crest and relating that back to what Dr. Pratt talked about concerning the turning on of the N-CAM, it appears that the first wave of neural crest cells is migrating. The second group should have come down more caudally but remained at the neural tube. That would be consistent with what Dr. Pratt said, had there been a failure of differentiation in the N-CAM at that time. The real effect is not necessarily on migration; it is turning on the differentiation of the cells. The downstream effects would be seen in thymus, in the aortic arch patterns, and in the ear. Am I right in that?

Lammer: You may be partially right. However, I think there is convincing evidence from in vitro studies that retinoids have an effect on cell migration. My suspicion is that several mechanistic phenomena may contribute; as Dr. Pratt pointed out, an abnormality of differentiation may prevent cranial neural crest cells from initiating their migration. Based on the amount of evidence from the in vitro studies, it is reasonable to assume that other mechanisms may act on these cells' migratory activity as well; for example, the membrane blebbing that was found would have a very deleterious effect on the recycling of plasma membrane components that is necessary during the process of migration. In addition, there is evidence that there are two time periods during which ear development is susceptible, the second being at day 9. In association with the ear malformations, extensive cell necrosis was found in the area of the developing epibranchial placodes. Although the cell death appeared to be limited to mesenchymal cells, often the externally visible phenotypic abnormality appeared as "holes" in the closing membranes.

Brent: One of the other explanations from your slide is differential death. If you noticed, the day that the incidence of ear anomalies dropped, you still had a high mortality for your embryos. When the incidence of ear anomalies rose, there was a sudden drop in mortality.

Lammer: That's correct.

Scott: You mentioned that the ear malformations resemble those from thalidomide children, also seen by Dr. Lois Newman, and reported very clearly in the nonhuman primate embryo. However, it is worthwhile to point out that the similarity essentially ends with the ear defects. The other pattern of malformations is, in fact, very different between the two. It is also worth mentioning that there is speculation that thalidomide works by the inhibition of neural crest migration.

Lammer: Your point is well taken. We have hypothesized that isotretinoin may be affecting the cranial portion of the neural crest. We have no evidence that it affects truncal neural crest cells, which were hypothesized as the susceptible cell population for the induction of limb abnormalities after thalidomide exposure.

Pedersen: I would like someone who knows the history of Accutane to tell me how the FDA could have approved such a compound for use in women of child-bearing age.

Brent: The FDA knew that Accutane was a teratogen. The warning on the drug said that it was a teratogen and should not be used by pregnant women. The same is true with methotrexate and many drugs that are teratogens.

Lammer: But a major difference is that isotretinoin is not nearly as toxic to the mother as most teratogenic drugs, like methotrexate. Isotretinoin is one of the first of a major new class of drugs, and it is probable that applications to market other synthetic retinoids in North America will continue. At this time, only isotretinoin and etretinate have been approved for systematic use in the United States.

Pedersen: Why are we wasting our time?

Lammer: I think that's an excellent question. A forum is urgently needed to discuss the public health policies related to the marketing of medications, like synthetic retinoids, which are likely to be human teratogens but which have relatively low maternal toxicity and therapeutic efficacy. The criteria that allow one teratogenic compound, isotretinoin, to be approved for marketing, but disallow others (thalidomide) are unclear and need to be discussed thoroughly. Yesterday, Dr. Braun reported that potential therapeutic benefits in the treatment of leprosy were attributed to thalidomide. Yet, it is apparent that it will not be marketed because of its teratogenicity.

Pedersen: Isn't it true that there are more retinoid-produced malformed babies in North America than babies with malformations produced by thalidomide?

Lammer: Probably.

Hanson: The concept of controlled release of teratogens into the human population is an interesting one.

Brent: Recently, I received a letter from a pregnant obstetrician who has leprosy, asking me to write a letter. She was about 56 days into gestation, and her leprosy was getting worse. The physician at Lexington wouldn't let her use thalidomide because this drug can't be used in pregnancy. We will never be free of the dilemmas of these insoluble problems. We strongly recommend that women continue using their oral contraceptives, but a small percentage of them do get pregnant.

Haney: I want to comment on the phenomena of teratogens. Cyproterone acetate is an antiestrogen used in Europe. The simple solution is combining this drug with 50 μg of ethynylestradiol, which inhibits ovulation, controls bleeding, induces hair loss, and increases antigen-binding protein. The combination is such that it virtually can't be a teratogen because it prevents ovulation. One could make a very strong case for a drug that is used for treating cystic acne, one that has an androgen-related component. You could combine it as described above, and virtually guarantee that it wouldn't be a teratogen.

Lammer: Cystic acne is largely a disease of males. Most of the patients who are appropriately prescribed isotretinoin should be male. This makes it impractical to formulate isotretinoin with oral contraceptives.

Haney: But you could take that approach with the women who took it because it's very easy to do.

Brent: You could have a male pill and a female pill.

Dencker: I have a more mechanistic comment. If there is a retinoic-acid-binding protein in the neural crest cells, it is probably also there when these cells have reached their site in the visceral arches, because we have seen a strong accumulation of retinoic acid in those areas. Therefore, retinoic acid might not only affect neural crest cell migration but also the future fate of the cells when they have reached their final location.

Lammer: That is a good point. Work needs to be done in the experimental models to determine whether or not the distribution of cytoplasmic retinoic-acid-binding protein changes during cranial neural cell migration and among those cells when they have completed their migration.

Iannaccone: In the case of the children that had thymic abnormalities, is there any evidence concerning the immunological status of the surviving children, particularly the distribution of Thy-1 and Thy-2 markers and their circulation?

Lammer: No. Currently, we are doing a follow-up study of more than 50 children exposed to isotretinoin in the United States, Canada, and Puerto Rico. We are only performing quantitative lymphocyte studies, using samples from the exposed children; we are not measuring levels of thymic hormones.

Iannaccone: Can that be done?

Lammer: Yes, perhaps in the future. We are doing total B-lymphocyte counts, total T-lymphocyte counts, and T4 and T8 antigen percentages. This part of the study has been very difficult because the participants live in more than 25 states, and the transportation problems with lymphocytes are a limitation to this work.

Pratt: In your talk, you mentioned that etretinate will probably be the next human teratogen to reach the market. What is known about the effects of etretinate in animal models? Does it produce the same sort of spectrum that you see with Accutane?

Lammer: There is some overlap in the types of malformations associated with human exposure to etretinate and those caused by isotretinoin. There are eight case reports from Europe of children with birth defects associated with exposure to etretinate. Some of the birth defects that were reported are different from those that we have found. Several fetuses exposed to etretinate had neural tube defects. To my knowledge, not a single child in the United States has been identified with a neural tube defect following exposure to isotretinoin. Also, several infants exposed to etretinate had limb reduction defects; we have not seen that among fetuses and infants exposed to isotretinoin. So, I think that the spectrum of defects is quite similar but also includes neural tube defects and limb abnormalities. These malformations can be induced by isotretinoin in experimental animals, but we have not seen them among exposed humans. The reason is unclear but may be related to the dose dependency of their induction. In experimental models, higher doses of isotretinoin are required to induce limb abnormalities than to induce the other malformations.

Brent: If you saw a patient with a limb defect born to a mother on Accutane, would you postulate that it was an isolated malformation unrelated to the exposure?

Lammer: First, it would be important to look at the child's X-rays in order to view the pattern of bony abnormalities and to correlate the findings with the timing of exposure. Dr. Kochhar and others have shown that the type of limb abnormality depends on the timing of the exposure; early exposures tend to produce more proximal abnormalities (Kochhar 1977). The pattern of abnormalities should be related to the timing of medication use, if the mechanism of retinoid-induced limb abnormalities is comparable between the experimental models and humans.

Kochhar: My impression was that the animal embryos you showed were growth retarded, but the newborn babies that you presented did not seem to be growth retarded.

Lammer: That is largely correct. The scanning electron microscopy images that I showed were taken 12 hr after the single isotretinoin exposure. Prenatal growth retardation is uncommon among malformed infants exposed to isotretinoin; postnatal growth retardation seems to be much more common. Only 2 of the first 21 malformed infants exposed to isotretinoin were growth retarded.

Kochhar: Perhaps human embryos are also growth retarded during the initial exposures to isotretinoin, but is there a catch-up growth by birth?

Lammer: I have no information about first trimester growth of the exposed human embryos.

Hanson: I would like you to speculate on the consumption of megadoses of vitamin A in the human population.

Lammer: There are no epidemiologic studies of the benefits or risks of the use of vitamin A, dietary or supplementary, during pregnancy. A handful of case reports of suspected vitamin A teratogenicity have been published. Some of the children described in these reports have malformations similar to those that we have seen among children exposed to isotretinoin, but some of the children exposed to vitamin A have abnormalities that are quite different. At best, I think that we can say that there are some similarities. I think that unsupervised supplementation of vitamin A during pregnancy is a concern. I don't think that anyone knows a daily dosage at which a clinically significant increased risk exists, but women should be encouraged not to take large amounts of vitamin A during pregnancy. I would define "large" as more than 10,000 units per day. That, I think, is a reasonable recommendation.

Oakley: It would help if there were no capsules containing more than 5000 units of vitamin A. You can go to most every grocery store in this country and buy capsules containing 25,000 units. If you go to the health food store around the corner, you can buy capsules containing 50,000 units. If capsule containment was limited to 5000 units, the possible risk might be decreased.

Extrapolation of Teratogenic Responses Observed in Laboratory Animals to Humans: DES as an Illustrative Example

MICHAEL D. HOGAN,* RETHA R. NEWBOLD,† AND JOHN A. MCLACHLAN†
*Division of Biometry and Risk Assessment
†Developmental Endocrinology and Pharmacology Section
Laboratory of Reproductive and Developmental Toxicology
Division of Intramural Research
National Institute of Environmental Health Sciences
Research Triangle Park, North Carolina 27709

OVERVIEW

This paper is concerned with the selection of the most appropriate dosage scale for extrapolating results observed in animal teratology studies to humans. The synthetic estrogen, diethylstilbestrol (DES), is used as a model compound to make interspecies comparisons of potency estimates for different reproductive tract anomalies found in the prenatally exposed offspring of humans and various laboratory animals. Potency is expressed as the (excess) risk or probability of developing a given anomaly per unit dose of exposure and is computed for a number of different dosage scales. Comparison of these potency calculations suggests that agreement between human and experimental animal risk projections is greatest when the selected dosage scale reflects exposure during comparable species developmental periods for the specific anomaly of interest.

INTRODUCTION

The extrapolation of toxicological outcomes observed in laboratory animals to humans is a very complex process since a variety of factors can cause differential responses among species exposed to the same hazardous agent. Included among these factors are interspecies differences in life span, body size and weight, xenobiotic absorption and metabolism, genetic homogeneity, concurrent disease(s), inherent susceptibility, and exposure regimen. In the case of a teratogenic agent, the process of species extrapolation is further complicated by the fact that the exposed organism is actually the entire maternal-placental-conceptus complex. Thus, a particular adverse response in a conceptus may result solely from some maternal action, may be due to

Banbury Report 26: Developmental Toxicology: Mechanisms and Risk

the modifying effect of the placental transfer mechanism, may be attributable to the direct action of the conceptus itself, or may involve some combination of these modes of teratogenic activity. Other factors unique to the field of teratology, such as interspecies differences in the timing of developmental events, may also bear on the extrapolation of laboratory results to humans.

Because of the complexity of the progress of teratogenesis, the traditional regulatory approach to species extrapolation in this area has been to rely on the use of safety factors, usually applied to either an experimentally determined no-observed or minimum-effect level of exposure (EPA 1979). Safety factors are often selected to reflect both the relative sensitivity of the test animal, as compared with humans, and the presumed increased variability in sensitivity within the human population. However, the safety factor approach to the determination of "acceptable" exposure levels may be too simplistic and nonspecific for use in teratogenesis (Gaylor and Shapiro 1979). As a result, there is a growing interest in approaching both teratogenic dose-response modeling and species extrapolation in a more analytical manner than, for example, that associated with carcinogenic risk estimation.

In carcinogenesis, interspecies comparisons are usually made in terms of a standardized dosage scale (e.g., mg/kg/day, parts per million (ppm) in the diet, $mg/m^2/day$, or mg/kg/lifetime) chosen to account, at least crudely, for some of the various factors that contribute to differential responses among species (Hogan and Hoel 1982). Comparisons of experimental and epidemiologic data for the limited set of compounds that are carcinogenic in both laboratory animals and humans, and for which sufficient data exist to make interspecies comparisons, indicate that when cancer risk or potency is expressed in terms of an average daily lifetime dose either in mg/kg or in mg/m^2, agreement between animals and humans within an order of magnitude is often observed (Crouch and Wilson 1978; Hoel 1979).

By analogy to carcinogenesis, the most obvious dosage scale to employ in the extrapolation of laboratory-based teratogenic responses to humans would be the average daily maternal dose administered throughout gestation. However, because of the differential timing of various developmental events, it is also reasonable to consider standardizing risk or potency estimates in terms of exposure or administered dose during the appropriate developmental periods with which specific anomalies may be associated.

In the present study, these proposed dosage scales are evaluated by comparing "standardized" human and animal potency estimates for the synthetic estrogen DES. DES was picked to serve as a model compound because it has one of the most extensive human data bases of any known chemical teratogen. Furthermore, DES produces a spectrum of relatively rare reproductive tract anomalies in both humans and various animal models, which may imply a commonality of action across species (Herbst and Bern 1981).

MATERIALS AND METHODS

Data for potency comparisons were abstracted from a number of different epidemiologic and experimental investigations. One of the primary sources of information on in utero DES exposure of humans is the University of Chicago DES efficacy trial (Dieckmann et al. 1953) and subsequent follow-up studies (Gill et al. 1979; Schumacher et al. 1981). The efficacy trial was a double-blind study that originally involved 2162 women who were enrolled in the University's prenatal clinic between September 1950 and November 1952. After deletion of subjects for such reasons as failure to be pregnant, non-compliance, change of address, and so forth, the final study sample was reduced to 840 women exposed to DES and 806 controls. On the average, study participants began their DES exposure at the end of week 12 of their pregnancy and followed a systematic dosing regimen in which the daily dose was increased from 5 mg/day during weeks 7–8 of pregnancy to 150 mg/day by weeks 34–35.

Another study of the potential teratogenic effects of prenatal DES exposure was conducted by Herbst and his colleagues (Herbst et al. 1975), who followed up the female offspring of mothers treated at the Boston Lying-In Hospital Diethylstilbestrol Clinic between 1947 and 1958. The study cohort consisted of 110 DES-exposed daughters, aged 18 or older, and 82 unexposed controls. As was the case in the Chicago efficacy trial, mothers followed a standardized treatment regimen during which exposure was increased from 2.5 mg/day through week 6 of pregnancy to 150 mg/day by week 35, after which treatment was discontinued. Nearly half of the study subjects' mothers had initiated their DES therapy by weeks 9–12 of their pregnancies.

The remaining human data employed in this paper were taken from a Seattle study (Stenchever et al. 1981) that used the sperm penetration assay to evaluate the risk of infertility among males whose mothers had been treated with DES during pregnancy. After elimination of study volunteers who had previously engaged in unprotected intercourse and, therefore, might have some knowledge about their fertility potential, 13 DES-exposed males and 11 nonexposed controls remained. All study participants underwent a sperm penetration assay and a physical examination during which various reproductive tract anomalies were scored. Although information on the duration of maternal exposure to DES was obtained for each study subject, the actual maternal dosage was not always known, and no dosage data were reported in the paper.

The observed results in humans were compared to laboratory findings in the CD-1 mouse, the ICR mouse, and the rhesus monkey. The experimental protocol for the CD-1 mouse studies (McLachlan et al. 1975; McLachlan 1981; Newbold and McLachlan 1982), which were performed to assess the association between intrauterine DES exposure and the onset of reproductive tract anomalies in male and female offspring, specified treatment of pregnant mice with 100 μg/kg/day on days 9–16 of gestation (where vaginal plug detection signaled day 0 of pregnancy). In the ICR mouse study (Nomura and Kanzaki 1977), interest centered on the induction of urogenital anomalies and tumors in the offspring of female mice who received a single 10-μg/g injection of DES between days 6 and 18 of gestation. The study protocol for the evaluation of the embryotoxic and fetotoxic effects of prenatal DES exposure in rhesus monkeys (Hendrickx et al. 1979; Thompson et al. 1981) called for the administration of 1 mg/day of DES to 19 pregnant females, beginning on day 21, 100, or 130 of gestation and continuing until delivery. Male and female offspring were examined over time for a number of reproductive tract anomalies.

RESULTS

Selected results from the various epidemiologic and laboratory studies considered in this paper are summarized in Table 1, along with an indication of the dosing regimen employed in each investigation. In the case of the Chicago and Boston studies, data on average duration or individual duration of treatment were combined with information about the standard dosing schedule prescribed at the respective hospitals to generate cohort-specific average daily dose estimates for both the entire gestational period and particular developmental periods of interest. In the absence of other knowledge, the dosing regimen of Smith and Smith (Boston Lying-In Hospital), which was a standard treatment regimen employed by many physicians throughout the country, was assumed to be applicable in the Seattle fertility study.

The most critical part of the potency comparison process was the determination of the corresponding developmental periods in humans and the various animal models that, when perturbed by DES exposure, might be associated with one of the anomalies or reproductive effects under consideration, that is, (1) vaginal adenosis, (2) cryptorchidism, (3) epididymal cysts, or (4) potentially impaired male fertility. On the basis of a detailed review of the published human and animal literature relevant to the issue, the following developmental periods associated with each of the anomalies mentioned above were defined:

1. The differentiation of the vaginal epithelium in humans follows the regression of the vaginal plate and canalization of the vagina. The transformation of vaginal epithelial cells from pseudostratified columnar cells to squamous cells begins after this regression and is complete by the end of the fourth

Table 1
Summary Data on Reproductive Tract Anomalies

Study	Treatment or dosing regimen	Anomaly			
		vaginal adenosis	cryptorchidism	epididymal cysts	impaired fertility[a]
Chicago	5–150 mg/day (week 7–35)[b]	153/229(5/136)[c]	17/308(1/307)	64/308(15/307)	
Boston	2.5–150 mg/day (≤week 6–35)[b]	35%(1%)			
Seattle	No dose data available[b]			5/13(0/11)	10/13(1/11)
CD-1 mouse	100 μg/kg/day (day 9–16)	3/20(0/20)[d]	6/24(0/14)[e]	8/24(0/14)[e]	6/10(0/10)[f]
ICR mouse	10 μg/g[g] (day 6–18)		71.9%(0%)[h]		
Rhesus monkey	1 mg/day (day 21, 100, or 130 to term)	2/8(0%)[i]			

[a]Low sperm penetration assay scores in humans; sterility in male mice.
[b]Variable subject entry dates.
[c]Corresponding control value in parentheses.
[d]At age 1 month.
[e]At age 9–10 months.
[f]At age 20–25 weeks.
[g]Single dose on 1 day in designated exposure period.
[h]Average of 11/15 on day 16 and 19/27 on day 18; response observed at age 12 months.
[i]At age 3.5 years.

month. Interruption of this process can lead to the persistence of columnar epithelial cells within the squamous cell epithelium of the vagina, that is, to vaginal adenosis. The transformation of vaginal epithelium in the mouse starts in the latter part of gestation and continues through the first neonatal week. In the rhesus monkeys, the gestational time periods of day 100 through day 130 and day 130 through delivery are associated with vaginal plate breakdown and the latter stages of vaginal development and differentiation, respectively.

2. In the development of cryptorchidism, or abnormal descent of the testes, two phases of testicular descent can be recognized. The first, in which the testes migrate from their position adjacent to the lower pole of the kidney to the opening of the inguinal canal, occurs in the human between the second and the sixth month of pregnancy and in the mouse between days 14 and 16 of gestation. The second phase involves passage of the testes through the inguinal canal and final location in the scrotum. This occurs in man during the seventh and eighth months and in the mouse on days 17–21, that is, from day 17 of gestation through neonatal day 3.
3. In the male, Müllerian duct regression is seen in the human by the end of the second month and in the mouse by day 16 of gestation. The regression in the mouse occurs between days 14 and 16. The remaining Wolffian duct differentiates further and, in the human, epididymal differentiation is complete by the fourth month. The corresponding process in the mouse is complete between neonatal days 1 and 2.
4. In the case of male fertility, the explicit cause(s) of the observed low sperm penetration assay scores in humans and sterility in the CD-1 mice is not known. However, both species displayed cryptorchidism and epididymal cysts associated with their prenatal DES exposure. Therefore, it seems plausible to postulate that the appropriate developmental period related to "male fertility problems" might be one of those associated with these two anomalies. The estimated comparable developmental periods for each of four teratogenic/reproductive endpoints under consideration are summarized in Table 2.

Finally, since much of the available human and animal data were limited to or summarized to yield an observed response at a single level of exposure or following a given dosing regimen, we elected to estimate the unknown, underlying dose-response relationship by a straight line from the observed response to the origin after correcting for background. Clearly, this is a simplistic estimate of this relationship, necessitated by the paucity of available data. However, similar procedures have been employed in the comparison of human and animal carcinogenicity data (e.g., NAS 1975; Hoel 1979). Furthermore, the slope of such a line should provide a measure of potency, and a

Table 2
Comparable Species-specific Developmental Periods

Anomaly	Species	Developmental period (%)
Vaginal adenosis	Humans	Week 7–17 (27.5)[a]
	Mouse	Day 16–25 (38.5)[b]
	Monkey	Day 100–168 (40.5)
Cryptorchidism	Humans	
	phase 1	Week 7–27 (52.5)
	total	Week 7–36 (75.0)
	Mouse	
	phase 1	Day 14–16 (13.6)
	total	Day 14–21 (36.4)[b]
Epididymal cysts	Humans	Week 7–18 (30.0)
	Mouse	Day 14–20 (33.3)[b]
Male fertility	Humans	As for cryptorchidism and
	Mouse	epididymal cysts

[a]Percent of gestational period.
[b]Percent of "extended" gestational period.

relative potency index (RPI) can be generated merely by considering the ratio of species-specific slopes. The dosage scale that produces the greatest agreement among species would then be that which gives an RPI closest to one. The various dosage scales being considered in the present analysis include average mg/kg/day doses throughout gestation, throughout some specified developmental period, and during that portion of a given developmental period in which exposure occurred.

Given the estimated "comparable" developmental periods and the incidence and exposure data from the epidemiologic and laboratory studies described previously, relative potency indices can be computed for each of the dosage scales and anomalies/reproductive effects of interest. These values are given in Table 3.

The results in Table 3 indicate that there is good overall agreement among the various epidemiologic studies, with maximum differences in relative potency ranging from approximately two- to fourfold, depending on the anomaly. Excellent agreement between the CD-1 mouse and the rhesus monkey is also observed when comparisons are made using either the first or the third dosage scales under consideration (i.e., RPIs of 1.19 and 1.07, respectively). The near order-of-magnitude difference in relative potency seen when the dose is standardized in terms of mg/kg/day average exposure throughout the developmental period may be, in part, a reflection of the fact that the mouse was exposed for only a small portion (10%) of that period, whereas the monkey's average exposure covered most (>80%) of the same period.

Table 3
Relative Potency Ratios for DES-Associated Urogenital Tract Anomalies and Reproductive Problems

	RPI (animal vs. human)			
Comparison	dosage scale[a]	gestation	DP	DPE
	Vaginal adenosis			
CD-1 mouse vs. Chicago[b]		4.19	4.67	1.00
vs. Boston		7.26	6.66	1.23
Monkey vs. Chicago		3.45	0.52	0.93
vs. Boston		6.13	0.74	1.15
	Cryptorchidism			
CD-1 mouse vs. Chicago[c]		82.78	29.18	46.47
ICR mouse vs. Chicago		19.04	9.35	1.73
	Epididymal cysts			
CD-1 mouse vs. Chicago		36.06	11.33	9.50
vs. Seattle		9.98	3.93	2.23
	Male fertility			
CD-1 mouse vs. Seattle				
crypt. DP[d]		10.21	4.07	5/69
EC DP[e]		10.21	4.48	2.54

[a]Dosage scale: gestation, average dose of a mg/kg/day throughout gestation; DP, average dose of a mg/kg/day throughout the comparable developmental period; DPE, average dose of a mg/kg/day during that portion of the developmental period in which exposure occurred.
[b]That is, versus appropriate offspring from the Chicago cohort.
[c]Assumes that Chicago anomalies are primarily first-stage effects.
[d]Assuming a DP similar to that for cryptorchidism.
[e]Assuming a DP similar to that for epididymal cysts.

The closest agreement between animal and human potency estimates occurs for vaginal adenosis. On the other hand, the animal model seems to be particularly sensitive, relative to humans, when the teratogenic endpoint is cryptorchidism. Limited data on the rhesus monkey, that is, one affected out of five exposed (Thompson et al. 1981), also seem to support this contention, indicating a relative potency for the monkey of 20- to 30-fold, regardless of the dosage scale employed. The small relative potency observed when excess risk for the ICR mouse and man are compared, using the dosage scale that reflects the average daily exposure during that portion of the developmental period during which exposure actually occurred (i.e., DPE), may be an artifact attributable to the large pulse dose used in treating these mice. (The single or one-time dose administered to the ICR mouse was 100 times higher than the corresponding daily dose given to the CD-1 mouse over an 8-day period, even though the related excess risk for cryptorchidism was increased by less than a factor of 3.)

DISCUSSION

In many respects, DES represents an almost "ideal" compound for comparing animal and human teratogenic responses. There is a significant body of data in the published literature indicating that DES exposure produces a variety of rare, reproductive tract anomalies in the offspring of both humans and various animal species, which strongly underscores the relevance of the animal model in this instance. Moreover, the close quantitative, as well as qualitative, agreement between the animal and human experience following DES exposure described in this paper is further supported by other reproductive tract anomalies seen in animals but not yet documented in humans to any significant degree. For example, female offspring of CD-1 mice exposed to 100 μg/kg/day of DES on days 9–16 of gestation displayed a high degree of infertility ($>70\%$), as well as a 100% incidence of oviductal malformations characterized in terms of arrested development (Newbold et al. 1983). In fact, DeCherney and his colleagues (DeCherney et al. 1981) reported results for 16 DES daughters who were evaluated for infertility problems and found to have "...unique (fallopian) tubal morphologic features." A histological feature of DES-induced oviductal dysmorphogenesis in mice is the presence of glandular elements in the oviducts of virtually all the exposed female offspring. These histological changes in the mouse share morphologic similarities with a human condition known as salpingitis isthmica nodosa (SIN) (Newbold et al. 1984), which is inconsistent with normal fertility (Woodruff and Pauerstein 1969; Honore 1978). (It is noteworthy that SIN has been reported in a case study of a woman exposed to DES [Shen et al. 1983].) The importance of embryological mechanisms was further underscored by the recent finding of paraovarian cysts in humans and mice exposed to DES prenatally (Haney et al. 1986); the ovarian cysts in females of both species were determined to derive from the male embryonic rudiment, the mesonephric duct.

The initial relative potency comparisons for the DES-related anomalies under consideration in this paper (i.e., those based on exposure over the entire gestational period) suggest that animals are more sensitive than humans to the teratogenic effects of the synthetic estrogen. However, review of the DES treatment regimen employed in the case of threatened pregnancies indicates that the heaviest exposure occurred in the later stages of pregnancy, whereas the developmental periods, typically of interest in the case of humans, fell in the earlier portion of the gestational period. Thus, if attention is restricted to exposure during comparable developmental periods, the observed differences between animal and human responses are reduced.

When evaluating the specific DES potency comparisons developed in this paper, it is important to bear in mind that these comparisons are obviously dependent on the observed outcomes, the assumptions of dose-response linearity and additive background, and the estimates of the appropriate developmental periods used to make interspecies comparisons. For simplicity, it was also assumed that the exposed pregnant female/conceptus was equally sensitive throughout the estimated developmental period, and no attempt was made when estimating potency to adjust for the possibility that DES may have accumulated in the pregnant female/conceptus prior to the beginning of a specified developmental period. Nevertheless, it is clear that in the case of DES, comparison of relative species potency on a dosage scale that reflects exposure during comparable developmental periods increases the agreement between human and animal estimates. Before any general conclusions can be reached about the most appropriate dosage scale for interspecies extrapolation of teratogenic effects, additional animal/animal and, if possible, animal/human comparisons need to be made for a variety of compounds. However, the results of this study certainly suggest that rigorous attention to the temporal details of organ development among different species and the perturbation of these processes by chemicals should be a component of any attempt at cross-species extrapolation of teratogenic events.

REFERENCES

Crouch, E. and R. Wilson. 1978. Interspecies comparison of carcinogenic potency. *J. Toxicol. Environ. Health* **5:** 1095.

DeCherney, A.H., I. Cholst, and F. Naftoliu. 1981. Structure and function of the fallopian tubes following exposure to diethylstilbestrol (DES) during gestation. *Fertil. Steril* **36:** 741.

Dieckmann, W.J., M.E. Davis, L.M. Rynkiewicz, and R.E. Pottiger. 1953. Does the administration of diethylstilbestrol during pregnancy have therapeutic value? *Am. J. Obstet. Gynecol.* **66:** 1062.

Environmental Protection Agency (EPA), Office of Pesticides Program. 1979. Endrin: Intent to cancel registrations and denial of applications for registration of pesticide products containing endrin, and statement of reasons. *Fed. Register* **44:** 43632.

Gaylor, G.W. and R.E. Shapiro. 1979. Extrapolation and risk estimation for carcinogenesis. In *Advances in modern toxicology,* part 2: *New concepts in safety evaluation* (ed. M.A. Mehlman et al.), vol. 1, p. 65. Wiley, New York.

Gill, W.B., G.F.B. Schumacher, M. Bibbo, F.H. Straus, and H.W. Schoenberg. 1979. Association of diethylstilbestrol exposure in utero with cryptorchidism, testicular hypoplasia, and semen abnormalities. *J. Urol.* **122:** 36.

Haney, A.F., R.R. Newbold, B.F. Fetter, and J.A. McLachlan. 1986. Paraovarian cysts associated with prenatal diethylstilbestrol exposure: Comparison of the human with a mouse model. *Am. J. Pathol.* **124:** 405.

Hendrickx, G., K. Benirschke, S. Thompson, K. Ahern, E. Lucas, and H. Oi. 1979. The effects of prenatal diethylstilbestrol (DES) exposure on the genitalia of pubertal Macaca Mulatta. II. Female offspring. *J. Reprod. Med.* **22:** 233.

Herbst, A.L. and H.A. Bern, eds. 1981. *Developmental effects of diethylstilbestrol (DES) in pregnancy*. Thieme-Stratton, New York.

Herbst, A.L., D.C. Poskanzer, S.J. Robboy, L. Friedlander, and R.E. Scully. 1975. Prenatal exposure to stilbestrol. A prospective comparison of exposed female offspring with unexposed controls. *N. Engl. J. Med.* **292:** 334.

Hoel, D.G. 1979. Low-dose and species-to-species extrapolation for chemically induced carcinogenesis. *Banbury Rep.* **1:** 135.

Hogan, M.D. and D.G. Hoel. 1982. Extrapolation to man. In *Principles and methods of toxicology* (ed. A. Wallace Hayes), p. 711. Raven Press, New York.

Honore, L.H. 1978. Salpingitis isthmica nodosa in female infertility and ectopic tubal pregnancy. *Fertil. Steril.* **29:** 164.

McLachlan, J.A. 1981. Rodent models for perinatal exposure to diethylstilbestrol and their relation to human disease in the male. In *Developmental effects of diethylstilbestrol (DES) in pregnancy* (ed. A.L. Herbst and H.A. Bern), p. 148. Thieme-Stratton, New York.

McLachlan, J.A., R.R. Newbold, and B. Bullock. 1975. Reproductive tract lesions in male mice exposed prenatally to diethylstilbestrol. *Science* **190:** 991.

National Academy of Sciences (NAS), Consultative Panel on Health Hazards of Chemical Pesticides. 1975. *Pest control,* vol. 1. *An assessment of present and alternative technologies*. National Academy of Sciences, Washington, D.C.

Newbold, R.R. and J.A. McLachlan. 1982. Vaginal adenosis and adenocarcinoma in mice exposed prenatally or neonatally to diethylstilbestrol. *Cancer Res.* **42:** 2003.

Newbold, R.R., B.C. Bullock, and J.A. McLachlan. 1984. Diverticulosis and salpingitis isthmica nodosa (SIN) of the fallopian tube. Estrogen-induced diverticulosis and SIN of the mouse oviduct. *Am. J. Pathol.* **117:** 167.

Newbold, R.R., S. Tyrey, A.F. Haney, and J.A. McLachlan. 1983. Developmentally arrested oviduct: A structural and functional defect in mice following prenatal exposure to diethylstilbestrol. *Teratology* **27:** 417.

Nomura, T. and T. Kanzaki. 1977. Induction of urogenital anomalies and some tumors in the progeny of mice receiving diethylstilbestrol during pregnancy. *Cancer Res.* **37:** 1099.

Schumacher, G.F.B., W.B. Gill, M.M. Hubby, and R.R. Blough. 1981. Semen analysis in males exposed in utero to diethylstilbestrol (DES) or placebo. *Obstet. Gynecol.* **9:** 100.

Shen, S.C., M. Bansal, R. Purrazzella, V. Malviya, and L. Stauss. 1983. Benign glandular inclusions in lymph nodes, endosalpingiosis, and salpingitis isthmica nodosa in a young girl with clear cell adenocarcinoma of the cervix. *Am. J. Surg. Pathol.* **7:** 293.

Stenchever, M.A., R.A. Williamson, J. Leonard, L.E. Karp, B. Ley, K. Shy, and D. Smith. 1981. Possible relationship between in utero diethylstilbestrol exposure and male fertility. *J. Obstet. Gynecol.* **140:** 186.

Thompson, R.S., D.L. Hess, P.E. Binkerd, and A.G. Hendrickx. 1981. The effects of prenatal diethylstilbestrol exposure on the genitalia of pubertal Macaca Mulatta. I. Male offspring. *J. Reprod. Med.* **26:** 309.
Woodruff, J.D. and C.J. Pauerstein. 1969. In *The fallopian tube*, p. 117. Williams and Wilkins, Baltimore, Maryland.

COMMENTS

Iannaccone: The data imply that there is some permanent or heritable change that is induced by DES treatment perinatally. Is this correct? Is there a construct into which this fits? The compounds are not mutagenic, for example; isn't that correct?

McLachlan: In all the short-term assays that have been done with DES, there is no mutagenic effect of which I am aware. I look at it as a permanent alteration in a program of differentiation that does not require a mutational event. What we have not sorted out yet is whether we are looking at an alteration in the same cell in terms of what it expresses or whether we have changed the population of cells so that we have a different contribution of mesonephric and paramesonephric duct epithelial cells in the same population. It is a very difficult problem.

Miller: To follow up, one issue that you seem to have perhaps laid to rest, but deserves a question—DES being both an estrogen and a carcinogen—that for many years, many have looked at the issue of metabolism here. Could you comment on whether you feel that the effects that you are observing now are related to anything other than the estrogenic actions of DES?

McLachlan: As you know, we have tried to address this problem for over 10 years. One of the ways we initially did that was to take known target cells, cells that did not have an estrogen receptor (Syrian hamster embryo fibroblasts). We could show that DES and other estrogens, based on their pharmacology, could transform these cells neoplastically, in a predictable way. I think that this has demonstrated that hormones have a capacity to alter a program of differentiation or induce a permanent differentiation defect that is consistent with neoplastic transformation in a diploid primary cell culture.

We have approached that same problem in the fetal uterus and can show that DES accumulates there. Organ cultures of this fetal material metabolize radioactive DES via the same peroxidase-mediated pathways that the transforming cells use. Then metabolites have been found in humans and every other species we have looked at.

So far, all of the experiments that we have done have not shown that metabolism does not play a role, but there are no definitive studies that say that metabolism of estrogens is involved in the induction of these differentiation defects. The more I look at the biology of this system, the more I think there are many things that we can see, many molecular lesions that really don't require bioactivation. Cancer, on the other hand, may be the one that does. But these things that I have shown you are just seeing some hormone at the wrong time or in the wrong amount and, normally, the fetal genital tract would not see this much unconjugated estrogen.

Brent: Dr. McLachlan, you know more about DES and its effects on reproduction than anybody in the world. With your expertise, you mentioned that in one of your models you are able to get 90% tumor formation. Is that so?

McLachlan: Right, that is true.

Brent: Now, with this knowledge, when you look back, do you have an explanation as to why cancer is so rare in the human? In view of the fact that really massive doses of DES over long periods of time were given to 3.5 million people, the incidence is believed to be about 1 in 1000 to 1 in 10,000 for the tumor. From a quantitative biochemical standpoint, do you have an explanation for that, outside of just species difference?

McLachlan: The doses we gave prenatally to mice were less than or equal to what was given to humans. We went over a 5-log dose range. The tumor you are talking about, vaginal adenocarcinoma, which is the cancer that has been seen in humans, is, as you mention, rare—some 1 in 1000 or 1 in 10,000. In our hands, vaginal adenocarcinoma in mice is also very rare; we looked at 3 cases out of some 248 animals. I think in every other model system vaginal adenocarcinoma is a very rare event.

The cancers that arise in high prevalence in mice are uterine cancers, which occur at a time that would be equivalent to reproductive senescence in the mouse. I think one concern that many clinicians express is that the total story on DES-induced changes in humans has yet to be played out and that things that we can see in terms of the reproductive life span of the mouse in 18 months perhaps will take 20–30 years to be seen in humans.

I don't think there is any evidence for preneoplastic lesions in the uteri of women. Clearly, the uterus is affected. Everything that we can find in a mouse so far has been found in a human. I think that the biology of the systems is very similar, and the metabolism is very similar. I think it is still an open issue whether these lesions are there.

Structural and Functional Consequences of Prenatal Exposure to Diethylstilbestrol in Women

A.F. HANEY
Division of Reproductive Endocrinology and Infertility
Department of Obstetrics and Gynecology
Duke University Medical Center
Durham, North Carolina 27710

OVERVIEW

Diethylstilbestrol (DES) is a human reproductive tract teratogen. Abnormalities in women exposed in utero include (1) vaginal adenosis, (2) unique cervical deformities (cockscomb, hood, collar, and pseudopolyp), (3) a hypoplastic uterus with an irregular "T-shaped" endometrial cavity, (4) abnormal development of the fallopian tubes, and (5) an increased number of paraovarian cysts, some with unusual histology. The genital tract abnormalities result in a variety of clinical problems, including an unusual vaginal clear-cell adenocarcinoma and increased rates of spontaneous abortion, ectopic pregnancy, and premature labor. The healing response of the cervix is altered, with cervical stenosis occurring frequently after surgical therapy of the cervix. A mouse paradigm has proved useful in studying this problem, with the pattern of anomalies paralleling those in women. Additional insight into genital tract development should be forthcoming, with a better understanding of the mechanism of action of this potent genital tract teratogen.

INTRODUCTION

DES was the first orally active synthetic estrogen available for clinical use. DES is not a steroid but rather a stilbene derivative with estrogen-receptor-association kinetics and biologic activity equal to the most potent naturally occurring estrogen, estradiol-17-beta. In the 1940s, a variety of clinical problems of pregnancy, such as abortion, toxemia, and gestational diabetes, were attributed to a deficiency of placental estrogen production. DES was given to pregnant women from the mid-1940s through the 1960s in an attempt to alleviate these problems.

Despite no evidence of efficacy, prenatal DES exposure was not thought harmful until the early 1970s, when an unusual cluster of a rare clear-cell adenocarcinoma of the vagina in young women was associated with prenatal DES exposure (Herbst et al. 1971). Subsequently, DES-induced abnor-

Banbury Report 26: Developmental Toxicology: Mechanisms and Risk

malities of virtually the entire female reproductive tract have been described (Fig. 1), including the vagina (Herbst et al. 1972; Jefferies et al. 1984), the cervix (Herbst et al. 1972; Haney et al. 1979; Jefferies et al. 1984), the uterus (Kaufman et al. 1977; Haney et al. 1979), and the fallopian tubes (DeCherney et al. 1981). DES-associated anomalies are apparently unique as they have not been described previously despite extensive pathologic evaluation of the genital tract. These anomalies have biologic importance, as a series of reproductive problems has been encountered correlating with these anomalies (Table 1).

Male offspring have also been noted to be affected with anomalies of organs of Wolffian or mesonephric duct origin (Bibbo et al. 1975; Henderson et al. 1976; Whitehead and Leiter 1981). Analogously, mesonephric remnants, that is, paraovarian cysts, in women exposed to DES are more frequent and occasionally have unusual histology (Haney et al. 1986). Remnants of the Mullerian ducts have not been noted to be present in men.

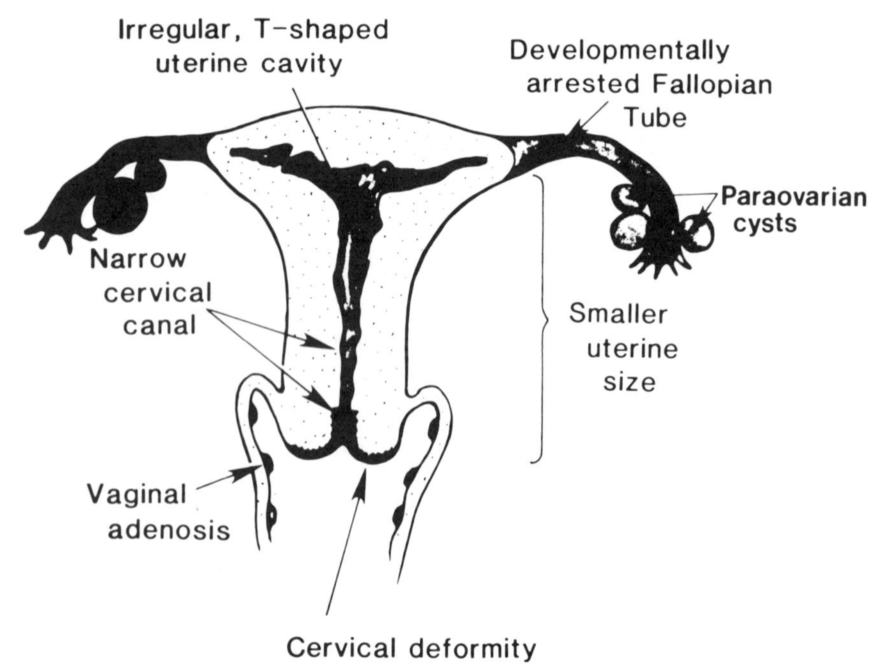

Figure 1

A schematic diagram of DES-associated abnormalities in the adult uterus. Every genital tract structure derived from the Mullerian duct can be affected by prenatal exposure to DES. The timing of administration and dose of DES is likely critical to the pattern of anomalies in each individual patient.

Table 1
Correlation of Anomalies Induced by Prenatal DES Exposure and Clinical Problems

Location	Anomaly	Clinical problem
Vagina	Adenosis	Clear-cell adenocarcinoma
Cervix	Cervical anomalies (cockscomb, collar, hood, pseudopolyp)	Cervical mucus production (?) incompetent cervix (?)
	Narrow endocervical canal	Posttherapy cervical stenosis, infertility (?)
Uterus	Hypoplastic uterus (irregular, small, T-shaped endometrial cavity	Premature labor, spontaneous abortion, infertility (?)
Fallopian tubes	Hypoplastic oviducts	Ectopic pregnancy, Infertility (?)
Mesonephric remnants	Multiple paraovarian cysts (unusual histology)	(?)

A mouse paradigm of prenatal DES exposure has been exploited to study these genital tract anomalies (McLachlan et al. 1980; Newbold et al. 1983; Haney et al. 1984), and murine teratogenicity parallels that observed in humans. This model should prove useful in future efforts to study the mechanism of action of DES and assist in predicting any subsequent clinical problems. What follows is a compilation of the various genital tract abnormalities observed in exposed women and the subsequent reproductive problems encountered.

RESULTS

Vagina

Vaginal adenosis, the presence of mucus-secreting columnar epithelium in the vagina, is well-described in association with prenatal DES exposure (Herbst et al. 1972; Jefferies et al. 1984) and is the presumed cell of origin of the clear-cell vaginal adenocarcinoma. Current estimates suggest a risk of developing this neoplasia between 1 in 2000 and 10,000 exposed women. The mechanism of the development of the adenocarcinoma is unclear, but as it typically occurs after puberty, reproductive hormones from the ovary may be involved. Squamous metaplasia often covers these lesions after puberty. Palpable nodules, analogous to the Nabothian cysts seen on the cervix, usually can be appreciated, and these areas fail to take up iodine stains. No effect of adenosis on reproduction has been described.

Cervix

Unique structural changes of the cervix (cockscomb, collar, hood, and pseudopolyp) have been described in association with prenatal DES exposure (Herbst et al. 1972; Haney et al. 1979; Jefferies et al. 1984). The endocervical canal is approximately one-third the caliber of the canal in nonexposed women. No histologic abnormalities of the cervix, however, have been described. Anecdotal reports suggest a lower volume and poorer quality cervical mucus, but no objective data are available for comparisons. Furthermore, no increase in the rate of cervical factor infertility has been identified in a group of infertile women with prenatal DES exposure (Haney and Hammond 1984), suggesting no biologically significant effects on cervical secretion.

As the endocervical canal is significantly narrower in women exposed to DES, it is not surprising that any surgical manipulation of the cervix (cold-knife conization, cauterization, cryotherapy, or laser vaporization) is associated with a higher rate of cervical stenosis (~75%) than similar therapy in nonexposed women (Schmidt and Fowler 1980). It is unclear whether this is because the narrow endocervical canal cannot withstand surgical trauma without circumferential scar contraction and stenosis or because the multifocal nature of the DES-associated intraepithelial neoplasia requires more aggressive therapy.

Several isolated clinical reports have suggested that an incompetent cervix is more common in women exposed to DES (Goldstein 1978; Singer and Hockman 1978; Nunley and Kitchin 1979; Rosenfeld and Bronson 1980; Mangan et al. 1982). This clinical diagnosis is based on a history of painless effacement and dilatation of the cervix in the second-trimester pregnancy with terminal labor and delivery of previable fetus. In contrast, the majority of the case reports are composed of women whose incompetent cervix was treated successfully by cervical cerclage in the first pregnancy. Interestingly, many of these women required induction of labor at term when removal of the cerclage failed to precipitate delivery. The cervix has a unique structure in women exposed to DES, and on examination, this may "feel" differently during pregnancy, specifically with regard to effacement. This confusion is understandable, as premature labor is extremely common in women exposed to DES and is the terminal event in women with an incompetent cervix. Until more is known about this clinical problem, uniform application of cervical cerclage to pregnant women exposed to DES is not justified on the basis of the clinical data available.

Uterus

A unique anomaly of the uterus has been identified by hysterosalpingography in women exposed to DES (Kaufman et al. 1977; Haney et al. 1979). The endometrial cavity is small, with an irregular contour and a prominent

T-shape (Fig. 1). As the endometrium has not been reported to be abnormal histologically, the fundamental anatomic change appears to be in the underlying structural (fibromuscular) development of the uterus.

Virtually all investigators have noted an increase in the spontaneous abortion rate in women exposed to DES prenatally (Barnes et al. 1980; Kaufman et al. 1980; Schmidt et al. 1980; Sandberg et al. 1981; Veridiano et al. 1981; Mangan et al. 1982). Clinically, the abortions appear indistinguishable from those in the nonexposed women, but they are not habitual. No etiology for the increased abortion rate has been proved, but because of the irregular endometrial contour and overall hypoplastic appearance of the uterus, a uterine factor seems to be the likely explanation. In the mouse model, altered secretory uterine luminal proteins may provide a possible explanation for failure to maintain early gestations (Maier et al. 1985). No data on uterine luminal secretions are yet available in humans.

Many investigators have noted an increase in premature labor, with a decrease in term deliveries and higher perinatal morbidity and mortality, attributed primarily to prematurity (Barnes et al. 1980; Berger and Goldstein 1980; Cousins et al. 1980; Kaufman et al. 1980; Veridiano et al. 1981; Sandberg et al. 1981; Mangan et al. 1982). The mechanism of premature labor remains unclear, but these events have been correlated with the uterine anomalies observed by hysterosalpingography (Kaufman et al. 1980). The risk of premature labor begins in the second trimester and is similar to that observed in nonexposed women at high risk for premature labor, that is, those with spontaneously occurring uterine anomalies, a history of premature labor, and leiomyomata uteri.

Fallopian Tubes

Although no characteristic anomalies have been noted on hysterosalpingography, a "withered" or hypoplastic fallopian tube with inadequate development of the fimbrial portion and infertility has been described (DeCherney et al. 1981). The frequency of this finding apparently is low and probably relates to the timing and dose of DES administration. Few human specimens have been available for pathologic study, as the age of the population exposed to DES is relatively low.

A higher rate of ectopic pregnancy in women exposed to DES has been observed repeatedly, representing approximately 3–5% of all clinical pregnancies (Barnes et al. 1980; Kaufman et al. 1980; Schmidt et al. 1980; Sandberg et al. 1981; Veridiano et al. 1981; Mangan et al. 1982). The mechanism of development of these ectopic pregnancies is likely related to the aforementioned alteration in tubal development, as the usual association with pelvic inflammatory disease is absent (Haney and Hammond 1984). Interestingly, marked alterations in the oviducts of mice exposed to DES prenatally are observed (Newbold et al. 1983), and the pattern is reminiscent of the human pathologic progress of salpingitis nodosa, which is also associated with a high rate of ectopic pregnancy. Similarly, uterotubal junction obstruction has been observed in an infertile population of women exposed to DES (Haney and Hammond 1984), and uterotubal junction abnormalities are also noted in the mouse model (Newbold et al. 1983).

Paraovarian Cysts

Recently, it has become apparent that persistence of paraovarian cysts, some with unusual histology, is more common in women exposed to DES prenatally (Haney et al. 1986). The origin of paraovarian cysts in humans remains controversial, but based on similar changes in the mouse model, these are likely to be of mesonephric tubule origin. The clinical significance of these lesions, if any, is not apparent at this time.

Ovary

Too few surgical specimens of ovaries from women exposed to DES prenatally are available for adequate comparison to those of nonexposed women. To date, no obvious clinical problems attributable to the ovary have been noted. The reason for concern, however, is the finding in the mouse model of functional and histologic changes in the interstitial compartment of the mouse ovary (Haney et al. 1984). Elevated synthesis of testosterone in vitro has been demonstrated in this system. Interestingly, in women with ovulatory cycles exposed to DES, serum levels of testosterone are elevated (Wu et al. 1980) and hirsutism has been reported (Peress et al. 1982), suggesting that an effect similar to that in mice may be present in women. Whether this will be reflected in alterations in human ovaries anatomically remains to be determined.

Infertility

The frequency of infertility in women exposed to DES remains controversial, with several large, well-done epidemiologic studies coming to contradictory conclusions (Bibbo et al. 1977; Barnes et al. 1980; Herbst et al. 1980). In the mouse paradigm, a clear dose-related reduction in fertility has been observed (McLachlan et al. 1982). A recent study evaluating the pattern of fertility in the population of women exposed to DES found no unique infertility problems (Haney and Hammond 1984). As virtually the entire reproductive tract is affected, there may be multiple factors operative to reduce fecundity. Candidates include (1) a higher rate of pelvic adhesions secondary to surgery for ectopic pregnancies, (2) changes in the cervical mucus, (3) uterine abnormalities, (4) functional tubal abnormalities, and (5) an alteration in ovarian function.

DISCUSSION

A clear understanding of the reproductive consequences of these genital tract anomalies has been difficult to obtain because of the problems inherent in studying human reproductive parameters. Furthermore, evaluation of the uniformly exposed population of women has been impossible due to the difficulties related to retrospective analysis and variations in dose and timing of administration. The single population given a standardized dosage schedule prospectively is no longer accessible for study. Accordingly, clinical information regarding any potential future problems encountered by this group of women will be obtained from case reports and retrospective case control analysis.

Despite the obvious importance of the genital tract anomalies caused by prenatal DES exposure, the mechanism of action of this teratogen remains uncertain. It is not clear whether this effect is a direct consequence of estrogen action or whether it is due to the unique teratogenic action of DES or its metabolites. Because it is not a steroid, substantially higher quantities of DES may be able to reach fetal tissues in an active form, resisting protein binding and metabolic inactivation to which steroids are susceptible. Alternatively, DES is a xenobiotic and may have unique teratogenic properties expressed in the genital tract because of the presence of hormone receptor sites. Further effort will be necessary to clarify the mechanism(s) of action.

There has not been anything comparable to the impact of prenatal DES exposure both in the scientific community and with the public. In contrast to human teratogens known previously, the alterations in the genital tract only became apparent a generation after the exposure, with the disturbing discovery of a rare vaginal malignancy in young women. As the scrutiny became more focused, reproductive tract anomalies throughout the entire genital duct system became apparent and were associated with clinical reproductive problems. This problem has undermined public confidence in both the scientific and regulatory communities responsible for protecting the public from teratogenic hazards. Hopefully, another human tragedy will not be necessary for a comparable expansion of our understanding of the impact of xenobiotics on human development.

REFERENCES

Barnes, A.F., T. Colton, J. Gundersen, K.L. Nolter, B.C. Tilley, T. Strama, D.E. Townsend, P. Hatab, and P.C. O'Brien. 1980. Fertility and outcome of pregnancy in women exposed in utero to diethylstilbestrol. *N. Engl. J. Med.* **302:** 609.

Berger, M.L. and D.P. Goldstein. 1980. Impaired reproductive performance in DES-exposed women. *Obstet. Gynecol.* **55:** 25.

Bibbo, M., M. Al-Naqeeb, I. Baccarini, W. Gill, M. Newton, K.M. Sleeper, M.

Sonek, and G.L. Wied. 1975. Follow-up-study of male and female offspring of DES-treated mothers: A preliminary report. *J. Reprod. Med.* **15:** 29.

Bibbo, M., W.B. Gill, F. Azizi, R. Blough, V.S. Fang, R.C. Rosenfield, G.F.B. Schumacher, K. Sleeper, M.G. Sonek, and G.L. Wied. 1977. Follow-up study of male and female offspring of DES-exposed mothers. *Obstet. Gynecol.* **49:** 1.

Cousins, L., W. Karp, C. Lacey, and W.E. Lucas. 1980. Reproductive outcome of women exposed to diethylstilbestrol in utero. *Obstet. Gynecol.* **140:** 194.

DeCherney, A.H., I. Cholst, and F. Naftolin. 1981. Structure and function of the fallopian tubes following exposure to diethylstilbestrol (DES) during gestation. *Fertil. Steril.* **37:** 741.

Goldstein, D.P. 1978. Incompetent cervix in offspring exposed to diethylstilbestrol in utero. *Obstet. Gynecol.* **52:** 735.

Haney, A.F. and M.G. Hammond. 1984. Infertility in women exposed to diethylstilbestrol *in utero. J. Reprod. Med.* **28:** 851.

Haney, A.F., R.R. Newbold, and J.A. McLachlan. 1984. Prenatal DES exposure in the mouse: Effects on ovarian histology and steriodogenesis *in vitro. Biol. Reprod.* **30:** 471.

Haney, A.F., C.B. Hammond, M.R. Soules, and W.T. Creasman. 1979. Diethylstilbestrol-induced upper genital tract abnormalities. *Fertil. Steril.* **31:** 142.

Haney, A.F., R.R. Newbold, B.F. Fetter, and J.A. McLachlan. 1986. Paraovarian cysts associated with prenatal diethylstilbestrol exposure: Comparison of the human with a mouse model. *Am. J. Pathol.* **124:** 405.

Henderson, B.E., B. Benton, M. Cosgrove, J. Baptista, J. Aldrich, D. Townsend, W. Hart, and T.M. Mack. 1976. Urogenital tract abnormalities in sons of women treated with diethylstilbestrol. *Pediatrics* **58:** 505.

Herbst, A.L., R.J. Kurman, and R.E. Scully. 1972. Vaginal and cervical abnormalities after exposure to stilbestrol in utero. *Obstet. Gynecol.* **40:** 287.

Herbst, A.L., H. Ulfelder, and D.C. Poskanzer. 1971. Adenocarcinoma of the vagina. Association of maternal stilbestrol therapy with tumor appearance in young women. *N. Engl. J. Med.* **282:** 878.

Herbst, A.L., M.M. Hubby, R.R. Blough, and F. Azizi. 1980. A comparison of pregnancy experience in DES-exposed and DES-unexposed daughters. *J. Reprod. Med.* **24:** 62.

Jefferies, J.A., S.J. Robboy, P.C. O'Brien, E. Berstralh, D.R. Labarthe, A.B. Barnes, K.L. Noller, P.A. Hatab, R.H. Kaufman, and D.E. Townsend. 1984. Structural anomalies of the cervix and vagina in women enrolled in the Diethylstilbestrol adenosis (DESAD) Project. *Am. J. Obstet. Gynecol.* **148:** 59.

Kaufman, R.H., E. Adam, G.L. Binder, and E. Gerthoffer. 1980. Upper genital tract changes and pregnancy outcome in offspring exposed in utero to diethylstilbestrol. *Am. J. Obstet. Gynecol.* **137:** 299.

Kaufman, R.H., G.L. Binder, P.M. Gray, and E. Adam. 1977. Upper genital tract changes associated with exposure in utero to diethylstilbestrol. *Am. J. Obstet. Gynecol.* **128:** 51.

Maier, D.B., R.R. Newbold, and J.A. McLachlan. 1985. Prenatal diethylstilbestrol exposure alters uterine responses to prepubertal estrogen stimulation. *Endocrinology* **116:** 1878.

Mangan, C.E., L. Borow, M.M. Burtnett-Rubin, V. Egan, R.L. Giuntoli, and J.J. Mikuta. 1982. Pregnancy outcome in women exposed to diethylstilbestrol in utero, their mothers, and unexposed siblings. *Obstet. Gynecol.* **59:** 315.

McLachlan, J.A., R.R. Newbold, and B.C. Bullock. 1980. Long-term effects on the female mouse genital tract associated with prenatal exposure to diethylstilbestrol. *Cancer Res.* **40:** 3988.

McLachlan, J.A., R.R. Newbold, H.C. Shah, M. Hogan, and R.C. Dixon. 1982. Reduced fertility in female mice exposed transplacentally to diethylstilbestrol (DES). *Fertil. Steril.* **38:** 364.

Newbold, R.R., A.F. Haney, and J.A. McLachlan. 1983. The developmentally arrested oviduct: A structural and functional defect in mice following prenatal exposure to diethylstilbestrol. *Teratology* **27:** 417.

Nunley, W.C., Jr. and J.D. Kitchin. 1979. Successful management of incompetent cervix in a primigravida exposed to diethylstilbestrol in utero. *Fertil. Steril.* **31:** 27.

Peress, M.R., C.C. Tsai, R.S. Mathur, and H.O. Williamson. 1982. Hirsutism and menstrual patterns in women exposed to diethylstilbestrol in utero. *Am. J. Obstet. Gynecol.* **144:** 135.

Rosenfeld, D.L. and R.A. Bronson. 1980. Reproductive problems in the DES-exposed female. *Obstet. Gynecol.* **55:** 453.

Sandberg, E.C., N.C. Riffle, J.V. Higdon, and C.E. Getman. 1981. Pregnancy outcome in women exposed to diethylstilbestrol in utero. *Am. J. Obstet. Gynecol.* **140:** 194.

Schmidt, G. and W.C. Fowler, Jr. 1980. Cervical stenosis following minor gynecologic procedures on DES-exposed women. *Obstet. Gynecol.* **56:** 333.

Schmidt, G., W.C. Fowler, Jr., L.M. Talbert, and D.A. Edelman. 1980. Reproductive history of women exposed to diethylstilbestrol in utero. *Fertil. Steril.* **33:** 21.

Singer, M.S. and M. Hochman. 1978. Incompetent cervix in a hormone-exposed offspring. *Obstet. Gynecol.* **51:** 625.

Veridiano, N.P., I. Delk, J. Rogers, and M.L. Tancer. 1981. Reproductive performance of DES-exposed female progeny. *Obstet. Gynecol.* **58:** 58.

Whitehead, E.D. and E. Leiter. 1981. Genital abnormalities and abnormal semen analyses in males exposed to diethylstilbestrol (DES) in utero. *J. Urol.* **125:** 47.

Wu, C.H., C.E. Mangan, M.M. Burtnett, and G. Mikhail. 1980. Plasma hormones in DES-exposed females. *Obstet. Gynecol.* **55:** 157.

COMMENTS

Hanson: Would you comment on men exposed to DES in terms of infertility and anomalies, and then malignancy?

Haney: My experience is relatively limited, but my general impression is that they do have a slightly higher rate of genital tract anomalies, everything from hypospadias to retained testes, and epidydimal cysts, but there is certainly not a dramatic alteration in their reproduction. My interpretation is that the spectrum of "normal" is so wide that quantitative reductions in semen production have to be incredibly severe to alter the fertility potential of a male significantly.

Hanson: What about tumors?

Haney: There have been several malignant testicular tumors reported anecdotally in men.

McLachlan: We showed epidermidosis in male mice 10 years ago. The Chicago group showed a higher prevalence of retained testes and epidermidosis, but those data have not been followed up. As far as I know, no studies have been published in the last 6 years that looked at human males exposed to DES. There are three, four, or five case reports of seminomas of the testes, but these are just case reports and a single report here and there.

We found that 5–7% of the 12 to 18-month-old male mice treated with DES prenatally have adenocarcinoma, which is a very rare lesion in men and male mice. It may be the same kind of single lesion as vaginal adenocarcinoma in females, and it is permanent.

Haney: One has to look for a very unique tumor like adenocarcinoma to be able to say that it is related to prenatal DES exposure because retained testes result in a high rate of malignancy.

Lammer: Are the vaginal lesions restricted to the upper part of the vagina that is derived from the Mullerian structures?

Haney: The lesions are in the upper two-thirds to one-half of the vagina.

Miller: Do you think that structural defects induced by DES occurred at one instant in time and that set the stage for a chain of events?

Haney: There are two dramatic effects. One is that this lesion was only identified a generation later. Nothing like this has ever occurred before, but it changed the perspective of evaluation of prenatal exposure entirely. The second effect is that puberty may have an influence. The vast majority of people affected are between the ages of 17 and 25. It may be, potentially, a hormonally triggered phenomenon.

Johnson: In light of what we know today, what general protocol would you recommend when designing a safety evaluation study in experimental animals?

Haney: Does anybody else want to answer that?

Manson: Yes. I think one would see this immediately in a 2-week to 1-month developmental or general adult toxicity study. Hyperplasia would be easy to pick up.

Haney: The level of scrutiny is so much more intense today that genital tract changes in the offspring would be detected much more certainly now than in the 1960s. Whether or not we will be able to screen for agents that affect the offspring but will only be noticeable later on, I can't say, but I presume that we are going to have to do multiple-generation animal studies to have any hope of answering that question.

McLachlan: I don't think there is a study in which one can give a compound to pregnant animals and do a totality study of immune or reproductive dysfunction in the offspring that would be meaningful. I don't think it would even be appropriate.

Similar tested compounds are still being used. Dr. Howard Berns's group at the University of California at Berkeley induced the first transplantable cervical carcinoma by treating neonatal mice with progesterone. I would guess that there are still many obstetricians who give progesterone to pregnant women for pregnancy symptoms of one sort or another. Is anyone following up on progesterone-treated pregnancies?

Haney: There was an article in the late 1970s suggesting that hypospadias was associated with prenatal progesterone, but it is the only one that even hinted at it. There have been several follow-up studies, all of which have been negative.

Brent: We have extensive animal models for both DES and progesterone. Your animal model is positive, and the progestational animal model isn't with regard to sexual function.

McLachlan: It depends when you give the compound.

Brent: It also depends on the dose, because you can give massive doses of these compounds.

Oakley: You had difficulty finding any adenocarcinoma of the vagina in mice. That is exactly the same experience that occurred in the follow-up of the large human series. They did not have a single adenocarcinoma among the exposed daughters. Isn't that right?

Haney: Correct.

Oakley: All of the follow-ups of the clinical trials came up with no cases of adenocarcinoma. The reason is that although this is a horrible disease, it doesn't happen very often after exposure.

Kimmel: Apparently, there were a lot of specimens available that showed abnormalities of the genital tract. If those had been heeded as abnormalities of concern, attention would have been drawn to the whole issue of DES sooner.

Haney: No, I don't think that is true. I don't think you can blame the pathologists. They had no idea that this lesion even existed before. Their job in those selected circumstances is to determine whether or not the patient has cancer and how far it has spread, and recommend therapy.

Oakley: Regarding Dr. Kimmel's comment on what could have been done to prevent this problem from occurring, the tragedy would have been prevented if treatment recommendations had not been made until the results from randomized clinical trials were available. Each of the randomized trials showed that the drug was not effective. We should encourage randomized clinical trials as a very important method, of even more importance than screening in animals, to prevent the risk of low incidence but serious adverse effects of new drugs.

Haney: They were done and, in actuality, showed a detriment in using DES.

Oakley: But somehow the obstetric community didn't pay any attention to that.

Haney: Exactly. There was an article in the *American Journal of Obstetrics* saying that it wasn't efficacious. These workers even put dye into the compound that came out in the urine to prove that the women took the drug. I think the data were available, but the physicians' scrutiny of the literature and acting upon it was not adequate.

Oakley: But there were a variety of uncontrolled trials that claimed it worked.

Lammer: I don't think we learned much from this lesson. The same legacy is continuing today with progestational agents. Physicians continue to prescribe progestational agents during pregnancy when there is very little evidence that they benefit women who have a history of miscarriage or bleeding early in their pregnancy.

Hanson: Dr. Rice laid the groundwork for suggesting that some of these agents may promote tumors later in life in exposed children and may set the children up to be more susceptible to subsequent exposure to the same agent or other agents. I think this is an issue that has not been examined. It really is something that agencies should be evaluting prospectively or retrospectively.

Manson: I think the biggest problem is not so much that the animal models aren't there or aren't predicting problems, but that people don't believe them. Dr. Druckery had models for transport cell carcinogenesis 30 years ago. In 1966, there were federal regulations to do segment-I female fertility studies, which included exposures throughout pregnancy and following the offspring until their reproductive maturity. The uterotropic effects of estrogenic compounds were certainly known long before DES. I think for virtually every agent, drug, or environmental agent known to cause disease in humans, that information can be found in the animal literature.

Haney: Retinoids are a good example.

Ianaccone: There are many examples. Why don't we believe the animal literature?

Brent: Science is not only a discovery by one person but belief by the scientific community. It is a shame that science is a slow-moving process, but we are not being objective in thinking that, in hindsight, all of these things were obvious at the times that the discoveries were made and that the implications were clear.

Haney: I think there are two points concerning that. Look at the thickness of Dr. Shepard's book, go to page 5 or 6 and see the list of known human teratogens, and then look through all the other animal teratogens.

I agree with you; it is impossible to put oneself back in the era when a drug first was used, in the mind-set and understanding of the time. A lot of things are obvious retrospectively but not nearly as obvious prospectively.

Iannaccone: I wanted to ask about clomiphene. It was tried as a contraceptive for a while, wasn't it?

Haney: I think it was designed and developed clinically to be a contraceptive, but it only made it to the phase-I and phase-II clinical trials as a contraceptive, when it started precipitating ovulations and conceptions.

Iannaccone: Is it estrogenic?

Haney: Yes, it is.

Iannaccone: Is there any evidence in the animal models that it has any effect?

Haney: It gives you the exact same pattern of anomalies that DES does.

Oakley: Is there evidence of adenocarcinoma of the vagina?

Haney: I don't believe that has been shown. It has a lot of advantages. One is that it is used prior to ovulation. The problems occur with the inadvertant or indiscriminate use in women who are amenorrheic without a pregnancy test.

Lammer: During the 1970s in Atlanta, 2% of women used clomiphene after conception, and there were a lot of conceptuses exposed to clomiphene.

Brent: Since clomiphene is taken in the early part of pregnancy, one is not really going to worry about these phenomena. One has to wait until the beginning of differentiation of the genitourinary system. Before some of the problems you are talking about here can be induced, a woman would have to continue clomiphene past the first, and possibly the second, missed menstrual period.

McLachlan: I would like to comment on the discussion concerning negative results after prenatal treatment with a chemical. There have been examples of mice treated prenatally or neonatally with DES, which are more sensitive to tumor induction than the adult animal. Especially worth noting is some published work that was done by Dr. Elizabeth Bolan in rats. She treated rats prenatally with DES and looked at mammary cancers. The numbers she saw in aging offspring were about the same in control and DES-treated litters. However, when she treated those rats with dimethylbenzanthracene, a mammary tumorigen in rats, the yield of tumors was much greater in those rats treated prenatally with DES. They looked like normal rats and had the same overall spontaneous tumor incidence, but they had about a four- to fivefold increase in the numbers of tumors per rat and the number of rats responding with tumors.

Brent: Since one of the major issues is to protect the public, would you suggest that pregnant women exposed to DES have their PAP smears done more frequently up until the time of menopause?

Haney: The Department of Health and Human Services spent several million dollars asking that question. Their recommendation was an annual PAP smear.

Hanson: I want to make a comment that relates to the clomiphene issue. We have tended to restrict our analysis of teratogenic effects to those that derange normal morphogenetic processes, that is, they produce malformations. We really haven't looked at the issue of either disruptional or deformational events. Clomiphene is an example. Regardless of its effect on the genital tract, it produces multiple fetuses in utero with some regularity. As a result, we should see the frequency of deformational events (which, in other studies, have always been correlated with the number of individuals in utero at any one time) increase in that population of individuals. By correcting for the number of individuals in utero, one ought to be able to account for that postulated increase, which I suspect is there but has never been documented.

Haney: From my knowledge of ovulation-induction agents, the offender is perganol, not clomiphene.

Hanson: The concept should apply to any agent that produces extra babies in utero.

Haney: In vitro fertilization produces multiples.

Hanson: That would be another one. You saw a growth-stunting effect, which is a third-trimester phenomenon in those cases.

Etiology of Human Birth Defects: What Are the Causes of the Large Group of Birth Defects of Unknown Etiology?

ROBERT L. BRENT
Jefferson Medical College
Thomas Jefferson University
Philadelphia, Pennsylvania 19107

INTRODUCTION

Although congenital malformations and mental retardation were reported in the earliest recordings of history, teratology did not become a modern science until the middle of the 20th century (Warkany 1977). Malformed and retarded offspring have been the focus of ridicule and superstition for thousands of years. It was the development of modern genetics that influenced scientists and physicians during the first half of the 20th century, since it provided a simple explanation for the etiology of human malformations.

With the discovery and rediscovery in the 1930s and 1940s that environmental factors could also cause congenital malformations, the expansion of experimental teratology began (Hale 1937; Warkany and Schraffenberger 1943). During the evolution of experimental teratology and clinical teratology, the emphasis was on discovering human teratogens and expanding the basic science foundation of experimental teratology. Teratology was quite isolated in the mid-20th century: (1) There was little impact of teratological research on clinical medicine and minimal interaction of teratologists with society; (2) there was only a small group of scientists and physicians interested in the problems of congenital malformations; and (3) although congenital malformations were considered a serious and debilitating group of diseases, they were also considered to be untreatable and unpreventable and were dwarfed by the morbidity and mortality of infectious diseases.

With the founding of the Teratology Society and the occurrence of the thalidomide tragedy, the field of teratology no longer was isolated from the mainstream of science and clinical medicine. Progress in medicine in the past 20 years has resulted in improvement in the prevention and treatment of congenital malformations. Experimental teratologists have been joined by epidemiologists, clinicians, geneticists, and obstetricians with specific interests in the problem of congenital malformations. Regulatory agencies, politicians, lawyers, unions, and the lay public all have become interested in the problem of congenital malformations.

Banbury Report 26: Developmental Toxicology: Mechanisms and Risk

DISCUSSION

Human Reproductive Potential and Risks

The perception of human reproductive potential is frequently misrepresented by the lay press or even scientists who are not familiar with human epidemiological studies. The concept that pregnancy is *not* without significant risk comes as a surprise to those not familiar with reproductive risks. A substantial number of conceptions end in early or late spontaneous abortion. As many as 50–60% of these spontaneous abortions have chromosomal abnormalities that result in severe malformations and embryonic death. A substantial number of spontaneous abortions that have a normal karyotype are also structurally abnormal; although Table 1 indicates that as many as 25% of conceptions may end in spontaneous abortion (15% are clinically diagnosed), the actual rate of abortion may be higher.

Second, 30 of the 1000 surviving embryos that reach viability will manifest major malformations. Third, approximately 100 of every 1000 live births will manifest either early- or late-onset genetic disease, such as cystic fibrosis, diabetes, Huntington's chorea, hyperlipidemia, and so forth. Thus, at least one in three conceptions will result in either spontaneous abortion, malformation, or genetic disease.

A frequent response to this information is that it is untenable to think that the human species has greater reproductive risks than most laboratory animals with which we are familiar. As an example, the resorption rate in the rat is 5%, and the malformation rate is only 2%. Actually, there is a simple explanation for this discrepancy. Most animals are bred for reproductive perfection, and strains or individual animals with repetitive pregnancy loss or birth defects are eliminated. In the human species, we do exactly the opposite. We provide the opportunity for all women and families to reproduce by treating diseases of infertility or diseases that interfered with reproduction previously, such as diabetes. Actually, since the discovery of insulin and the "new-found" reproductive potential of diabetics, we have been increasing the

Table 1
Human Reproductive Potential and Risks: Results of 1350 human conceptions

Live births	1000
Spontaneous abortions	350
Major congenital malformations	30
Early- and late-onset genetic disease	100

Data from Brent and Harris (1976).

human pool of diabetic genes gradually, which will result in an increase in the incidence of diabetes in future generations.

With this as a background, one can readily understand why it is inappropriate for obstetricians to reassure patients uncritically about the outcome of their pregnancies. Human pregnancy is a high-risk experience.

Etiology of Congenital Malformations

Various authors have estimated the etiology of human congenital malformations (Table 2) (Wilson 1973; Brent 1976, 1982a; Heinonen et al. 1977; Holmes 1985). It is interesting that most analyses conclude that a definite etiology for 60–70% of human malformations is unknown. All agree that drugs and chemicals account for a very small percentage of malformations. At the conclusion of the 50,000-patient collaborative perinatal project, Heinonen et al. (1977) found that there was "no common drug analogous to thalidomide." They suggested that drugs with less potent teratogenic potential are the concern because their effects can go undetected. But they found their results "generally reassuring." Their results say, in a very sophisticated

Table 2
Relative Contribution of Various Causes to the Frequency of Human Malformations

Etiology	Malformed live births (%)
Genetics	20–25
Cytogenic	5
Mendelian inheritance	15–20
Spontaneous mutations	
Unknown	65–70
Polygenic	
Multifactorial	
Synergism	
Spontaneous errors of development	
Environmental (exposure of the embryo)	10
Maternal infections (rubella, toxoplasmosis, herpes simplex, C.I.D., syphilis, etc.)	
Maternal disease states (diabetes, phenylketonuria, endocrinopathies and endocrine tumors, alcoholism, smoking, nutritional problems)	
Problems of constraint	
Drugs, chemicals, irradiation, hyperthermia	1

Data from Brent and Harris (1976).

manner, that we will never be able to prove with the certainty of laboratory experiments that a drug is absolutely safe in the human population. That degree of risk reduction is not available for any potential environmental hazard.

The unknown category has several hypothetical models that have both experimental and clinical evidence in their favor. There are many congenital malformations that occur sporadically, such as Poland syndrome, and also exhibit familial occurrence. This common phenomenon for rare syndromes raises the question as to whether there are multiple etiologies for many syndromes or whether complicated genetic transmission is responsible for a large percentage of human malformations. No one can deny the great importance of genetics to this problem. McKusick (1975) reports the dramatic explosion in genetic information in the past 20 years. According to McKusick, the number of cataloged genetic diseases has increased sixfold between 1958 and 1975—from 412 to 2336. It is true that new diseases with classical Mendelian inheritance will not account for the unknown group, since they obviously will be extremely rare, but new genetic and embryologic concepts may explain a substantial portion of the unknown group of malformations.

There may be confusion between the terms polygenic and multifactorial. If one assumed that the diseases with an increased recurrence risk are polygenic, that is, solely due to two or more genetic loci, mathematical models then can be established for this group of malformations (neural tube defects, pyloric stenosis, cleft lip and palate, congenital dislocation of the hip, and certain congenital heart diseases). If one assumed that a polygenic disease was due to a homozygous recessive state at two loci, the probability that parents who were heterozygous for both loci would deliver an infant that is homozygous recessive for all four loci is 6.25%. Thus, the mathematics is consistent with the reported recurrent risk of polygenic disease at approximately 5%. The problem is that the epidemiological and clinical data partially refute a purely genetic explanation for these malformations, since there should not be discordance for these malformations in identical twins, and there are many case reports in the literature of diseases with an increased recurrent risk that have been reported in only one of two identical twins. Thus, some other influence must be modulating the genome.

The multifactorial/threshold hypothesis (Wright 1934; Fraser and Fainstat 1951; Fraser 1977) is consistent with experimental and even some clinical observations because it includes several alternatives, namely, multiple-loci/genetic disease and/or the interaction of intrinsic and extrinsic environmental factors with a continuum of genetic characteristics. Although there are experimental models to support the hypothesis, we are not certain which factors are important in producing this group of malformations in the human. Are the modulating factors (1) variations in placental blood flow, placental transport, or site implantation; (2) maternal disease states; (3) infections; (4) drugs and chemicals; (5) all the other known environmental factors; or (6) simply "chance"?

For an erudite discussion of the multifactorial/threshold hypothesis, one should read Fraser's analysis (1977).

A third suggestion is that some malformations are due to a multiplicity of environmental factors that might be acting synergistically. Although there are experimental models consistent with apparent synergism, Fraser (1977) warns that if one believes that environmentally produced malformations have a threshold, additive effects of teratogens easily could be misinterpreted to be synergistic. In the human situation, it is extremely difficult to evaluate this hypothesis. The hypothesis is anxiety provoking in itself, because it suggests that several nonteratogenic factors may, under certain circumstances, result in a malformation. The difficulty with the theory is that it is utilized occasionally even when the exposures of the alleged teratogenic agents are several orders of magnitude below the threshold dose and no biochemical or physiological effects can be produced with the combination of agents.

Another explanation for the etiology of this large group of malformations is that they are intrinsic spontaneous errors of development (Brent 1964, 1976)—"biologic errors inherent in the reproductive and developmental process, similar to the concept of spontaneous mutations." Some have referred to this as developmental noise (Adams and Niswander 1967). This pessimistic viewpoint suggests that even if the environment were optimal, embryopathology would occur because the reproductive and developmental processes have a built-in probability of going awry. If this interpretation is correct, embryopathology can never be reduced to zero unless we develop a technique that prevents abnormal embryos from reaching term. Nature does eliminate most abnormal embryos spontaneously (Nishimura and Yamamura 1969), and it is very likely that by the year 2000, biomedical science will have available electronic, biochemical, and genetic techniques to evaluate the status of every human embryo at very early stages of gestation. What society will do with this information is yet to be determined, since the controversy over the appropriateness and/or the appropriate use of therapeutic interruption of pregnancy has not been resolved in many societies (Brent 1976).

We have now spent many decades with these several hypotheses in an attempt to explain the etiology of a large group of human malformations. Although we have more information, and some of these hypotheses appear to be more likely to be correct, we must recognize the tremendous inertia that exists in the field of teratology with regard to these hypotheses. We may have to wait for breakthroughs in genetics, developmental biology, computer science, and biochemistry in order to determine the etiology of many malformations.

Mechanisms of Teratogenesis

Although most teratologists agree that teratogenesis resulting from environmental agents has a threshold below which an exposure would not result in an increase in malformations, one occasionally sees in print that all exposures to potential teratogens present a risk to the embryo (Epstein 1979). Frequently, mutagenesis, carcinogenesis, and teratogenesis are linked together as though all three had similar mechanisms and risks. In reality, mutagenesis and carcinogenesis caused by environmental agents are considered to be stochastic phenomena, and, therefore, although it decreases with exposure, the risk, from a theoretical standpoint, never disappears. It is therefore important to differentiate between stochastic and threshold phenomena (Table 3).

Although it is true that mutagens may be teratogenic, they do not produce malformations in the developing embryo by producing a mutation that leads to a clone of cells that results in a malformation. Exposures of mutagens (X-ray, chemical) that may produce a significant number of somatic mutations will produce multicellular and organ cytotoxic effects simultaneously that result in embryotoxicity. Therefore, an exposure of 1 Gy (100 rads) to the rat embryo on the tenth day of development will kill cells at a rate that is several orders of magnitude greater than its capacity to produce mutations in surviving cells. Furthermore, once the dose drops below a certain exposure, the incidence of embryotoxic effects is identical to that of the nonexposed embryos. This supports the threshold hypothesis and also emphasizes the fact that the embryo, although sensitive to cytotoxic agents, also has sophisticated recuperative mechanisms, especially when the exposure is low or administered at a low dose rate (Brent 1971).

There is one time during embryogenesis when the embryo may respond in a stochastic fashion (Brent 1980a,b), namely at the time of fertilization. In this circumstance, the risk of cell death to the one-cell embryo might be stochastic in nature. One can readily recognize that a methodology of drug and chemical regulations should have a much different approach, depending on whether one is dealing with a threshold or stochastic phenomenon.

Etiology of Malformations Resulting from Environmental Factors

It is well recognized that while there are literally hundreds of agents that can produce malformations in experimental animals, there are considerably fewer proven human teratogens or teratogenic milieus (Table 4). This apparent contradiction is explained readily by defining *teratogen*, *potential teratogen*, or *nonteratogen* as they relate to the human.

Table 3
Relationship of Diseases Produced by Environmental Agents and the Risk of Occurrence

Relationship	Pathology	Site	Diseases	Risk	Definition
Stochastic phenomena	Damage to a single cell may result in disease	DNA	Cancer, mutation	Exists at all exposures, although at low exposure, the risk is below the spontaneous risk	The incidence of disease increases with exposure, but the severity and the nature of the disease in the patient remain the same
Threshold phenomena	Multicellular injury	Great variation in etiology affecting many cell and organ processes	Malformations, growth retardation, death, and many other toxic phenomena	Completely disappears below a certain threshold dose	Both the severity and incidence of the disease increase with higher exposures

Table 4
Agents or Milieu That Can Result in Teratogenicity or Reproductive Toxicity and That Occur or Have the Likelihood of Occurring during Human Pregnancies

Thalidomide embryopathy
Androgenic masculinization
Diethylstilbestrol adenosis and adenocarcinoma of the vagina
Fluorescent tooth staining from tetracyclines
Goiter resulting from antithyroid medications and iodine deficiency
Syndromes resulting from infection with rubella virus, toxoplasmosis, cytomegalovirus infection, treponema palidum, herpes simplex, Venezuelan equine encephalitis, varicella; with isolation or identification of the organism
Documented drug-induced hemolytic anemia in the fetus
Mental retardation, cerebral palsy, associated with maternal methyl mercury exposure (Minamata disease)
Nasal hypoplasia following warfarin exposure
Virus infection of fetus with live virus vaccine
Retinoic acid and high doses of vitamin A
Excessive exposure to vitamin D (mental retardation, IUGR, aortic stenosis)
Ionizing radiation greater than the usual diagnostic exposure
Abnormal sex organ development following exposure to *high* dose of certain progestational agents (medroxyprogesterone norethisterone, norethindrone)
Growth retardation, embryolethality, and malformations following exposure to folic acid antagonists and certain other cancer chemotherapeutic drugs (methotrexate, amniopterin, busulfan, etc.)
All maternal disease states: Abortion, IUGR—smoking addiction, hyperthyroidism, IUD, collagen disease, renal vascular disease, hypertensive vascular disease
a. Diabetes
b. Phenylketonuria
c. Alcoholism
d. Epilepsy associated with anticonvulsant therapy
e. Mechanical or constraint problems causing limb malformations, cleft lip and palate, open neutral tube, emphalocele related to uterine malformations, oligohydramnios, amniotic rupture, bleeding, velamentous cord insertion, and multiple pregnancies
Deafness following the administration of streptomycin to a pregnant woman
Documented hypoglycemia—insulin induced or idiopathic
Fetal hydantoin syndrome (chronic drug administration)
Trimethadione syndrome (chronic drug administration)
PCB cola babies
Penicillamine
Valproic acid (chronic drug administration)
Lithium

From Beckman and Brent (1984); Brent and Beckman (1986).

Teratogen or Teratogenic Milieu

A teratogen or teratogenic milieu is a chemical, drug, metabolic state, physical agent, or physiologic alteration during development that has been demonstrated to produce a permanent pathologic alteration in human offspring at exposures or circumstances that commonly occur. Thus, the quantitative aspects of the exposure are an important part of the definition.

Nonhuman Teratogen with Teratogenic Potential

A nonhuman teratogen with teratogenic potential is a chemical, drug, metabolic state, physical agent, or physiologic alteration during development that has not been demonstrated to produce a permanent pathologic alteration in offspring at exposures or circumstances that commonly occur. As an example, vitamin A and vitamin D at maintenance doses are not teratogenic, but if given at much higher doses, they could result in specific malformations. The same example could be used for phenylalanine. Likewise, many drugs and chemicals are not teratogenic at doses to which the population is exposed but can produce some effects if doses of several orders of magnitude higher are given. Bendectin and Meclizine are examples, since they can produce reproductive loss or malformations in animals, if the exposure is increased 100- or 1000-fold. Thus many therapeutic agents or chemicals are safe for the human embryo at their usual exposure but can produce malformations in an animal model at much higher exposures (Szabo and Brent 1974, 1975; Fraser 1977; Brent 1980a,b; Brent and Beckman 1986.)

Nonteratogen

Nonteratogens in humans or animals refer to agents that have no teratogenic potential at any exposure or have greater maternal toxicity and therefore affect the mother before the embryo or fetus. Interestingly, many very toxic compounds are considered to be nonteratogenic, as are certain agents that have no potential for reproductive toxicity.

Timing

The stage of gestation when particular teratogens can have a deleterious or teratogenic effect is quite variable. One should refrain from assuming that all stages of the first trimester are equally vulnerable to teratogenic or embryopathic effects. Actually, each teratogenic agent has its own embryopathic or teratogenic profile. For short-acting drugs or chemicals and high-dose radiation, the first 2 weeks of human pregnancy are a stage that is much less vulnerable to teratogenesis or growth retardation, although it is susceptible to the lethal effects of these agents.

Each teratogenic agent has its unique period of gestation when it can affect the embryo in specific ways. Thalidomide has been demonstrated to produce limb defects during a very short period of gestation (Lenz and Knapp 1966). Therefore, counseling about the risk of each agent has to be tailored individually and should be based on all the knowledge relating to teratogenicity and timing, as well as the benefits of those drugs (Beckman and Brent 1984; Brent and Beckman 1986). As an example, ^{131}I can result in fetal thyroid destruction if the dose is high enough, but it will not affect the thyroid if given before the eighth week of pregnancy. Warfarin presents a similar situation, namely that it is less likely to be teratogenic early in pregnancy, because of the nature of the teratogenic effect. In some instances, the benefits of the drug may be such that the family and counselor may be unwilling to withdraw the drug, as in the case of lithium.

The evaluation of alleged teratogenicity cannot be based on the results of one human epidemiological study, since it is to be expected that a few significant associations are invariably found in epidemiological studies of the cohort type that test for associations of drug exposure and a large number of malformations (Holmes 1983; Brent 1985a,b). It has been suggested that a formal approach for evaluating alleged teratogenicity be followed, as outlined in Table 5 (Brent 1985c,d). It requires consistency in epidemiological results, uses of secular trends when appropriate, use of animal models, and the evaluation of basic science plausibility.

When two of the most common litogens (Bendectin, progestational drugs) are evaluated with this format, it can be concluded that they do not have a

Table 5
Proof of Teratogenesis

1. Controlled epidemiologic studies consistently demonstrating an increased incidence of a particular congenital malformation in exposed human populations.
2. Secular trends demonstrating a relationship between the incidence of a particular malformation and exposures in human population.
3. An animal model that mimics the human malformations at clinically comparable exposures:
   A. Without evidence of maternal toxicity.
   B. Without reduction in food and water ingestion (Szabo and Brent 1974, 1975; Brent and Szabo 1975).
   C. With careful interpretation of malformations that occur in isolation, such as anophthalmia in the rat, cleft palate in the mouse, vertebral and rib malformations in the rabbit, and ompholocele in the ferret.
4. The teratogenic effects are dose related.
5. The mechanisms of teratogenesis are understood and/or the results are biologically plausible.

Table 6
Agents Likely to be Alleged Teratogens (Litogens)

1. Used frequently during pregnancy.
2. Associated with sexually related or pregnancy-related problems.
3. A prescription item so that documentation of exposure is more likely.
4. Relatively nontoxic to mother and embryo, so that very high exposures can be utilized in animal experiments. Therefore, some type of embryotoxicity may be demonstrated in pregnant animals exposed at several orders of magnitude above the dose expected to be used in humans.

From Brent (1985c,d).

nongenital teratogenic effect in spite of their frequent appearance in congenital malformation litigation (FDA talk/paper 1979; Holmes 1983; Brent 1985c,d). Bendectin and progestational agents are drugs that are likely to be suspected of being teratogenic by the novice, because they were prescribed so frequently and therefore will be utilized in a large number of pregnancies that will happen to result in malformed fetuses for other reasons (litogens) (Table 6). Thus, a formal evaluation of this allegation allows one to readily conclude that Bendectin is not a human teratogen and that progestational agents do not affect the development of nongenital or nonsexually related organs.

SUMMARY

There is widespread belief that many congenital malformations are caused by drugs and chemicals in the environment. This belief has been popularized in the lay press and even by some physicians (Brent 1972, 1981, 1985a). The magnitude of the problem is represented in the United States by the exponential rise in negligence litigation involving malformed children (Brent 1967, 1977, 1982a). The vast majority of these lawsuits are nonmeritorious. Recent examples of lawsuits include malformed infants whose parents allege that the following environmental factors were responsible for the child's malformation: (1) animal flea collar, (2) mimeograph machine, (3) paint restoration materials, (4) electronic surveillance, and (5) routine pest control.

Most of these lawsuits would never be initiated if it were not for the fact that some scientists are willing to state with a "reasonable degree of certainty" that the child's malformation was due to the exposure in question. This type of irresponsible behavior occurs frequently because scientific experts have never been taught their proper roles. Many of these experts function as partisans, rather than scholars (Brent 1982b). The solution to this serious problem rests with the scientific community, but so far, organized science and medicine have ignored the problem of the irresponsible expert witness and the not-so-knowledgeable scientist.

The prevention of congenital malformations is a goal worth attaining and is likely to be accomplished in the next several decades. It is quite clear that the solution to the birth defect problems will not occur if we eliminate all known and future environmental teratogens. This is due to the fact that only a small percentage of human malformations are due to environmental agents. The two approaches are both related to prevention. Advances in genetics will permit us to diagnose or alter the genetic constitution of the fertilized ovum, or diagnostic tests pertaining to the status of the embryo or fetus will become so sophisticated that the exact condition of each embryo will be known. In any case, society will have a difficult choice to make and either of these medical opportunities is likely to stimulate controversy before the matter is resolved. It will be just another example of scientific advances having outdistanced societal adaptability.

ACKNOWLEDGMENT

This work was supported by the National Institutes of Health grants HD18396 and HD19165.

REFERENCES

Adams, M.S. and J.D. Niswander. 1967. Developmental noise and a congenital malformation. *Genet. Res.* **10:** 313.

Beckman, D.A. and R.L. Brent. 1984. Mechanisms of teratogenesis. *Annu. Rev. Pharmacol. Toxicol.* **24:** 483.

Brent, R.L. 1964. Drug testing in animals for teratogenic effects: Thalidomide in the pregnant rat. *J. Pediatr.* **64:** 762.

———. 1967. Medicolegal aspects of teratology. *J. Pediatr.* **71:** 288.

———. 1971. Response of the 9-1/2 day-old embryo to variations in dose rate of 150R X-irradiation. *Radiat. Res.* **45:** 127.

———. 1972. Protecting the public from teratogenic and mutagenic hazards. *J. Clin. Pharmacol.* **12:** 61.

———. 1976. Environmental factors: Miscellaneous. In *Prevention of embryonic fetal and perinatal disease* (ed. R.L. Brent and M. Harris), publ. no. 76-853. Department of Health, Education, and Welfare, Washington, D.C.

———. 1977. Litigation produced pain, disease, and suffering in experience with congenital malformation lawsuits. *Teratology* **16:** 1.

———. 1980a. Radiation teratogenesis. *Teratology* **21:** 281.

———. 1980b. Radiation-induced embryonic and fetal loss from conception to birth. In *Human embryonic and fetal death* (ed. I.H. Porter and E.B. Hook), p. 177. Academic Press, New York.

———. 1981. Drugs as teratogens. The Bendectin saga, another American tragedy. *Teratology* **23:** 28A.

———. 1982a. Drugs and pregnancy. Are the insert warnings too dire? *Contemp. Obstet. Gynecol.* **20:** 42.

———. 1982b. The irresponsible expert witness: A failure of biomedical graduate education and professional accountability. *Pediatrics* **70:** 754. [Congenital Malformations Surveillance Report. 1980. Center for Disease Control, Atlanta, Georgia.]

———. 1985a. The magnitude of the problem of congenital malformations. In *Prevention of physical and mental congenital defects*, Part A (ed. M. Marois), p. 55. A.R. Liss, New York.

———. 1985b. Methods of evaluating the alleged teratogenicity of environmental agents. In *Prevention of physical and mental congenital defects*, Part C: *Basic and medical science, education and future strategies* (ed. M. Marois), p. 199. A.R. Liss, New York.

———. 1985c. Editorial comment on comments on "Teratogen update: Bendectin." *Teratology* **31:** 429.

———. 1985d. Bendectin and interventricular spetal defects, editorial. *Teratology* **32:** 317.

Brent, R.L. and D.A. Beckman. 1986. *Teratology, clinics in perinatology*. W.B. Saunders, Philadelphia. (In press.)

Brent, R.L. and M. Harris, eds. 1976. Prevention of embryonic fetal and perinatal disease. In *Fogarty international series on preventive medicine,* vol. 3, p. 461. DHEW publication (NIH) 76-853, Washington, D.C.

Brent, R.L. and K.T. Szabo. 1975. Nutritional supplementation of the pregnant mouse as a method of reducing the incidence of drug induced cleft palate. *Pediatr. Res.* **9:** 358.

Epstein, S.S. 1979. *The politics of cancer.* Anchor Press, New York.

Food and Drug Administration (FDA). 1979. Pregnancy labeling. *Drug Inform. Bull.* **9:** 23.

Food and Drug Administration (FDA) talk/paper. 1979. *Bendectin*, T79. Food and Drug Administration, Washington, D.C.

Fraser, F.C. 1977. Relationship of animal studies to man. In *Handbook of teratology* (ed. J.G. Wilson and F.C. Fraser), vol. 1, p. 445. Plenum Press, New York.

Fraser, F.C. and T.D. Fainstat. 1951. The production of congenital defects in the offspring of pregnant mice treated with cortisone. A progress report. *Pediatrics* **8:** 527.

Hale, F. 1937. The relation of maternal vitamin A deficiency to microphthalmia in pigs. *Tex. State J. Med.* **33:** 37.

Heinonen, O.P., D. Slone, and S. Schapiro. 1977. *Birth defects and drugs in pregnancy*, p. 516. Publishing Sciences Group, Littleton, Massachusetts.

Holmes, L.B. 1983. Teratogen update: Bendectin. *Teratology* **27:** 277.

———. 1985. Malformation attributed to multifactorial inheritance. *Pediatr. Rev.* **6:** 269.

Lenz, W. and K. Knapp. 1966. Thalidomide embryopathy. *Arch. Environ. Health* **5:** 100.

McKusick, V.A. 1975. *Mendelian inheritance in man*, 4th edition. Johns Hopkins University Press, Baltimore.

Nishimura, H. and H. Yamamura. 1969. Comparison between man and some other mammals of normal and abnormal developmental processes. In *Methods for teratological studies in experimental animals and man* (ed N. Nishimura and J.R. Miller), p. 223. Igaku Shoin, Tokyo.

Szabo, K.T. and R.L. Brent. 1974. Species differences in experimental teratogenesis by tranquilizing agents. *Lancet* **I:** 565.

———. 1975. Reduction of drug induced cleft palate in mice. *Lancet* **I:** 1296.

Warkany, J. 1977. History of teratology. In *Handbook of teratology* (ed. J.G. Wilson and F.C. Fraser), vol. 1, p. 3. Plenum Press, New York.

Warkany, J. and E. Schraffenberger. 1943. Congenital malformations induced in rats by maternal nutritional deficiency. V. Effects of a purified diet lacking riboflavin. *Proc. Soc. Exp. Biol. Med.* **54:** 92.

Wilson, J.G. 1973. *Environment and birth defects.* Academic Press, New York.

Wright, S. 1934. The results of crosses between inbred strains of guinea pigs, differing in number of digits. *Genetics* **19:** 537.

COMMENTS

Pedersen: On the basis of what I recall about population genetics, I would take issue with your statement that we are increasing the number of "defective" genes in the human gene pool by letting homozygotes survive and bear young; the Hardy-Weinberg equation states that you must have a much greater number of carriers in order to get a few affected homozygotes. Affected people bearing children should not have a dramatic effect on the incidence of genetic disease in the human population.

Brent: How can you say that? If a diabetic ends up with the genes that lead to diabetes and they cannot reproduce, those genes are lost.

Pedersen: Yes, but you don't lose the carriers, which greatly outnumber the homozygotes.

Brent: That is true if you examine the problem from the perspective of where future genetic diseases will be derived. Most recessive diseases will continue to be derived from parents that are carriers, especially if the carrier state is a lethal one. In the case of diabetes, you are doing two things. First you permit the patient to survive, and second, you can treat the disease so that the patient will be able to reproduce. Over many generations, the addition of these extra genes to the pool has to increase the incidence of the disease. Furthermore, the phenomenon that I described—permitting or supporting patients with diseases to reproduce—increases the incidence of malformations in more ways than increasing the incidence of homozygous recessive disease. There are dominant diseases that decrease fertility or reproductive potential, and there are genetic diseases that lead to birth defects when treated, not because of the genetic disease in the offspring but because of the metabolic disease in the mother. Diabetes happens to work both ways, as does phenylketonuria. So I believe the medical therapy, while having many positive effects, is also decreasing the reproductive fitness of the population.

Hanson: I think there is a variety of different population genetic models that show either kind of effect, even multigenerational effects, depending on the conditions that are set up in the model. But I would acknowledge that data from sickle cell studies suggest that there may be some transgenerational phenomena. I was surprised that you didn't address the issue of genetically susceptible subpopulations. I'm thinking more of low-frequency polymorphisms that are not single-gene phenomena, which might put people at strikingly different risks.

Brent: The phenomenon of demonstrating increased susceptibility to environmental teratogens in susceptible populations can be demonstrated in animal species. Certainly, the cleft palate situation in the mouse is a well-known example. It is a very complex medical problem to hypothesize on and evaluate because "susceptible" populations frequently have a higher "spontaneous" incidence of the malformation, and the genetic component may be related only indirectly to the frequency of the disease when that population is exposed to a particular teratogen. There is a possibility that genetic variations could change the threshold dose for a particular teratogen, as well as render some agents teratogenic for one genotype and not for another. The magnitude of these problems in the human is unknown. Although it is of practical and theoretical importance to clarify your concept, I do not believe that it is a major factor in the human malformation problem.

Scandalios: I think one aspect that is genetic and frequently underestimated is the fact that you don't need to have structural gene mutations. You can have a normal gene in terms of its output, and many chemicals or environmental signals may increase or decrease the quantitative output of that gene, and that can affect normal or abnormal development. So, among the causes of birth defects that you listed, more things may be genetic in an indirect sense, in that you are affecting the rate of production of the gene.

Brent: Are you saying that a teratogen might manifest its action by modulating a gene rather than by altering it?

Scandalios: Yes. I'm saying that's what it might do in some cases.

Brent: But it is still going to have to follow the general principles of teratology, namely, that it has to be at the right time and dose and that there is going to be a dose at which it is not going to have an effect. Even gene modulation during embryonic development is going to be a multicellular phenomenon and therefore have a threshold effect.

Kimmel: The issue of threshold is one that I have grappled with a lot and I would like to hear more from you about it. One of the problems I have is that the evidence for thresholds in this area is somewhat soft, especially if you look at the total spectrum of possible outcomes, including death. What do you think about that and, in particular, about thresholds in relation to radiation? Dr. Schull's recent publications suggest that there is no threshold for severe mental retardation as a result of radiation.

Brent: I firmly believe that the threshold phenomenon is an important part of environmental teratogenic effects, although I can demonstrate certain exceptions, which actually make sense. Toxic exposures very early in gestation could have a stochastic effect. Our studies on radiation of the one-cell fertilized ovum may not have a threshold, with regard to cell killing and therefore embryonic death. With regard to Drs. Otake and Schull's work, I happen to disagree with their interpretation of the data. While Dr. Schull admits that the data seem to fit a linear relationship better, there are a few patients in the low-exposure group that are mentally retarded. More importantly, those patients with mental retardation following high radiation dose in utero have a specific neuropathological phenomenon, namely failure in migration of neurons and decrease in central nervous system neurons. There is no way that 1 rad (0.01 Gy) of radiation could produce this phenomenon. Furthermore, animal studies do not support this concept. Drs. Jensh and Norton have demonstrated a threshold for all neurophysiological and neurological effects they have studied. Furthermore, the other factors that were present at Hiroshima cannot be eliminated from the list of etiological factors (infection, starvation, physical injury, etc.), as well as other explanations for the very few cases of mental retardation observed in the low-exposure groups. Interestingly, in Nagasaki there was no mental retardation in the small number of patients exposed in utero to less than 150 rads (1.5 Gy).

Johnson: I think that maybe we could have a definition of human teratogen, which states that it is either an agent with a large A/D ratio or something that is experienced by mothers near the maternally toxic dose. To think that the latter is not happening is unrealistic. I think we are exposing pregnant women to agents with very low A/D ratios but doing so at the maternally toxic dose.

Brent: Do you want to mention such a drug?

Johnson: Alcohol is one.

Oakley: If one drinks heavily throughout pregnancy, one's infant may have the fetal alcohol syndrome, and that is an effect on the mother and embryo.

Johnson: The other point I would like to make is that in one of your tables you said synergism was difficult to prove in the human. Dr. Scott has done some elegant studies involving interactions in mammals. They are very difficult studies, yet, to some extent, he has had some success. I would suggest and propose that we need to ask a different kind of question. In looking for synergism, maybe we should approach it from the viewpoint of trying to find two agents that affect the same kind of developmental phenomenon adversely.

Brent: I don't think I said that a teratogen did not affect the mother. I said that in order for an agent to be a teratogen, it has to produce malformations at the exposure at which you usually use that agent. Alcohol is an interesting example because it is a "drug" of abuse. No pharmaceutical firm or chemical company could market alcohol to be used at the level that an alcoholic ingests. Most drugs that produce teratogenic effects at the level of maternal toxicity are used at doses several orders of magnitude lower. With regard to synergisms, I said that there were some very good studies demonstrating interaction and synergism in animals, but it is very difficult to evaluate this phenomenon in the human because we cannot control exposures. It is very likely that in the vast majority of exposures to environmental contaminants, waste products and industrial chemicals, the combined exposures are too low for additive or synergistic effects. Of course, we must still entertain that hypothesis.

Risk Assessment

Biological Considerations for Risk Assessment in Developmental Toxicology

JEANNE M. MANSON
Smith Kline & French Laboratories
Philadelphia, Pennsylvania 19101

OVERVIEW

The application of the risk assessment process to developmental toxicology studies is relatively new. This presentation focuses on the first stage of the risk assessment process, the identification of hazard from laboratory animal studies. A lack of concordance in dose-response relationships between human and laboratory animal studies is evident and may be related to incomplete ascertainment of effects other than malformations. Embryolethality is a major outcome measured routinely in animal studies, and its selective occurrence warrants identification of an agent as a developmental hazard. Spontaneous abortion is a frequent yet rarely measured adverse pregnancy outcome in humans, and a far greater percentage of abortuses have abnormalities than are found in live births. Underestimation of adverse pregnancy outcome, and thus the true risk and pattern of response, is unavoidable in human studies whenever measurements are made only from the time of birth onward. Human studies must be designed to measure developmental toxicity and not just teratogenicity before adequate cross-species comparisons can be made.

Another factor may be lack of adequate control and interpretation of maternal toxicity and how it impacts on development of the conceptus in animal studies. Although models have been proposed for ranking agents according to relative maternal versus developmental toxicity, none of the models have yet been extensively applied to results from in vivo studies. Identification of a uniform method for ranking agents according to their selective toxicity to the conceptus is a key issue in reliable hazard assessment in developmental toxicology. Additionally, for those few agents that appear to cause selective and severe developmental toxicity, a more thorough examination of perturbations in the maternal system is needed than routine measurements of clinical signs, food consumption, and body weight. Otherwise, the lack of precision in identifying maternal toxicity will necessarily weight the ratio toward overestimation of developmental toxicity.

Banbury Report 26: Developmental Toxicology: Mechanisms and Risk

INTRODUCTION

The topic of risk assessment in developmental toxicology has received a great deal of recent attention (Kimmel et al. 1984; Schardein 1985; Schardein et al. 1985; Environmental Protection Agency 1986; Hart et al. 1986; National Research Council 1986). Regulatory requirements have mandated that developmental toxicology studies be carried out on new drugs for the past 20 years, and a substantial body of data exists for cross-species extrapolation of animal data to humans. This, in conjunction with recent efforts to expand the risk assessment process to noncancer areas of health effects (National Research Council 1986), has led to a reexamination of the reliability of experimental studies in predicting risk to human development.

Regulatory agencies have developed a systematic scientific and administrative framework to assess risks associated with exposure to chemical and physical agents (National Research Council 1983). The process usually begins with the *identification of a hazard,* which is defined as the potential of a given exposure to be toxic. Laboratory animal studies and occasionally human case reports are used to identify dose-response relationships between exposure to a substance and associated adverse responses. The next step is the *identification of risk*, which is the likelihood of the adverse effect occurring under real conditions of exposure. Information is obtained on the range of dose levels to which human populations are likely to be exposed. Finally, the dose-response model from animal studies is applied to the expected human exposure levels to produce a quantitative estimate of risk. To date, quantitative risk assessment has been used largely for estimating the risk of developing or dying from cancer.

This presentation will focus on the first stage of risk assessment, the identification of hazard. The following questions are considered: Are the patterns of dose-response seen in animal studies generally predictive of what occurs in humans? How much and what kind of animal data are needed to identify agents as potential developmental toxicants? What is the relationship between maternal and developmental toxicity? What models exist for ranking agents according to relative developmental toxicity across species? Subsequent presentations in this session will address the use of safety factors and mathematical models for quantitative assessment of risk.

Patterns of Dose-Response in Animal Studies: Concordance with Humans

Major outcomes measured in Segment-II developmental toxicology studies are embryo (fetal) lethality, growth retardation, and dysmorphogenesis (major and minor malformations and variations). Three general patterns of dose-response have been identified in animal studies for each of these outcomes (Fig. 1) (Neubert et al. 1980). To identify these patterns clearly,

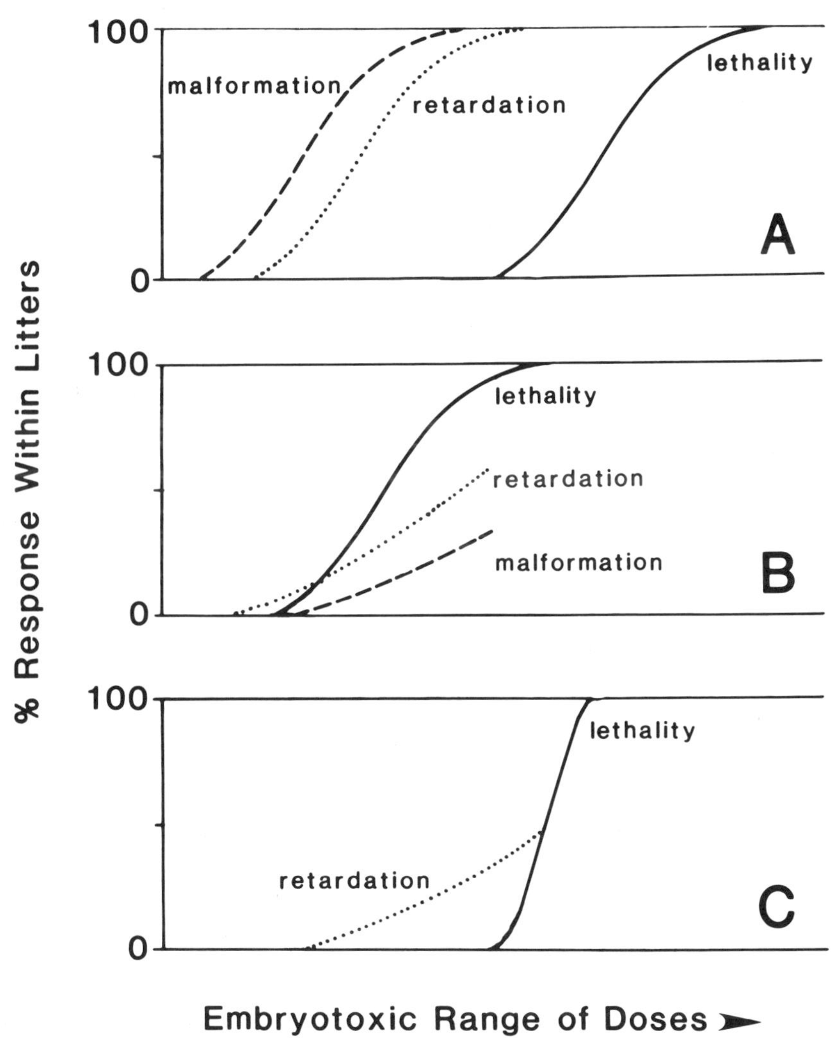

Figure 1
Patterns of dose-response for developmental toxicants in laboratory animal studies. (Adapted from Neubert et al. 1980.)

exposure must be at levels below those causing frank maternal toxicity. One pattern of response is seen with agents that cause malformation of the entire litter at exposure levels not causing embryo death (Fig. 1A). If the dose is increased beyond that causing malformations of the entire litter, embryo death can occur but often in conjunction with maternal toxicity. Fetal malformations are usually accompanied by growth retardation, and the curves for these two endpoints are parallel and slightly displaced from one another in Fig. 1A. This pattern of response is rare, indicating that the agents have high and selective teratogenic potency. Generalities about the mode of action of these agents cannot be made except that they are likely to cause perturbations in processes unique or highly selective for developing/differentiating systems.

A more common dose-response pattern involves a combination of resorptions, malformations, growth retardation, and apparently unaffected fetuses (Fig. 1B). Lower doses may cause predominantly resorptions or malformations, depending on the teratogenic potency of the agent. As the dosage increases, however, embryo death predominates until the entire litter is resorbed. Growth retardation can precede both outcomes or parallel the malformation curve. This response pattern is typical of agents that are cytotoxic to replicating cells by altering replication, transcription, translation, or cell division. These agents include alkylating, antineoplastic, and many mutagenic substances. Exposure to cytotoxic agents during organogenesis can produce all three outcomes both within and among litters. Some litters may be totally resorbed, others may include only growth-retarded fetuses at term, and still others may include a mixture of malformed or growth-retarded fetuses and resorption sites.

A third dose-response pattern consists of growth retardation and embryo death without malformations (Fig. 1C). The dose-response curve for embryo death in this case is usually steep, which may imply a dose threshold for survival of the embryo. Growth retardation of surviving fetuses usually precedes significant increase in lethality. Agents producing this pattern of response would be considered embryotoxic or embryolethal but not teratogenic. When such a pattern is observed, it is necessary to conduct additional studies at doses within the range causing growth retardation and embryo death. Results obtained at these intermediate doses can indicate whether teratogenicity has been masked by embryolethality. Agents inducing this response pattern typically affect fundamental cellular processes such as mitochondrial function or membrane integrity. There is no basis for target organ susceptibility with such agents, and all tissues appear to be equally affected. An early sign of perturbation is overall growth retardation, which progresses to death of the entire litter once a threshold for cellular energy supply or survival is exceeded. These conditions are incompatible with the production of malformations in which irreversible lesions are produced in some tissues while others are spared, permitting survival of abnormal embryos to term.

These patterns of dose-response indicate that for some agents, that is, those cytotoxic to replicating cells (Fig. 1B), growth retardation, embryolethality, and malformations are different degrees of manifestation of the same primary insult, cytotoxicity. For others, there is a qualitative difference in response, and prenatal exposure leads primarily to embryolethality (Fig. 1C) or to malformations alone (Fig. 1A).

These patterns of dose-response from animal studies have important implications in extrapolation of developmental toxicity data to humans. The major implication is that a spectrum of endpoints can be produced, even under the controlled conditions of timing and exposure that can be achieved in animal studies. In estimating hazard to humans, all exposure-specific adverse outcomes identified in animal studies must be taken into consideration, not just malformations. A similar spectrum of growth retardation, embryolethality, and malformations has been observed in humans after prenatal exposure to developmental toxicants, although clear patterns of dose-response have not been identified with individual agents. Consequently, manifestations of developmental toxicity cannot be presumed to be constant or specific across species. *Any* manifestation of exposure-related developmental toxicity in animal studies can be indicative of a variety of responses in humans (Kimmel et al. 1984).

An important factor to be considered in cross-species extrapolation is that the most common manifestation of developmental toxicity in humans is spontaneous abortion or early fetal loss (before the 28th week of pregnancies), occurring in at least 10–20% of all recognized pregnancies (Table 1). Estimates from prospective studies range even higher: between 20–25% of all conceptions spontaneously abort (Edmonds et al. 1981). Approximately one third of specimens obtained from spontaneous abortions occurring between 8 and 28 weeks of gestation contain chromosomal aberrations. The frequency of such aberrations is at least 60-fold higher among spontaneous abortions than among term births. Of those abortuses without chromosome aberrations, approximately half have structural malformations (Edmonds et al. 1981). The frequency of these malformations is not as well documented as that for chromosome aberrations because they are difficult to observe in specimens that are often macerated or incomplete.

These observations suggest that the majority of human embryos bearing chromosome aberrations or morphological abnormalities are lost through early miscarriage. Epidemiological approaches to monitoring the frequency of early fetal loss and detecting such fetal abnormalities have only been used to a limited extent. Most studies of humans have focused on outcomes occurring at the time of birth or later, that is, major malformations, stillbirths, low birth weight, and neonatal deaths. Underestimation of adverse pregnancy outcome and, thus, true risk and pattern of response are unavoidable in human studies whenever measurements are made only from the time of birth onward.

Developmental toxicants with dose-response patterns resembling those in Figure 1A could be detected by monitoring malformations at the time of

Table 1
Frequency of Selected Reproductive Failures

Event	Frequency per 100	Unit
Failure to conceive after 1 year	10–15	Couples
Spontaneous abortion (8–28 weeks)	10–20	Pregnancies or women
Chromosomal anomalies in spontaneous abortions (8–28 weeks)	30–40	Spontaneous abortion
Chromosomal anomalies from amniocentesis (>35 years)	2	Amniocentesis specimens
Stillbirths	2–4	Stillbirths and live births
Birth weight (<2500 g)	7	Live births
Birth defects	2–3	Live births
Chromosomal anomalies (live births)	0.2	Live births
Severe mental retardation	0.4	Children to age 15 years

Adapted from Edmonds et al. (1981).

birth, especially if the malformations were rare (such as those resulting from thalidomide), or if the exposed populations were large (such as those with rubella infections). Agents with patterns B and C (Fig. 1) would probably be missed, because early fetal loss is not monitored routinely in human populations even though it has been done successfully in isolated groups (Kline et al. 1977).

The sensitivity (ability to detect a true positive response in humans) and specificity (ability to detect a true negative response in humans) of laboratory animal studies have been evaluated recently by the FDA (Table 2) (Frankos 1985). Of 38 compounds having demonstrated or suspected teratogenic activity in humans, all except one (tobramycin, which causes otological deficits in humans) tested positive in at least one animal species. Furthermore, over 80% were positive in multiple species. A positive response was elicited 85% in the mouse, 80% in the rat, 60% in the rabbit, 45% in the hamster, and only 30% in the monkey. Overall, these findings indicate that conventional animal species have high sensitivity for detecting human teratogens.

Evaluation of specificity (Table 2) has indicated that of 165 adequately tested compounds with no evidence of being human teratogens, 29% ap-

Table 2
Sensitivity and Specificity of Laboratory Animal Studies for Predicting Teratogenesis (FDA Review)

	Number	Percent
Sensitivity		
Compounds with positive teratologic findings in humans	38	100
Positive in at least one laboratory animal test species	37	97
Positive in more than one laboratory animal test species	29	76
Positive in all laboratory animal species tested	8	21
Specificity		
Compounds studied in humans with no teratologic findings	165	100
Negative in at least one laboratory animal test species	130	79
Negative in more than one laboratory animal test species	84	51
Negative in all laboratory animal species tested	47	29
Positive in more than a single laboratory animal test species	68	41

Adapted from Frankos (1985).

peared negative in all animal species tested and 51% appeared negative in multiple species. However, 41% of these 165 compounds were positive in more than a single animal species. The nonhuman primate and the rabbit had the highest specificity of laboratory animal species, testing "negative" for substances reported to be negative in humans 80% and 70% of the time, respectively (Frankos 1985).

These findings indicate that laboratory animal species have high sensitivity but low specificity for predicting human teratogens. Schardein et al. (1985) have reviewed the literature on drugs and environmental chemicals and have found that of the 2800 agents now reported to have been tested in laboratory animals, approximately 1000 have demonstrated some measure of teratogenicity for which there is no evidence of positivity in humans. Consequently, the major concern in hazard assessment today is that far more agents have been shown to be positive in animal studies than have been identified in human studies. There are at least three possible explanations for this: No studies or inadequate studies have been conducted in humans with the animal teratogen; the animal studies have yielded false-positive results because of test conditions and interpretation; or the true risk for adverse pregnancy outcome is underestimated in human studies that only measure outcomes from the time of birth onward. Human studies must be designed to measure developmental toxicity and not just teratogenicity before adequate cross-species comparisons can be made.

Quality and Quantity of Animal Data Needed to Identify Agents as Potential Developmental Toxicants

There are substantial problems in the quality, quantity and interpretation of animal data that can complicate the hazard assessment process. Unequivocal identification of an agent as a developmental toxicant from animal studies is rarely straightforward. A major consideration is whether there are adequate data to perform a hazard assessment. Results from Segment-II studies in two species, preferably a rodent and a nonrodent, carried out according to proscribed guidelines (Interagency Regulatory Liaison Group 1981), should be available. Treatment should be via the likely human route and the highest dose of sufficient magnitude to cause measurable maternal toxicity. Results should permit identification of the no-observable effect level (NOEL) for maternal and developmental toxicity, as well as the lowest observable effect level (LOEL) for maternal and/or developmental toxicity. These represent minimal requirements that must be met before hazard assessment can be carried out.

Relationship between Maternal and Developmental Toxicity

Regulatory guidelines for identifying developmental toxicity call for dose-response studies in pregnant animals, the highest dose being of sufficient magnitude to cause measurable maternal toxicity. There is much controversy over what constitutes measurable maternal toxicity. In some cases, exhibition of frank clinical signs such as sedation, hyperactivity, and convulsions of the dam are considered sufficient. In other cases, a marginal but statistically significant depression in maternal weight and weight gain throughout treatment during organogenesis, or on individual days, is considered adequate for demonstration of maternal toxicity. In general, measurement of effects on the maternal system rarely exceeds identification of survival, clinical signs, food consumption, or body weight. In general toxicology studies, the definition of maximum tolerated dose includes a statistically significant weight loss and not more than 10% deaths. The rationale for using a maternally toxic dose is to maximize the potential to detect adverse effects in the fetus (Palmer 1981). Effects observed on the conceptus at maternally toxic doses are used as landmarks to focus attention on outcomes at lower doses. If a statistically significant incidence of a particular adverse effect is found in the high-dose group, the biological significance of a lower and perhaps nonsignificant incidence at lower doses is magnified. It can be difficult, however, to interpret some effects observed only at maternally toxic dose levels. Are they indicative of unique and selective developmental toxicity, or are they a function of nonspecific alterations in maternal homeostasis?

An initial factor to consider is that the state of pregnancy itself confers altered susceptibility to chemical insult. Table 3 details physiological changes in pregnancy related to drug handling. The general trend is that pregnancy-related physiological changes favor increased absorption of drugs. The one exception to this is increased renal function, which would result in elevated urinary excretion of free drug (Hytten 1984). Increased susceptibility due to these pregnancy-related changes can be addressed by comparing LOELs for adverse effects in the maternal system in pregnant versus nonpregnant animals. This comparison would be limited to those few measurements of maternal toxicity routinely made in Segment-II studies, that is, clinical signs, food consumption, and body weight. Given that adult toxicity studies routinely include a more complete assessment of health status (histology, hematology, clinical chemistry, etc.), it is unlikely that LOELs for maternal toxicity, as measured conventionally, would ever be lower than LOELs for adult toxicity.

It is generally accepted that developmental toxicity in the form of increased resorptions and decreased fetal body weight can occur at maternally toxic dose levels. The role of maternal toxicity in the production of congenital malformations is not clear, however. Khera (1984) reviewed more than 85 published studies in mice to examine the relationship between maternal

Table 3
Physiological Changes in Pregnancy Related to Drug Handling

Absorption	Decreased gastric and intestinal motility
Distributon	40% increase in blood and plasma volume
Protein binding	60% reduction in plasma albumin levels; increase in proportion of free drug in pregnancy plasma
Metabolism	Decreased hepatic drug metabolism (?)
Excretion	Renal function increased; 50% elevation in renal plasma flow and glomerular filtration rate

Adapted from Hytten (1984).

toxicity, embryo toxicity, and birth defects. He noted that doses of test agents that caused maternal toxicity, as indicated by reduced maternal body weight, clinical signs of toxicity, or deaths, commonly caused reduction in fetal body weight, increased resorptions and, rarely, fetal deaths. The structural variants/malformations found to be associated with maternal toxicity in mice are listed in Table 4; this pattern also includes findings from more recent analyses in rats, rabbits, and hamsters (Khera 1985). The severity and incidence of these defects could be directly related to the degree of maternal toxicity. They were absent or rare at doses that were nontoxic to the dam. Khera (1984, 1985) concluded that these defects resulted from maternal toxicity and did not reflect the teratogenic potential of the compounds.

A number of other investigators (Palmer 1972; Kimmel and Wilson 1973; Nishimura 1982; Kavlock et al. 1985) have addressed the association between maternal toxicity and the occurrence of retarded sternebral and vertebral ossification and supernumerary/wavy ribs in experimental animals. There is general agreement that these are variations and not malformations that are reversible and secondary to maternal toxicity. Those major malformations attributed to maternal toxicity in the studies done by Khera (exencephaly, open eyes, encephalocele, micro- or anophthalmia) require more detailed examination to determine whether they are representative of maternal toxicity or teratogenicity.

Table 4
Structural Variants/Malformations Associated with Maternal Toxicity

Species	Variants/malformations associated with maternal toxicity
Mice	Exencephaly
	Open eyes
	Hemivertebrae
	Fused arches or centra thoracic/lumbar vertebrae
	Missing or supernumerary ribs
	Fused or scrambled sternebrae
Rats and rabbits	Fused, extra, missing, or wavy ribs
	Fused, retarded, missing, or split vertebrae
	Fused, missing, or nonaligned sternebrae
Hamsters	Exencephaly
	Encephalocele (cranial blister)
	Microphthalmia or anophthalmia
	Fused ribs

Adapted from Khera (1984, 1985).

Models for Ranking Agents According to Relative Developmental Toxicity

Investigators concerned with the regulatory aspects of risk assessment have focused on the development of a quantitative index for comparing developmental toxicity across species, taking into account concurrent maternal toxicity. Underlying this approach is the perceived need to distinguish between compounds that are uniquely toxic to the embryo and those that induce developmental toxicity at exposure levels that are also toxic to the mother. Agents in the latter category should be regulated on the basis of their maternal toxicity, whereas those in the former would be regulated on the basis of their unique toxicity to the embryo.

Johnson (1980) has developed a testing system for addressing this issue quantitatively. He defined a teratogenic hazard potential as the ratio of adult to developmental toxicity, or the A/D ratio, that is,

$$\text{Log} \quad \frac{\text{Lowest adult toxic (lethal) dose}}{\text{Lowest developmental toxic dose}}$$

He has calculated this ratio for more than 70 compounds, using data from an in vitro system of *Hydra attenuata* adult and embryonic tissues. The A/D ratio from the hydra assay has been 1/10 to 10 times greater than the mammalian A/D ratio. Most compounds had ratios near 1, several had ratios greater than 5, and very few had ratios greater than 10 (Johnson and Gabel 1983). This system has been proposed for setting priorities for further testing of agents in mammalian developmental toxicity studies.

Fabro et al. (1982) have explored the quantitative characteristics of a similar type of index in mammalian studies. Dose-response data for adult mortality and fetal malformations were fitted (probit of response against log of dose) for eight compounds. The observed log-probit dose-response lines for lethality and teratogenicity were not parallel, and there was not a constant ratio between the slopes for the two lines. Consequently, a simple ratio between the median lethal dose and the median effective dose (i.e., LD_{50}:ED_{50}) could not be used. To calculate a relative teratogenic index (RTI), Fabro and colleagues established a ratio between one point on each dose-response line. The LD_{01} value was chosen to represent adult mortality on the basis that a low LD value is necessary to guard against compounds with a shallow dose-response curve for adult mortality. The tD_{05} was chosen to represent teratogenicity, that is, the dose causing a 5% elevation of the malformation rate above background. The tD_{05} can be estimated with confidence for most teratogens because induced malformations occur often in a frequency between 1% and 20% in animal studies.

This ranking system was developed to evaluate structure-teratogenicity relationships between structurally related compounds. For this purpose, the

RTI seems adequate. The potential usefulness of this index for interspecies comparisons and risk estimation, however, has not been established. In their evaluation of the RTI, Hogan and Hoel (1982) argued that because of the lack of parallelism between the probit lines for lethality and teratogenicity, the index will not be invariant in the selection of other LD and tD values; for example, if a ratio of LD_{10}:tD_{05} were chosen instead of LD_{01}:tD_{05}, a different ranking order for the RTI would be obtained. In addition, the index would be subject to the established deficiencies of the probit model, which tends to be insensitive in the low-dose region near the origin of the dose-response curve and requires use of the fetus rather than the litter as the unit of analysis.

Therefore, until the RTI has been applied and evaluated more extensively, it should not be used for hazard assessment. It is apparent, however, that a uniform method for ranking agents according to their selective toxicity to the conceptus needs to be established. Such a method would provide a yardstick against which all agents could be compared and would standardize the selection of the NOEL and LOEL for risk assessment. Selection of the safety factor could then be based on the severity of the endpoint.

A key consideration in obtaining a reliable ratio is that data of comparable quality and quantity need to be provided for identification of maternal and developmental toxicity. At present, far more detailed observations are made on the fetus than on the maternal system. The lack of precision in identifying maternal toxicity will necessarily weight the equation toward overestimation of developmental toxicity. For those few agents that appear to cause selective and severe developmental toxicity, a more thorough examination of perturbations in the maternal system is needed than routine measurements of clinical signs, food consumption, and body weight.

REFERENCES

Edmonds, L., M. Hatch, L. Holmes, J. Kline, C. Letz, B. Levine, R. Miller, P. Shrout, Z. Stein, D. Warburton, M. Weinstock, R.D. Whorton, and A. Wyrobek. 1981. Report of panel II: Guidelines for reproductive studies in exposed human populations. In *Guidelines for studies of human populations exposed to mutagenic and reproductive hazards* (ed. A.D. Bloom), p. 37. March of Dimes Birth Defects Foundation, White Plains, New York.

Environmental Protection Agency (EPA). 1986. Guidelines for health assessment of suspect developmental toxicants. *Fed. Regul.* **51(185):** 34026.

Fabro, S., G. Shull, and N.A. Brown. 1982. The relative teratogenic index and teratogenic potency: Proposed components of developmental toxicology risk assessment. *Teratog. Carcinog. Mutagen.* **2:** 61.

Frankos, V.H. 1985. FDA perspectives on the use of teratology data for humans risk assessment. *Fundam. Appl. Toxicol.* **5:** 615.

Hart, W.L., R.C. Reynolds, W.J. Krasavage, T.S. Ely, R.H. Bell, and R.L. Raleigh.

1986. Evaluation of developmental toxicity data: A discussion of some pertinent factors and a proposal. *J. Risk Anal.* (in press).
Hogan, M.D. and D.G. Hoel. 1982. Extrapolation to man. In *Principles and methods of toxicology* (ed. A.W. Hayes), p. 724. Raven Press, New York.
Hytten, F.E. 1984. Physiological changes in the mother related to drug handling. In *Drugs and pregnancy* (ed. B. Krauer et al.), p. 7. Academic Press, New York.
Interagency Regulatory Liaison Group (IRLG). 1981. Testing standards and guidelines work group. Recommended guidelines for teratogenicity studies in the rat, mouse, hamster or rabbit. National Technical Information Service Publ. no. PB-82-119488. Springfield, Virginia.
Johnson, E.M. 1980. A subvertebrate system for rapid determination of potential teratogenic hazards. *J. Environ. Pathol. Toxicol.* **4(5):** 153.
Johnson, E.M. and B.E.G. Gabel. 1983. An artificial "embryo" for detection of abnormal developmental biology. *Fundam. Appl. Toxicol.* **3:** 243.
Kavlock, R.J., N. Chernoff, and E.H. Rogers. 1985. The effect of acute maternal toxicity on fetal development in the mouse. *Teratog. Carcinog. Mutagen.* **5:** 3.
Khera, K.S. 1984. Maternal toxicity—A possible factor in fetal malformations in mice. *Teratology* **29:** 411.
———. 1985. Maternal toxicity: A possible etiologic factor in embryo-fetal deaths and fetal malformations of rodent-rabbit species. *Teratology* **31:** 129.
Kimmel, C.A. and J.G. Wilson. 1973. Skeletal deviations in rats: Malformations or variations? *Teratology* **8:** 309.
Kimmel, C.A., J.F. Holson, C.J. Hogue, and G.L. Carlo. 1984. Reliability of experimental studies for predicting hazard to human development. *NCTR Technical Report no.* 6015. National Cancer Center for Toxicological Research, Jefferson, Arkansas.
Kline, J., Z. Stein, B. Strobino, M. Susser, and D. Warburton. 1977. Surveillance of spontaneous abortions: Power in environmental monitoring. *Am. J. Epidemiol.* **106:** 345.
National Research Council (NCR). 1983. *Risk assessment in the federal government: Managing the process.* National Academy Press, Washington, D.C.
———. 1986. Developmental effects of chemical contaminants. In *Drinking water and health,* vol. 6, p. 11. National Academy Press, Washington, D.C.
Neubert, D., H.J. Barrach, and H.-J. Merker. 1980. Drug-induced damage to embryo or fetus: Molecular and multilateral approach to prenatal toxicology. *Curr. Top. Pathol.* **69:** 241.
Nishimura, M. 1982. Repairability of drug-induced "wavy ribs" in rat offspring. *Drug. Res.* **32(12):** 1518.
Palmer, A.K. 1972. Specific malformations in laboratory animals and their influence on drug testing. In *Drugs and fetal development* (ed. M.A. Klingsbury et al.), vol. 7, p. 45. Plenum Press, New York.
———. 1981. Regulatory requirements for reproductive toxicology: Theory and practice. In *Developmental toxicology* (ed. C.A. Kimmel and J. Buelke-Sam), p. 259. Raven Press, New York.
Schardein, J.L. 1985. Symposium: Risk assessment for developmental toxicology. *Fundam. Appl. Toxicol.* **70:** 607.

Schardein, J.L., B.A. Schwetz, and M.F. Kenel. 1985. Species sensitivities and prediction of teratogenic potential. *Environ. Health Perspect.* **61:** 55.

COMMENTS

Hanson: It seems to me that of some teratogens seen in animals and humans were identified after the fact; that is, until somebody knew what to look for, they did not seek those things. To suggest that this process is predictive is somewhat of an anomaly in terms. It can hardly be predictive after the fact, if prediction implies that one looked and then predicted something, not that one saw something, and then hunted until you found a system that would predict it.

Manson: I am not proposing predictability for a particular outcome. What I would like to emphasize is that any manifestation of treatment-related developmental toxicity in animal studies can be indicative of a variety of responses in humans.

Hanson: Yes, but I think the whole point is that the value of it, other than to look at mechanisms, is to suggest that there is a real risk. If one uses systems and they fail, and then one designs more systems until they pick something up after one knows that it is there, this should not suggest that they are predictive.

Oakley: To review what the words sensitivity and specificity mean would be helpful. The words mean we start out knowing that the compound is either bad or good, and then you test it and you see how it did. It is predictive in that way, not necessarily in a time-related phenomenon. I think pathologists and physicians confused that issue when they are doing clinical tests; something looks very good for either sensitivity or specificity, but when it is actually put into play in the unknown situation, how good it is not only is a function of those two issues but the a priori risk of a bad condition among the people or the drugs that the physicians are screening. That is where it becomes much, much more difficult.

Hanson: In that case, a lot of these compounds do not meet that criterion because for the nonteratogenic ones, at least one doesn't really know if they were good or bad in the final analysis. One can only say that it has not yet been found at this point in time, so that strict application of terms does not work.

Kimmel: Looking at the known human developmental toxicants, we used the term "comparability." In fact, in many cases there were animal studies that predated the human recognition of the problem. I do not know that "predictability" is a good word; perhaps comparability is more appropriate in our situation.

Oakley: It really gets down to behavior, doesn't it?

Kimmel: Yes, exactly. I would like to comment on the consideration of the dose/response curves for maternal and developmental toxicity that have been looked at in various cases. I guess the biggest controversy was over Fabro's attempt to make some comparison of those dose/response curves. One comment is that whether or not these dose/response curves are parallel is irrelevant. That is because the dose is given to the maternal animal and we don't really know the dose to the offspring. So, although we can plot a so-called dose/response curve, the dose is actually a maternal dose; therefore, that curve may not be really representative of the actual dose/response curve in the fetus.

Markert: What is the basis for selecting a given animal species as test objects? Is it simply economics?

Manson: At this point, it's historical.

Markert: Hasn't any thought been given from a biological point of view as to what would really be the most rational choice?

Manson: I think there has been a tremendous amount of thought given to that; and a spectrum of species has been tried. But in the end, if there are major malformations occurring spontaneously in the frequency of 1 or 2 per 1000, it is particularly important to understand what the background rate of the malformation is in the particular strain that is under consideration. Comparison with the concurrent control is always the most important first step. But, for any given study, where there are 20 litters in the group and 1000 fetuses, the investigator is going to spontaneously pick up a certain level of malformation, which may or may not, by chance, also have been present in the concurrent control. So, having some idea of the historical incidence of outcomes is particularly important. Right now, 20 years have gone into studies of the rat, rabbit, and the mouse, and a tremendous amount of data exists on the spontaneous incidence of adverse pregnancy outcomes in these species.

Regulatory Perspectives on Risk Assessment for Developmental Toxicity

JANET A. SPRINGER
Division of Mathematics
Food and Drug Administration
Washington, D.C. 20204

OVERVIEW

The thalidomide tragedy prompted passage of the 1962 Kefauver-Harris Amendments of the 1938 Food, Drug, and Cosmetic Act. These amendments required performance of certain animal studies prior to clinical testing of a drug (Frankos 1985). Specific guidelines for the conduct of such studies came later (United States Food and Drug Administration 1966, 1970; Kelsey 1974; Collins 1978).

The 1966 guidelines outlined protocols for the now familiar segment-I, -II, and -III studies. Segment I addresses fertility and general reproductive performances, segment II addresses classic teratology animal testing, and segment III limits treatment to perinatal and postnatal period. Although three-generation reproductive studies were required in these early years of animal testing, the present guidelines call for a minimum two-generation reproductive study design, with a teratology phase as outlined below:

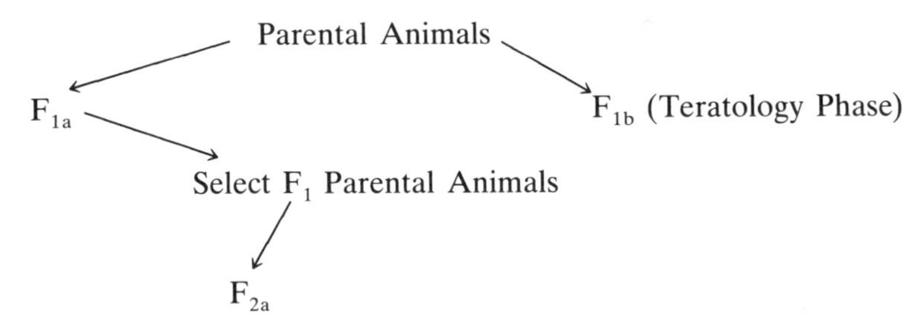

DISCUSSION

Recent Agency Activities

For food additives, color additives, and human animal drugs, internal agency guidelines (Bureau of Foods, unpubl.) were recommended in 1982 by the Teratology Guideline Committee of the Bureau of Foods. Chaired by Dr. Benjamin Jackson, this committee emphasized all endpoints of abnormal intrauterine development, including death, malformations, growth defects, and functional deficits. They attempted to classify these endpoints of developmental toxicity according to severity and reversibility. Effects were separated into two types. Type-I effects were perceived as permanent (irreversible), possibly life-threatening, and frequently associated with gross malformations. Type-II changes were perceived as nonpermanent (irreversible), non-life-threatening, and not necessarily associated with induced malformations. Table 1 lists specific endpoints associated with type-I and type-II outcomes.

The table lists type-II as type-I/type-II responses because type-II responses are often associated with type-I gross malformations at high doses of the test agent. The committee recommended treating possible type-II responses as type I unless other information is available that clarifies the data. For example, if a compound at the highest dose tested in preliminary studies caused decreased birth weights and minor skeletal effects, these responses may be classified as type-I changes until other data at other doses, indicating the absence of gross malformations or resolving the issue of maternal toxicity or postnatal recovery of the offspring, clarify the true nature and type of response.

The committee recommended a hazard assessment scheme using the results of endpoint effects in the two studies, multigeneration and teratology, as shown in Table 2. Safety factors were recommended for the various outcomes. For instance, a 1000-fold safety factor would be applied in those cases where there is evidence of serious developmental effects in two species. A 100-fold safety factor will be applied to the no-observed effect level for a type-II effect. When results of the two tests are discordant, consideration

Table 1
Developmental Toxicity Endpoints

Type I	
Decreased no. of live births (litter size)	Decreased no. of live fetuses (litter size)
Increase no. of stillbirths	Increased no. of resorptions
	Increased no. of fetuses with malformations, structural changes
Type I/type II	
Decreased birth weights	
Decreased postnatal survival	
Decreased postnatal growth, reproductive capacity	Increased no. of fetuses with retarded development (skeletal, soft tissue)

Table 2
Safety Factors Applied to Outcomes of Developmental Toxicity Tests

Teratology test	Findings from multigeneration test		
	type I	type II	no effect
Type I	Type I (1000×)	Type I[a]	Type I[a]
Type II	Type I[a]	Type II (100×)	Type II (100×)
No effect	Type I[a]	Type II (100×)	No effect (100×)

[a]An appropriate safety factor will be determined after consideration of additional information on metabolism, pharmacokinetics, natural toxicity, mechanism, etc.

would be given to information on maternal toxicity, metabolic differences, and mechanism in selecting appropriate safety factors.

This committee report was unpublished, although used as a guideline internally in Bureau of Foods. An agency-wide FDA Task Force was established in 1984 to review each FDA center's current recommendations and requirements for reproductive and developmental toxicity studies and to report on existing similarities and differences. The report of this Task Force (United States Food and Drug Administration, unpubl.) was essentially a listing of the various tests required by each center in FDA and the endpoints evaluated in these tests. There was little guidance on assessing risks, although there was general discussion of the precision and sensitivity of the various endpoints and the effect of maternal toxicity on risk assessment.

The recently published Environmental Protection Agency (EPA) guidelines on the Health Assessment of Suspect Developmental Toxicants (United States Environmental Protection Agency 1986) discuss sensitivity of various endpoints for risk assessment and the use of uncertainty factors for setting safe levels. This document does not, however, recommend any specific safety factors.

Safety Factors—Comparison of Animal and Human Effects

The use of 50- to 1000-fold safety factors to derive an allowable human exposure based on animal no-observed effect levels has been recommended in various documents. How well these factors are expected to protect humans can be assessed only by examining the concordance between responses of animals and humans to a representative group of chemicals. Although there is not a great deal of information on comparison of animal and human effect or no-effect levels, several comparisons have been made. The Council on Environmental Quality report, "Chemical Hazards to Human Reproduction,"

Table 3
Comparisons of Lowest Effective Doses of Eight Teratogens in Humans and Animals

Chemical	Lowest effect dose humans (per day)	Lowest effect dose animals (per day)	Species	Ratio of animal dose/ human dose
Thalidomide	0.5–1.0 mg/kg	2.5 mg/kg	Rabbit	5–2.5
Polychlorinated biphenyls[a]	70 μg/kg	125 μg/kg	Rhesus monkey	1.8
		1,000 μg/kg	Dog	14.3
Alcohol	0.4 – 0.8 g/kg	1.5 g/kg	Rat	3.8–7.6
Aminopterin	50 μg/kg	100 μg/kg	Rat	2
Methotrexate	42 μg/kg	200 μg/kg	Rat	4.8
Methylmercury[b]	5 μg/kg	250 μg/kg	Cat, rat	50
Diethylstilbestrol	20–80 μg/kg	200 μg/kg	Rhesus monkey	10–2.5
Diphenylhydantoin	2 mg/kg	50 mg/kg	Mouse	25

[a]For blood levels of 200–500 ng/ml, the equivalent long-term daily intake of mercury as methylmercury is 3–7 μg/kg/day. (Reprinted, with permission, from National Research Council 1978.)
[b]Methylmercury and/or polychlorinated dibenzofurans.

made such comparisons (Nisbet and Karch 1983). They showed ratios of lowest effect levels for eight teratogens in humans and animals. Table 3 shows that these ratios, between the lowest effective dose in animals to the lowest effective dose in humans, range from 1.8 to 50 on a mg/kg weight basis. Table 4 shows ratios ranging from 0.2 to 10 for four teratogens, with a much higher ratio for halothane (Hemminki and Vineis 1985). Table 5 shows ratios of 0.5 to 150, with 150 being for thalidomide (NCTR 1984). Note that these ratios are usually greater than 1, indicating that humans are more sensitive than animals. Although these comparisons all suffer from the quality of the data bases, they all tend to support the use of 100- to 1000-fold safety factors recommended by the Jackson committee and others.

Other Quantitative Methods

Other methods of evaluating risk-of-development toxicants have been proposed. Johnson (1980) developed a teratogenic hazard potential as the ratio of adult to developmental toxicity, the A/D ratio:

$$\text{Log} \quad \frac{\text{Lowest adult toxic dose}}{\text{Lowest developmental toxic dose}}$$

Fabro et al. (1982) have proposed a relative teratogenic index (RTI) as the ratio of LD_{01}/tD_{05}, the ratio of lethal dose causing 1% death in the mother to

Table 4
Animal Tests on Chemicals with High and Limited Evidence of Teratogenicity in Man After a Quantitated Exposure

Substance	Human data: daily dose (mg/kg)	Human data: risk	Animal data: species	Animal data: daily dose (mg/kg)	Animal data: risk
Carbon monoxide	?	Central nervous system defects	Mouse, rat, rabbit	30–250 ppm	Resorptions
Ethanol	2000	33% (FAS)	Mouse	3000–5000	−30% (FAS-type)
Halothane[a]	60 ppm = 0.6 ml/kg	1.5 times	Mouse	1.10 l/kg	Threefold (skeletal)
Methylmercury	1.0[b]	25% microcephaly	Mouse, rat	5–10	50% malformations
			Cat	0.25	32% malformations
Nitrous oxide[c]	3000 ppm = 0.3 l/kg	5.5% vs. 3.6% 1.5 risk	Rat	300 l/kg	Malformations sixfold
			Rat	0.2 l/kg	Fetal deaths
Polychlorinated biphenyls	0.5[d]	100% (symptoms) 30% (small for date)	Monkey	0.10	13% (skin symptoms) 50% (low birth weight) 40% (lethality)

[a]It is assumed that halothane is the agent responsible in anesthetics; in fact, the evidence is inadequate.
[b]For methylmercury, 1 mg/kg is the total ingested dose; the daily doses were on the order of 80 μg/kg.
[c]It is assumed that nitrous oxide is the agent responsible in anesthetics; in fact, the evidence is low.
[d]Estimated intake of Kanechlor.
(Reprinted, with permission, from Hemminki and Vineis 1985.)

Table 5
Comparisons of Lowest Reported Effect Levels for Selected Developmental Toxicants in Man and Test Species

Aminopterin	Man	0.1 mg/kg/day	Death/malformations
	Rat	0.1 mg/kg/day	Death/malformations
Diethylstilbestrol	Man	0.8–1.0 mg/kg	Genital tract abnormalities/death
	Mouse	1 mg/kg	Genital tract abnormalities/death
Ionizing radiation	Man	20 rads/day	Malformations
	Rat/mouse	10–20 rads/day	Malformations
Cigarette smoking	Man	20 cigarettes/day	Growth retardation
	Rat	>20 cigarettes/day	Growth retardation
Thalidomide	Man	0.8–1.7 mg/kg	Malformations
	Monkey	1.25–20 mg/kg	Malformations
	Rabbit	150 mg/kg	Malformations

the teratogenic dose causing 5% elevation of malformation rate above background. A probit model was used for both endpoints to estimate the LD_{01} and tD_{05}. There has been some criticism of this index because of the nonparallelism between probit lines for lethality and teratogenicity.

EPA used a number of models commonly applied to carcinogenicity data to evaluate development toxicity data for PCBs and glycol ethers (United States Environmental Protection Agency 1983a,b). The risks associated with low-exposure levels showed good agreement between models when the data were linear and wide disagreement when data were nonlinear, with upward curvature.

The glycol ether quantitative risk assessment was circulated to scientists in the field of teratogenic risk assessment for review and was to be included in the FDA's decision document on glycol ethers. However, it was felt by the reviewers and scientists in the FDA that these models were not validated for the area of developmental toxicity and were not included in any final documents by the EPA.

Another method proposed by Kimmel and Gaylor (1986) is basically the Gaylor-Kodell method used by FDA for cancer risk assessment. This method fits a dose-response model to the proportion of affected fetuses per litter and obtains an upper confidence limit on that curve. A lower confidence limit on the dose that produces a 10% risk is then divided by a safety factor (F) to give a dose that is not expected to produce a risk greater than (0.1) ($1/F$). (Fig. 1).

The Rai and Van Ryzin model is a probabilistic mechanistic model that has been proposed for developmental toxicity data to evaluate risk at low doses

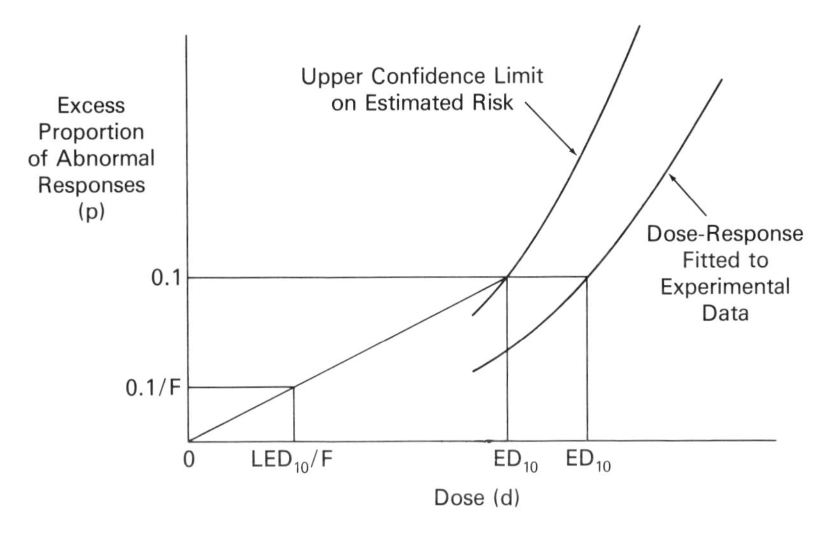

Figure 1
Graphic illustration of proposed low dose risk estimation for the proportion of abnormal responses in developmental toxicity. (Figure printed, with permission, from C. Kimmel and D. Gaylor, unpubl.).

(Rai and Van Ryzin 1985). This model includes a probability of an effect of dose on the litter and a probability of effect of dose on individual pups in the litter, given a certain litter effect. An assumption of the model was that the probability of an abnormal response at a given dose decreases as litter size increases. Both effects are modeled essentially as linear functions of dose in the exponent. In the published paper, the model was applied to dominant lethal data, which is atypical of most standard teratology studies.

EPA (Faustman-Watts 1986) has applied the Rai and Van Ryzin model to three different endpoints for each of six studies of developmental toxicity data. The fits obtained were reasonably good, accounting for over 80% of the variation. This model has the advantage of taking account of the nonindependence of responses of pups within the same litter.

However, this method is illustrative of the general problem in low-dose extrapolation. For three of these data sets, the linear trend with dose was not significant, whereas for several data sets, there was considerable upward curvature with dose (Fig. 2). Yet both estimates of dose giving a specified risk and lower limits on that dose became linear with all data sets at risks less than 10^{-3}, since the risk is essentially linear with dose in the model at low doses. Van Ryzin is developing models other than the one-hit model for developmental toxicity; however, no results have been shown. It is expected

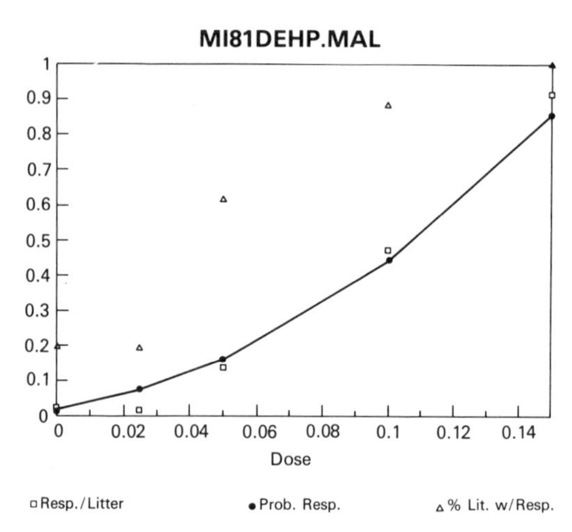

Figure 2

Dose-response curve for malformations in CD-1 mice treated with DHEP (diethylhexyl phthalate). (□) Malformations/litter; (●) probability of response; (△) % litters with malformed fetuses. (Reprinted, from Faustman-Watts 1986.)

that these models would show the same behavior as the cancer models, that is, estimates of risk from different models could vary greatly at low doses with data showing a high degree of upward curvature, yet upper confidence limits on these risks would all be essentially linear for any models that allow for a linear term in the exponent unless a threshold is explicitly incorporated into the model.

CONCLUSION

For the present, it appears that both FDA and EPA agree with the approach for risk assessment stated in the EPA Guidelines for the Health Assessment of Suspect Developmental Toxicants. The Gaylor method of combining use of mathematical models with the application of a safety factor is a reasonable method for controlling risk and is being examined along with other quantitative approaches. However, for the present, the agencies will continue to use uncertainty factors and margins of safety while evaluating other approaches.

ACKNOWLEDGMENTS

I would like to thank Dr. Dave Gaylor, Dr. Benjamin Jackson, and Dr. Carole Kimmel for their helpful comments and guidance and Mrs. Shirley Reed for her excellent secretarial assistance in preparation of this manuscript.

REFERENCES

Collins, T.F.X. 1978. Reproduction and teratology guidelines: Review of deliberations by the National Toxicology Advisory Committee's reproduction panel. *J. Environ. Pathol. Toxicol.* **2:** 141.

Fabro, S., G. Shull, and N.A. Brown. 1982. The relative teratogenic index and teratogenicity potency: Proposed components of developmental toxicity risk assessment. *Teratog. Carcinog. Mutagen.* **2:** 61.

Faustman-Watts, E. 1986. *Characterization of a developmental toxicity dose-response model report to EPA.* United States Environmental Protection Agency, Washington, D.C.

Frankos, V.H. 1985. Perspectives on the use of teratology data for human risk assessment. *Fundam. Appl. Toxicol.* **5:** 615.

Hemminki, K. and P. Vineis. 1985. Extrapolation of the evidence on teratogenicity of chemicals between humans and experimental animals: Chemicals other than drugs. *Teratog. Carcinog. Mutagen.* **5:** 251.

Johnson, E.M. 1980. A subvertebrate system for rapid determination of potential teratogenic hazards. *J. Environ. Pathol. Toxicol.* **4(5):** 153.

Kelsey, F.O. 1974. Present guidelines for teratogenic studies in experimental animals. In *Congenital defects—New guidelines in research* (ed. D.T. Janerich et al.), p. 195. Academic Press, New York.

Kimmel, C. and D. Gaylor. 1986. Issues in qualitative and quantitative risk analysis for developmental toxicology. *Risk Anal. Int. J.* (in press).

NCTR Technical Report. 1984. Reliability of experimental studies for predicting hazards to human development. National Center for Toxicological Research, Technical Report for Experiment 6015.

National Research Council (NRC). 1978. *An assessment of mercury in the environment.* National Academy of Sciences, Washington, D.C.

Nisbet, I.C.T. and N.J. Karch. 1983. *Chemical hazards to human reproduction.* Noyes Data, Park Ridge, New Jersey.

Rai, K. and J. Van Ryzin. 1985. A dose-reponse model for teratology experiments involving quantal responses. *Biometrics* **41:** 1.

United States Environmental Protection Agency (EPA). 1983a. *Quantitative risk assessment of reproductive risks associated with PCB exposure.* Office of Pesticides and Toxic Substances, United States Environmental Protection Agency, Washington, D.C.

———. 1983b. *Quantitative assessment of developmental and testicular toxicity: Ethylene glycol monomethyl ether and ethylene glycol monoethyl ether.* Office of Pesticides and Toxic Substances, United States Environmental Protection Agency, Washington, D.C.

———. 1986. Guidelines for the health assessment of suspect developmental toxicants. *Fed. Regist.* **51:** 34028.

United States Food and Drug Administration (USFDA). 1966. Guidelines for reproduction studies for safety evaluation of drugs for human use. United States Food and Drug Administration, Washington, D.C.

———. 1970. Advisory committee on protocols for safety evaluation: Panel on reproduction report on reproduction studies in safety evaluation of food additives and pesticide residues. *Toxicol. Appl. Pharmacol.* **16:** 264.

COMMENTS

Iannaccone: Does the FDA go through the review process you outlined for every new drug that comes on the market?

Springer: I am in the Bureau of Foods, which is probably more like EPA than the Bureau of Drugs. In the FDA the mindset for drugs compared with foods is different. I will read you a paragraph from Bill Frankos' paper:

"Drugs, radiation, medical devices and biologics have the latitude of benefit/risk considerations, always the case in drugs or biologics, which allow greater control over who, when, where, and how a woman of child-bearing age would be exposed to these products. Unfortunately, food additives, color additives, cosmetics, and food animal drugs do not have this leeway and must assume that women of child-bearing age will be exposed preconception, in utero, and postnatally."

Clearly, therefore, how the risks are evaluated and how they are looked at is completely different.

Iannaccone: But are there foods that we are not eating today because someone said that you cannot let an adolescent woman eat that?

Springer: Yes.

McLachlan: It seems that all of these mathematical models are based on having reasonable denominators in both the experimental models and humans. Part of the problem with human data is that we really don't know the denominator. The reason Dr. Haney said the risk for adenocarcinoma with diethylstilbestrol (DES) is somewhere between 1 in 10,000 and 1 in 1000 is because to this day, we still do not know how many people took DES, whether it was 2 million or 8 million. How can we make all of these models when we really do not have a denominator?

Springer: When we are using the models, we are applying them to animal data, so in that case, when we go down on this curve, we are just saying this is a point at which there would be a risk of 1 out of 1000 or 1 out of 10,000 or 1 out of 100,000. That is the difficult thing about the human data. I have seen a lot on the qualitative comparison. Statistically, we can hardly put any credence at all on it.

Brent: The more experimental work one does, the more one realizes the complexities of the relationship between exposure and effects. It is difficult to imagine how anyone—unless they have not done any experimental work—could simplify all of these into one curve. Furthermore, the mass of data shows that there is a threshold in teratogenesis and that to draw a line to zero is, in a sense, ignoring all the data that are there. In radiation research, there are no data at low dose to indicate cancer and mutation, either in Hiroshima, or Nagasaki, or in the human. In fact, the National Cancer Institute has put a moratorium on low-dose radiation research with regard to carcinogenesis because the risk is so small that millions of people would be needed to do the study. But the biological plausibility makes sense. The fact is that mutations do occur in the DNA, and the DNA one cell can be changed; therefore, to protect the public, the plausibility of a linear relationship or a quadrilinear relationship is assumed to be zero.

But to postulate that one change in one cell in a developing embryo, which has tremendous recuperative powers and the ability to replace, through organogenesis, is going to result in a major malformation, is not biologically plausible and does not fit in with all the data.

An Evaluation of the Safety Factor Approach in Risk Assessment

NORMAN KAPLAN, DAVID HOEL, CHRISTOPHER PORTIER, AND MICHAEL HOGAN
Department of Biometry and Risk Assessment
National Institute of Environmental Health Sciences
Research Triangle Park, North Carolina 27709

OVERVIEW

The safety factor approach is often used by regulatory agencies to determine acceptable exposure levels of developmental toxicants. A major assumption of this procedure is that the specified acceptable exposure level is below the threshold dose of most, if not all, exposed members of the population. The purpose of this paper is to study the validity of this assumption under a simple model that describes the variability of individual threshold doses in the human population. Our results show that the fraction of the population whose threshold doses are below an acceptable exposure level can be significant. This fraction is important since it is an upper bound on the added population risk resulting from exposure to the toxicant.

INTRODUCTION

In developmental toxicology, as in many other areas of toxicology, a safety factor approach is commonly used to regulate exposure levels of potential developmental toxicants. For a particular toxicant, the safety factor approach for calculating an acceptable human daily intake level (ADI) is to divide the highest experimentally determined dose for which there is no statistically significant adverse toxic effect (no observable effect level [NOEL]) by an appropriately chosen safety factor (SF), that is, ADI=NOEL/SF. In theory, the decision about the size of the safety factor should be based on scientific judgments about such issues as interspecies and intraspecies variability in response to the exposure of concern, the shape of the dose-response curve, the severity of the experimental outcome or effect, and the nature and extent of potential human exposure. However, in practice, a traditional safety factor of 100 is often employed. This safety factor is based on the dual assumptions that the average human may be an order of magnitude more sensitive than the test species used and that there may be an order of magnitude variability in sensitivity within the human population. Generally, the safety factor that is used in the calculation of an ADI is a multiple of 10, and values ranging from 10 to 5000 have been proposed in the literature (Weil 1972).

Banbury Report 26: Developmental Toxicology: Mechanisms and Risk
 0-87969-226-X/87. $1.00 + .00

A fundamental assumption of the safety factor approach is that each individual has a threshold dose. In view of the genetic variability of the human population, it is likely that these threshold doses, if they exist, vary among individuals. Therefore, it would seem that an important regulatory goal would be that the specified ADI be below the threshold doses of most, if not all, members of the exposed population. Unfortunately, there is no obvious way to determine whether or not this goal is achieved when a regulatory decision is made.

It is not known in general what the implications of failing to achieve this goal are for those individuals whose threshold doses are below the ADI. A worst-case scenario is that an individual's chance of having an adverse toxic response is certain. On the other hand, his/her incremental risk may be negligible when compared to the background risk. In any event, the fraction of the population whose threshold doses are below a specified ADI is always an upper bound on the corresponding added population risk.

In this paper we propose a simple model that describes the variability of individual threshold doses in the human population in order to study the behavior of the proportion of the exposed population that have thresholds below a specified ADI. We also study via simulation how the variability of the experimental NOEL affects this proportion.

RESULTS

To assess what fraction of the population has threshold doses below a specified ADI, it is necessary to make some assumptions. For simplicity, it is assumed that the test species used to determine the NOEL is genetically homogeneous and so the individual threshold dose is assumed to be the same for all animals. We denote this threshold dose by μ_A.

The next set of assumptions concerns the distribution of threshold doses in the human population. Let T denote the threshold dose of a randomly chosen individual and let μ_H denote the mean threshold dose in the human population. A representation commonly used for T is the additive model

$$T = \mu_H + \epsilon \tag{1}$$

where ϵ is a random variable with mean zero and variance σ_H^2. The quantity ϵ represents the random deviations of the individual threshold dose T from the population mean threshold dose μ_H. For those endpoints caused by a variety of genetic and/or environmental factors, a plausible assumption is that ϵ has a normal distribution. If, however, the endpoint is likely to have been caused by a single polymorphic gene or a single environmental factor, it may be more appropriate to assume that ϵ has a discrete distribution. For the purposes of this paper, we assume that the former case holds, that is, T has a normal distribution with mean μ_H and variance σ_H^2.

An assumption frequently employed in the safety factor approach is that a human may be more sensitive than the test species considered. We therefore let $\mu_H = \beta\mu_A$, where β is a number between 0 and 1. The value most commonly assumed for β is 1/10, but other values such as 1/50 have also been suggested (National Research Council 1986).

Our choice of a value for σ_H^2 depends on another common assumption of the safety factor approach, namely that individual threshold doses in the human population may vary by as much as an order of magnitude. One interpretation of this assumption is that only a very small fraction, η, of the population has threshold doses below $\mu_H/10$, that is,

$$P\left(T \leq \frac{\mu_H}{10}\right) = \eta \tag{2}$$

It follows from the properties of the normal distribution that

$$\eta = P\left(\frac{T - \mu_H}{\sigma_H} \leq \frac{-0.9\mu_H}{\sigma_H}\right) = \Phi\left(\frac{0.9\mu_H}{\sigma_H}\right)$$

where

$$\Phi(x) = \int_{-\infty}^{x} \frac{e^{-1/2y^2}}{\sqrt{2\pi}}\, dy \qquad -\infty < x < \infty$$

Thus

$$\sigma_H = \frac{-0.9\mu_H}{\Phi^{-1}(\eta)} \tag{3}$$

where Φ^{-1} is the inverse function of Φ. Hence, σ_H is a function of η, β, and μ_A (since $\mu_H = \beta\mu_A$).

We are now in a position to evaluate the probability $P(T < \text{ADI})$, which corresponds to the proportion of the population whose threshold doses lie below the specified ADI. It should be noted that after performing some algebra, we have

$$\begin{aligned} P(T < \text{ADI}) &= P\left(\frac{T - \mu_H}{\sigma_H} < \frac{\text{ADI} - \mu_H}{\sigma_H}\right) \\ &= \Phi\left(\left[1 - \left(\frac{\text{NOEL}}{\mu_A}\right)\left(\frac{1}{\beta \cdot \text{SF}}\right)\right] \frac{\Phi^{-1}(\eta)}{0.9}\right). \end{aligned} \tag{4}$$

There are a number of important implications of Eq. 4. First, if $\Phi^{-1}(\eta) < 0$, which is always true if η is small, (that is if $\eta < 1/2$) the probability $P(T < \text{ADI})$ is a decreasing function of the product $(\text{NOEL}/\mu_A)\,(\beta \cdot \text{SF})^{-1}$. Thus, decreasing the ratio NOEL/μ_A or increasing the product $\beta \cdot \text{SF}$ will decrease the probability $P(T < \text{ADI})$. A second consequence of Eq. 4 is that if the product $(\text{NOEL}/\mu_A)\,(\beta \cdot \text{SF})^{-1}$ is less than 1, the probability $P(T < \text{ADI})$ is an increasing function of η. However, because of the nonlinear nature of Φ, the proportion of the population with thresholds below the ADI increases at a slower rate than η (e.g., doubling η will increase the probability $P[T < \text{ADI}]$ by a factor <2).

Since decreasing the ratio NOEL/μ_A decreases the fraction of the population whose thresholds are below the specified ADI, this ratio is an interesting quantity to study. Furthermore, the value of the ratio is a measure of how well the experimentally determined NOEL estimates the unknown animal threshold dose μ_A.

In Table 1, values of the ratio NOEL/μ_A were determined for several different values of $P(T < \text{ADI})$ and η. Since the values commonly used for β and SF are 1/10 and 100, respectively, the product $\beta \cdot \text{SF}$ was set equal to 10. It is clear from Table 1 that a safety factor of 100 is adequate if the ratio NOEL/μ_A is not large, that is, if the NOEL does not overestimate μ_A significantly. For example, if one requires that the probability $P(T < \text{ADI})$ be less than or equal to 0.0001, the ratio NOEL/μ_A can be as large as 3 if $\eta = 10^{-6}$. However, if $\eta = 0.0001$, the ratio NOEL/μ_A cannot be larger than 1, that is, the estimated NOEL cannot be larger than the threshold dose of the test species employed. Table 1 also illustrates the sensitivity of $P(T < \text{ADI})$ to changes in the ratio of NOEL/μ_A. For example, when $\eta = 10^{-6}$, if the ratio NOEL/μ_A changes from 2.95 to 4.15, the probability $P(T < \text{ADI})$ increases from 10^{-4} to 10^{-3}, a factor of 10.

Because of the relationship between the ratio NOEL/μ_A and the product $\beta \cdot \text{SF}$ in Eq. 4, Table 1 can be modified easily for different values of β and SF.

Table 1
Value of NOEL/μ_A for Corresponding Value of the Probability $P(T < \text{ADI})$, Assuming T Has a Normal Distribution

			Probability			
η	10^{-5}	10^{-4}	10^{-3}	10^{-2}	10^{-1}	0.5
10^{-6}	1.92	2.95	4.15	5.59	7.57	10
10^{-5}	1.00	2.15	3.48	5.09	7.30	10
10^{-4}	—[a]	1.00	2.52	4.37	6.90	10

The product $\beta \cdot \text{SF} = 10$.
[a]Since $\text{NOEL}/\mu_A > 0$, $P(T < \text{ADI})$ cannot be $< \eta$.

If β and SF are changed so that their new product equals R, multiplying all the values of the ratio NOEL/μ_A in Table 1 by $R/10$ leads to the appropriate values of the ratio NOEL/μ_A for the new values of β and SF. For example, if β is kept at 1/10 but the safety factor is increased to 1000, $\beta \cdot \text{SF}=100$ and so all the values of the ratio NOEL/μ_A in Table 1 must be multiplied by 10. Thus, if $\eta=0.0001$, the ratio NOEL/μ_A can be as great as 10, and the probability $P(T < \text{ADI})$ will still be less than 0.0001.

It is also of interest to examine the dependency of these results on the assumption of Eq. 1 that the random deviations about μ_H are additive. A reasonable alternative model is to assume that the deviations are multiplicative, and so we propose that

$$\log T = \log \mu_H + \epsilon \tag{5}$$

where ϵ is normally distributed with mean zero and variance σ_H^2. It is not difficult to show that in this case,

$$P(T < \text{ADI}) = \Phi\left(\Phi^{-1}(\eta)\ \frac{\log\left[\left(\frac{\text{NOEL}}{\mu_A}\right)\left(\frac{1}{\beta \cdot \text{SF}}\right)\right]}{\log\,(0.1)}\right) \tag{6}$$

Calculations not presented here show that all previous conclusions about the ratio NOEL/μ_A are also valid for the alternative model in Eq. 5.

SIMULATIONS

The results of the previous section suggest that the ratio NOEL/μ_A is an important quantity to study. Since a NOEL is determined experimentally, it is of interest to investigate how its variability effects the ratio NOEL/μ_A. In this section, a simulation study is described that examines this issue.

In the area of teratology, a rodent assay commonly employed has an experimental design consisting of three doses and a control with 20 pregnant dams at each dose. We will adopt the frequently used convention that the low dose and middle dose are one-quarter and one-half the high dose, respectively. Also, all the doses are scaled in units of the high dose, that is, the high dose will be denoted by 1, the middle dose by 1/2 and the low dose by 1/4.

The approach of simulating the fetal response is similar to that used by Kupper et al. (1986). It is assumed that the number of implantation sites (n) for each dam is random, and its distribution is the same as that given in Kupper et al. (1986). Conditional on n, the number of fetuses exhibiting the particular endpoint in question is assumed to have a binomial distribution with parameters n and $p(D)$, where $p(D)$ is a function of dose whose general shape is given in Figure 1. The quantity $p(D)$ is interpreted as the probability that a fetus has the specified endpoint, assuming that the mother was exposed to a dose D. This method of simulating litter response assumes that there is no intralitter correlation; this is different from the procedure used by Kupper et al. (1986), which allows $p(D)$ to vary from mother to mother.

For each experiment an arc-sign t-test was used to determine which dose-related responses were significantly different from background. There are four possible outcomes to each experiment: Either one of the three doses is identified as the NOEL, or the responses for all three dose groups are significantly different from background and so a NOEL is not determined. In the latter case, another experiment is usually performed at lower doses to determine a NOEL. In this paper, however, we just note that a NOEL has not been identified. If the high dose is identified as the NOEL, the response for each of the three dose groups is not statistically different from background. For these kinds of data, use of the experimentally determined NOEL may be conservative since it could be significantly below the animal threshold. In such cases, additional experimentation at higher doses may be undertaken in the hope of observing a significant dose response.

For each set of parameter values, 1000 simulated experiments were performed, and the frequencies of the four outcomes are presented in Tables 2–7. The results in these tables should be interpreted as the probabilities of the four outcomes for a random experiment under a given set of parameters. The values in Figure 1 of $p(0)$, the background response (0.01, 0.05, and 0.15), and $p(1)$, the high dose response (0.15, 0.30, and 0.60), were chosen so that the behavior of the ratio NOEL/μ_A could be studied for a variety of slopes of the animal dose-response curve. For example, if the background response is 0.01 and the high dose response is 0.15, the dose-response curve in Figure 1 is relatively flat; but if the high dose response is 0.60, the dose response curve is quite steep. The values of the animal threshold dose, μ_A,

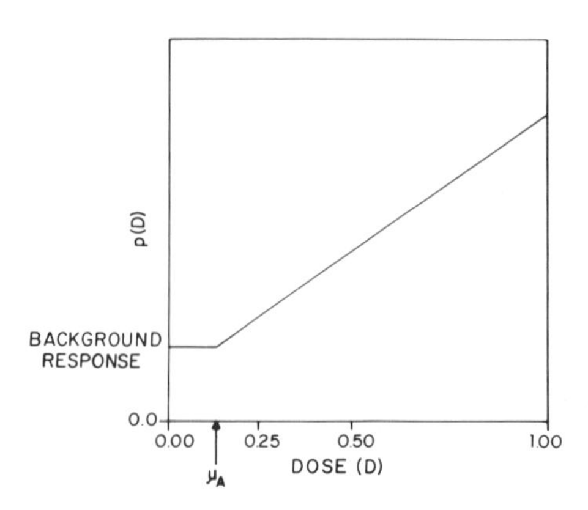

Figure 1

$p(D)$ is the probability that a live fetus will have a specified endpoint, assuming that the mother was exposed to dose D. Dose is expressed in units of the high dose.

Table 2
Simulation Results: $\mu_A = 0.01$

Parameters		Simulation results			
		NOEL	$NOEL/\mu_A$		
$p(0)$	$p(1)$	not identified	25	50	100
0.01	0.15	52.3	42.7	5.0	0.0
	0.30	97.4	2.6	0.0	0.0
	0.60	100.0	0.0	0.0	0.0
0.05	0.15	12.3	31.3	44.5	11.9
	0.30	57.7	39.3	3.0	0.0
	0.60	99.7	0.3	0.0	0.0
0.10	0.15	1.7	7.2	16.2	74.9
	0.30	23.2	51.2	25.0	0.6
	0.60	96.4	3.6	0.0	0.0

assumed in the simulations were 0.01, 0.05, 0.1, 0.15, 0.25, and 0.30.

It is clear from the results in Tables 2–7 that the slope of the animal dose-response curve is very important in determining the behavior of the ratio $NOEL/\mu_A$. In those cases where μ_A is near the lowest dose tested, for example, Tables 5–7, and the high dose response is 0.30 or 0.60, one either does not determine a NOEL, that is, the response at each dose is significantly higher than background, or the ratio $NOEL/\mu_A$ is not great (often <2). If a NOEL is not determined, additional experimentation at lower doses would be needed, and so the ratio $NOEL/\mu_A$ would be smaller than the lowest value in the table. The only case where there is a significant chance that the ratio $NOEL/\mu_A$ is large is when the high dose response is 0.15, that is, when the

Table 3
Simulation Results: $\mu_A = 0.05$

Parameters		Simulation results			
		NOEL	$NOEL/\mu_A$		
$p(0)$	$p(1)$	not identified	5	10	20
0.01	0.15	45.1	48.4	6.5	0.0
	0.30	94.3	5.5	0.2	0.0
	0.60	100.0	0.0	0.0	0.0
0.5	0.15	8.5	28.2	51.3	12.0
	0.30	47.4	49.7	2.9	0.0
	0.60	99.6	0.4	0.0	0.0
0.10	0.15	3.0	5.0	17.1	74.9
	0.30	22.9	47.4	29.0	0.7
	0.60	91.7	8.3	0.0	0.0

Table 4
Simulation Results: $\mu_A = 0.10$

Parameters			Simulation results		
		NOEL	$NOEL/\mu_A$		
$p(0)$	$p(1)$	not identified	2.5	5	10
0.01	0.15	32.5	59.6	7.9	0.0
	0.30	84.6	15.4	0.0	0.0
	0.60	99.9	0.1	0.0	0.0
0.05	0.15	6.2	25.5	55.1	13.2
	0.30	35.9	59.7	4.4	0.0
	0.60	95.1	4.9	0.0	0.0
0.10	0.15	1.9	5.2	15.6	77.3
	0.30	13.6	52.9	32.5	1.0
	0.60	78.4	21.6	0.0	0.0

animal dose-response curve has a flat slope. For example, in Table 5, the ratio $NOEL/\mu_A$ is almost always less than 1.25 if $p(1)=0.60$ and is almost always less than 2.5 if $p(1)=0.30$. If $p(1)=0.15$, the ratio $NOEL/\mu_A$ will be as large as 5, 12% of the time if $p(0)=0.05$ and 73% of the time if $p(0)=0.10$.

In contrast to the above cases, if μ_A is much smaller than the lowest dose tested, the chance that the ratio $NOEL/\mu_A$ is large may be appreciable even if the dose-response curve is not flat, as is indicated in Tables 2–4. For example, in Table 3, the ratio $NOEL/\mu_A$ will be as large as 10, 29% of the time if $p(0)=0.10$ and $p(1)=0.30$.

Table 5
Simulation Results: $\mu_A = 0.20$

Parameters			Simulation results		
		NOEL	$NOEL/\mu_A$		
$p(0)$	$p(1)$	not identified	1.25	2.5	5
0.01	0.15	5.8	77.2	17.0	0.0
	0.30	25.3	74.7	0.0	0.0
	0.60	83.7	16.3	0.0	0.0
0.05	0.15	3.2	21.7	63.2	11.9
	0.30	6.3	80.3	13.4	0.0
	0.60	35.8	64.2	0.0	0.0
0.10	0.15	2.0	5.1	19.7	73.2
	0.30	3.5	48.3	47.6	0.6
	0.60	18.7	81.0	0.3	0.0

Table 6
Simulation Results: $\mu_A = 0.25$

Parameters		Simulation results			
		NOEL	NOEL/μ_A		
$p(0)$	$p(1)$	not identified	1	2	4
0.01	0.15	0.7	70.8	28.5	0.0
	0.30	0.7	97.6	1.7	0.0
	0.60	0.6	99.4	0.0	0.0
0.05	0.15	1.9	17.2	68.3	12.6
	0.30	1.4	72.3	26.3	0.0
	0.60	1.2	98.2	0.6	0.0
0.10	0.15	1.3	4.7	18.1	75.9
	0.30	1.5	38.5	59.0	1.0
	0.60	1.4	94.9	3.7	0.0

DISCUSSION

Because of the long-standing use of the safety factor approach in most areas of toxicology and the relative ease with which it is applied, its use may be taken for granted in many instances. This complacency arises, in part, because it is a quantitative procedure that is essentially being used to produce qualitative results. That is, under a safety factor approach, "safe," "acceptable," or "minimal risk" levels of exposure are determined presumably, but little consideration, if any, is given to the incremental risk over background that may be associated with this procedure.

Table 7
Simulation Results: $\mu_A = 0.30$

Parameters		Simulation results			
		NOEL	NOEL/μ_A		
$p(0)$	$p(1)$	not identified	0.833	1.67	3.33
0.01	0.15	0.5	58.8	40.7	0.0
	0.30	0.7	97.2	2.1	0.0
	0.60	0.7	99.3	0.0	0.0
0.50	0.15	1.6	14.1	71.1	13.2
	0.30	1.6	61.1	37.3	0.0
	0.60	1.6	97.3	1.1	0.0
0.10	0.15	1.0	2.5	19.5	77.0
	0.30	1.4	28.1	69.7	0.8
	0.60	1.4	90.0	8.6	0.0

The purpose of this investigation is to study the behavior of an upper bound on this added risk, namely $P(T < \text{ADI})$, the fraction of the population whose threshold doses are below the specified ADI. It is shown that for a simple model describing the variability of threshold doses in the human population, the probability $P(T < \text{ADI})$ can be appreciable if the ratio NOEL/μ_A is large. For example, if η lies between 10^{-4} and 10^{-6}, β is 1/10, a safety factor of 100 is employed, and if the NOEL is at least an order of magnitude greater than the unknown animal threshold, (NOEL/$\mu_A \geqslant 10$), 50% or more of the population will have thresholds below the specified ADI (see Table 1). Furthermore, the simulations indicate that if the experimental dose-response curve in a developmental toxicology study is relatively flat, or if the unknown animal threshold lies well below the lowest dose employed in the experimental study used to determine the NOEL, there is a real chance that the ratio NOEL/μ_A will be large.

The upper bound $P(T < \text{ADI})$ may or may not be close to the actual added population risk. For example, if the teratogenic effect is sure to occur when individual thresholds are exceeded, the added population risk actually equals $P(T < \text{ADI})$. Alternatively, if the incremental risk over background is negligible at exposure levels near individual threshold values, the probability $P(T < \text{ADI})$ could significantly overestimate the additional population risk.

In conclusion, although the results obtained in this study are dependent on both the various assumptions that were made, such as the underlying normality of the human threshold distribution, and on the simulation parameters that were employed, they do suggest that there are instances when the traditional safety factor of 100 may be inadequate. If the slope of the experimental dose-response curve is very flat or if there is evidence that the animal threshold is well below the lowest experimental dose considered, a safety factor greater than the traditional value may be necessary. In other cases, the conclusions are more equivocal, and so the safety factor of 100 may or may not be adequate. Therefore, it might be advisable to also estimate the risk associated with the given ADI under some specified dose-response curve, such as a linear model, to provide an indication of the incremental risk that may be incurred if a threshold assumption is not appropriate.

REFERENCES

Kupper, L.L., C. Portier, M.D. Hogan, and E. Yamamoto. 1986. The impact of litter effects on dose-response modeling in teratology. *Biometrics* **42:** 85.

National Research Council. 1986. *Drinking water and health*, vol. 5. National Academy Press, Washington, D.C.

Weil, C.S. 1972. Statistics vs. safety factors and scientific judgment in the evaluation of safety for man. *Toxicol. Appl. Pharmacol.* **21:** 454.

COMMENTS

Oakley: Dr. Kaplan's paper is open for brief discussion.

Brent: I thought it was very interesting and provocative. I would venture to say that your next step would be to get some data and look at some of the parameters that you hypothesized about, namely variability of population. I would think that you would find that the variability is probably smaller with regard to dose/response curves both in the human and in the animal population than your ranges of hypothesis.

In the fields of cancer induction and teratogenesis there is the tremendous range over which we use dosages. For instance, in the Hiroshima and Nagasaki data, if you take all the cancer induction and average out the ability to double the incidence of tumor with cancer in the United States of having an incidence of around 20%, in order to double the incidence of cancer in a population, the ABCC or the Radiation Research Foundation data indicate that you need a dose of about 2700 rads, which means there is no population left; so, in order to double the cancer rate, you have to kill the population. That tells us that what we are dealing with in all of our studies is way down in the 100-rad range where the curve is undissectible, and that's why we have to use that maximum risk.

In spite of the fact that a lot of very good scientists say, "There is a threshold; cancer isn't related to linear," there is no way to defend that statement if you have the responsibility to protect a population.

But with the data that you are working on, most teratogenic agents, unless you are dealing with the receptor phenomena, have their range over 1 or 2 orders of magnitude. You either have 100% malformations, all dead embryos, or nothing. So, we are dealing with radiation and, with protraction, you can go over 4 or 5 orders of magnitude and do the studies. We cannot do that in teratogenesis.

Kochhar: I know that the usual safety factor of 100 is used to account for the species differences in sensitivity, and human is presumably the more sensitive. Is the decision about the safety factor influenced if the data are derived from several experimental animal species?

Morrissey: I can comment on that. Usually, in setting the safety levels, if you have information in humans, you can use a lower level, such as 10; whereas, if you have information, for example, in two experimental animal species, you can use a safety factor of perhaps 100; and where there are very scanty data, you might want to use a factor of 1000.

One of the factors of 10 in extrapolating from animal to human data relates to the fact that we express doses on a milligram-per-kilogram basis. In fact, if you look at those data, the best way to correlate the animal studies with humans is on a milligram-per-surface-area basis. So, to compensate for the differences in size between humans and rodents, there is about a 10-fold factor there that makes up one of those 10s that goes into the 100-fold factoring.

Oakley: Isn't it fair to say that those tenfolds that are picked are rather arbitrary. Is there any biology to back that?

Morrissey: For the animal-to-human dose comparison, there are dose and surface area considerations, which are one tenfold factor. The other one is, I think, a little more arbitrary and uncertain.

Kimmel: The other consideration that is taken into account is the exposure consideration. Often, the 1000-fold safety factor is used when you know that a large portion of the population may be exposed, as in the case of food additives. That is why we sometimes use a higher-fold safety factor. So, there are scientific and other kinds of considerations that are taken into account.

Brent: To pursue the point about the relationship with doses I think that the final thing is the pharmacokinetics. It is much better to know whether the blood levels are the same and if the blood is metabolized the same than to know some of these other parameters. There is a lot more that we could do to make these estimates more accurate.

Kaplan: The issue of dosimetry is starting to become a very big issue in carcinogenesis too. They are really starting to worry about what is the appropriate measure of dose.

Kimmel: All of those factors are important in better estimating what the real threshold is. We have got a long way to go in the kind of testing studies that we do now to estimate where that threshold is. That's part of the problem.

Teratogen Information Services: Developmental, Clinical, and Public Health Aspects

JAMES WILLIAM HANSON
Division of Medical Genetics
Department of Pediatrics
University of Iowa
Iowa City, Iowa 52242

OVERVIEW

In the mid-portion of the 20th century, the twin tragedies of rubella and thalidomide not only captured the attention of regulatory, public health, and scientific groups but of health practitioners and the general public as well. The past 25 years have seen a substantial growth of knowledge about risks to the fetus stemming from exposure to potentially hazardous environmental agents during pregnancy. Yet, despite this attention, information about this class of agents, collectively known as "teratogens," among laymen remains incomplete and is poorly understood.

Governmental agencies have attempted to provide access to information on these subjects. However, until recently, concerned individuals and their health care providers have had serious problems in obtaining comprehensive, authoritative information regarding such potentially hazardous exposures and interpreting the implications of such information for a given pregnancy. To remedy this deficiency in part, since 1979, "teratogen information systems," sometimes referred to as "hotlines," have begun to appear in the United States and elsewhere throughout the western world. These programs appear to be of uneven quality, have differing goals and operational methods, and have been poorly integrated into the existing health care services system. Furthermore, the sources of information on which they depend are still developed inadequately, raise serious medical/legal questions regarding appropriateness of interpretation, and have not identified adequate ongoing funding mechanisms.

Nonetheless, these programs have achieved substantial public attention and are increasingly popular sources of information that have the potential to alleviate a serious public health problem. Furthermore, they may play a vital role in appropriate education of the public and the health care provider community and may contribute significantly to the abatement of birth defects attributable to avoidable exposures during pregnancy.

Banbury Report 26: Developmental Toxicology: Mechanisms and Risk

INTRODUCTION

In the 1950s and 1960s the attention of the world was attracted by the tragedies of thalidomide and pandemics of rubella which, together, damaged tens of thousands of children. Not only were governmental public health and regulatory agencies stimulated to greater efforts to monitor the environment and avoid potentially hazardous environmental exposures during pregnancy, but scientists and health care professionals placed increased emphasis on understanding environmental causes of birth defects, collectively known as teratogens. Though the past 25 years has seen a substantial growth in our scientific knowledge of types of potentially hazardous agents, their mechanisms of action, other potentially contributing risk factors, and approaches to their control and treatment, substantial barriers remain to the widespread effective implementation of this information at the clinical level.

Partially in response to the continued need for access to information regarding these agents, various governmental agencies, such as the National Institutes of Health and the National Library of Medicine, have collaborated in developing information resources. Examples of existing services include (1) Environmental Teratology Information Center File at the National Institute of Environmental Health Sciences, a subfile of TOXLINE, which is a part of MEDLARS, the National Library of Medicine's computerized information system; (2) the Chemical Carcinogenesis Research Information System (CCRIS), a joint program of the National Cancer Institute and the National Library of Medicine; (3) the Information Retrieval and Analysis Section of the National Institute of Occupational Safety and Health; (4) the National Toxicology Program; and (5) the National Pesticide Telecommunications Network at the Texas Tech Health Science Center School of Medicine. Other sources of information include the U.S. Centers for Disease Control and the Food and Drug Administration. Whereas these agencies and programs have access to an immense amount of information, until relatively recently, even knowledgeable scientists and public health officials had serious problems accessing this information appropriately and obtaining comprehensive and authoritative information on which to base public policy decisions, to say nothing of interpreting the information in clinical management settings. For the average health practitioner or layman, the vast array of available information was both inaccessible and incomprehensible.

In the 1970s, increasing public interest in pollution of the environment and possible health consequences of related exposures, including reproductive toxicity, began to focus the attention of the health care community (in part, through medical tort liability actions) on ways of providing this information to potentially pregnant couples and their health care providers (Brent and Jackson 1985). In 1979, the first such program was developed at the University of California, San Diego, by Dr. Kenneth Jones (Jones et al. 1986). The California Teratogen Registry had, as an early objective, the identification of a cohort of individuals whose exposure to given environmental agents could be identified so that reproductive outcome could be prospectively assessed. However, the subsequent response of the public suggested more interest in the service aspects of this telephone service than in the research aspects. Subsequently, over 25 other such systems or programs have developed throughout the United States and Canada, most of which emphasize the service components (Vogt 1985). These systems are extremely heterogeneous in terms of their organizational base, operational methods, type and quality of personnel, information retrieval and interpretation, target audience, sponsorship, and program of services (Allen and Hoyme 1986; Biesecker et al. 1986; Manchester et al. 1986; Martinez et al. 1986; O'Brien et al. 1986; Peters and Garbis 1986; Quirk et al. 1986; Shepard et al. 1986; Shulman et al. 1986; Vogt and Librizzi 1986).

Relatively recently, more organized programs with a national perspective have evolved. The two principal such programs include REPROTOX (Brown and Scialli 1986), a computerized information service of the Reproductive Toxicology Center of the Research Foundation of the Columbia Hospital for Women in Washington, D.C., developed by Dr. Sergio Fabro, and the Teratogen Information System under development by Dr. Jan Friedman at the University of Texas Health Sciences Center at Dallas through a grant from the Federal Office of Maternal and Child Health.

The development and operation of these service programs have resulted in the identification of a number of problems that have been the basis for extensive discussion in recent years at national meetings of the Teratology Society (1986), the American Society of Human Genetics, and meetings sponsored by the March of Dimes Birth Defects Foundation.

DISCUSSION

The central question concerning exposure to a potentially teratogenic agent is "Has the risk to the baby been increased?" However, various parties to the communication process resulting from this question have somewhat different areas of emphasis in this risk assessment. A teratogen information service may be judged effective on the degree to which it meets these needs.

From a consumer perspective, the patient who requests information in general is asking "Is my baby well?" Families wish reassurance that *nothing* has gone wrong and are likely to be equally distressed by the presence of birth defects unrelated to a given exposure. Hence, such information needs to be given in the context of other pregnancy risk assessment data. Furthermore, it should be understood that attitudes regarding the desirability of pregnancy and acceptable risks vary widely. In addition, when told of possible increases in risk, patients may infer that they (or someone) are guilty of some misconduct.

A different question arises from the perspective of the physician or other health care provider. "What should be done?" Thus, the physician's risk assessment must evaluate the interplay of many factors and must allow negotiation of a complex interpersonal communication process, resulting in appropriate management steps. In this process, the physician must confront information that is often inadequate, incomplete, or even contradictory. Furthermore, the physician inevitably becomes aware that he/she may be held responsible for the outcome. In view of questions in the minds of many obstetricians as to who (mother or baby) is their patient, potentially serious conflicts of interest often arise.

Finally, from the standpoint of scientists, public health officials, and regulatory agencies, the risk assessment questions of interest are substantially different: "What is the nature and magnitude of reproductive risks associated with exposure?" "How should these be reflected in public health policy?" "What regulatory actions are necessary to protect the public?" "Can teratogen information services be used effectively to collect scientific information to be utilized in risk assessment or risk management?"

It should also be noted that teratogen information services are asked to respond to a wide array of concerns. Agents of interest not only include drugs (prescription and nonprescription) and chemicals (environmental and in the home and workplace) but also a host of infectious agents, physical factors, and maternal factors such as health conditions, genetic disorders, and nutritional status. Questions about agent interactions and genetic susceptibility, as well as dosage and exposure timing, often arise. Thus, the information is often complex, raising a host of psychological, educational, and communication problems that cannot be resolved inexpensively or briefly.

Any consideration of the problems that such concerns engender identifies a wide range of issues to be addressed in the development of a teratogen information service. A partial listing of such issues is presented in Figure 1. A complete discussion of these issues is beyond the scope of this paper; however, a few observations are in order:

1. Integration of this information into the existing health care system seems highly desirable. The alternative would be unacceptably inefficient and expensive.
2. Follow-up on identified exposed pregnancies is essential to any proposed scientific utilization of information collected. Furthermore, without appropriate follow-up activities, there can be no satisfactory quality assurance or evaluation of program effectiveness.

1. How is access assured?
2. Costs and financing
3. Who provides information to clients?
4. How can these services be integrated into existing health care system?
5. Should services be regionalized?
6. Quality assurance
7. Follow-up
8. Administrative authority
9. Legal/ethical issues
10. Research

Figure 1
Issues for teratogen information services.

3. A unified authoritative information resource is highly desirable and could be shared effectively by many programs. This would reduce costs and improve the quality and accessibility of information.
4. Appropriate administrative mechanisms are needed at state and/or national levels. This could help to assure the development of adequate public policy, sufficient financing, and a rational process for the resolution of legal and ethical concerns.
5. Finally, it must be noted that education of both health providers and the public is essential to the success of any such program. This must lead to a basic understanding of the health risk/benefit comparison process and should be sufficient to assure appropriate and timely access to information for decision making.

CONCLUSIONS AND RECOMMENDATIONS

From the foregoing considerations it is obvious that there is clear indication of a need for, and interest in, improved access to reliable information on potentially teratogenic exposures. This need is currently unmet, and developing systems are not yet equipped to meet this task. An orderly consideration of numerous issues will be necessary if this situation is to be remedied. Adequate program development will include consideration of several program components, which are summarized in Figure 2.

Furthermore, a planned evaluation system is essential to assure that existing or new programs attain their objectives and meet the needs of providers, government officials, scientists, and the general public.

Finally, there is a clear need for national level interagency commitment and planning if these needs are to be met. A variety of agencies and groups, including the Food and Drug Administration, National Institutes of Health, Centers for Disease Control, and the Environmental Protection Agency, as well as private and voluntary groups, such as the pharmaceutical and chemical

1. Administrative authority
   A. legal basis/issues/legislation
   B. policy formulation
   C. planning and evaluation
   D. development
   E. economic factors
   F. advisory panels
   G. ethical issues
2. Finances
3. Database
4. Communication system
5. Clinical services
   A. risk assessment
   B. counseling
6. Follow-up system
7. Education system
   A. provider
   B. consumer/advertising

Figure 2
Components of a teratogen information service.

industries and the March of Dimes Birth Defects Foundation, have a vested interest in ensuring that such information is collected, analyzed, and made accessible to those who need it in a manner that is economical, reliable, and comprehensive.

REFERENCES

Allen, E.F. and H.E. Hoyme. 1986. The Vermont teratogen information network. *Teratology* **33:** 55c.

Biesecker, B.B., P. Feldman, and L. Weik. 1986. Clinical teratology programs in Wisconsin. *Teratology* **33:** 52c.

Brent, R.L. and L. Jackson. 1985. The development of teratology hotline or teratology referral services: Their place in the medical care system. *Teratology* **31:** 40a.

Brown, N.A. and A.R. Scialli. 1986. REPROTOX: A computerized information system in reproductive toxicology. *Teratology* **33:** 50c.

Jones, K.L., K.A. Johnson, A.R. Aylor, and L.M. Dick. 1986. The California teratogen registry. *Teratology* **33:** 54c.

Manchester, D., B. Petterson, E. Sujansky, J. Capra, A. Davis, L. Golightly, K. Wruk, and B. Rumack. 1986. Teratogen information system: Use of existing resources to contain costs. *Teratology* **33:** 51c.

Martinez, L.P., S.A. Gunderson, J.C. Carey, N.K. Kochenour, M.G. Emery, C. Stock, M. McCormick, T. Wells, and P.C. Van Dyke. 1986. Pregnancy risk line: Teratology information service for the state of Utah. *Teratology* **33:** 53c.

O'Brien, J., S. Rosenwasser, and M. Feingold. 1986. Teratology information services: Preliminary analysis of data. *Teratology* **33:** 50c.
Peters, T.W.J. and J.M. Garbis. 1986. The operation of the Dutch information center. *Teratology* **33:** 55c.
Quirk, J.G., D. Hill, L. Kepper, F. Char, M. Brewster, B. Butler, F. Hawks, M. Hals, and D.R. Mattison. 1986. Teratology information and counseling in a rural state. *Teratology* **33:** 52c.
Shepard, T.H., A.G. Fantel, P.E. Mirkes, and D. Nelson. 1986. Teratogen information service: 25 years of experience by the Central Laboratory for Human Embryology. *Teratology* **33:** 53c.
Shulman, S.A., D.L. Quinn, and S.B. Cassidy. 1986. Connecticut pregnancy exposure information service. *Teratology* **33:** 53c.
Teratology Society. 1986. Public affairs forum of teratogen information services in honor of Sergio Fabro. *Teratology* **33:** 27c.
Vogt, B.L. 1985. Coordination of clinical teratogen information programs. *Teratology* **31:** 40a.
Vogt, B.L. and R.J. Librizzi. 1986. The pregnancy health line: Characteristics of callers, exposures, and utilization. *Teratology* **33:** 51c.

COMMENTS

Oakley: One point you didn't make with regard to safety factors is that you lean in the direction of suggesting that a compound is a problem. It may not, however, be a human teratogen. That same logic used in the physician/patient interaction moves in the direction of suggesting an abortion, meaning an induced abortion, where the risk might be very small or nonexistent. It is one of the side effects of this risk assessment game that we play.

Hanson: Yes. I think most of us who have worked in these areas have heard of families that have gone to a physician and said, "I had a dental x-ray. What should I do?" and the doctor says, "Well, I think you should have an abortion." That is quite safe from his perspective.

Johnson: I think the considerations which you described for disseminating information are very useful. There seems to be another consideration that could be added, and that is exposure level. If a woman said she ate a carrot and thus got some vitamin A, and if another woman took 100,000 units of vitamin A, one might give very different kinds of thoughts to the individual. Perhaps, you could add to Dr. Shepard's idea of what it takes to identify human teratogens, a definition of a developmental hazard or a human teratogen.

A human teratogen could be defined as something either uniquely targeted on embryonic development or something used by women at maternally toxic doses. If that kind of consideration were added on top, you might be able to give even better kinds of advice to people.

McLachlan: How does this word get to physicians, how are physicians alerted, why are physicians still giving humans teratogens and why do we have another human teratogen coming on the market? It's good to have this information on computers, but if physicians don't actually get the word, and if there isn't a public action, in a way it really doesn't make a lot of difference.

Hanson: That's exactly my reason for saying that the data resource is only one component of a whole system and for emphasizing the educational system. There has to be a timely and appropriate ongoing system. Not only do you need to know that there is something new out there but, in a more chronic sense, you need to know how to use that information and how to deliver it in a timely fashion—and not just to physicians. It also has to be given to the general public, because physicians may well not know when a woman has had a given kind of exposure. She herself may not know of specific effects; but if she doesn't even know to ask, the information will certainly not be applied in a timely fashion.

Pedersen: I would just like to make a point that the public awareness about risks of prenatal exposure to environmental hazards has been increasing now for years. I think, if anything, it might have overtaken the concern that teratologists have had for a long time, to the point that it may eventually compel legislation, like the Delaney Amendment. I would like to hear some comment by people of what impact that would have and how they would view it.

Oakley: Clearly, the Teratology Society is on record 3 or 4 years ago as having said that they thought it would be bad policy to have a Delaney-type amendment related to teratogens. There may be other comments on that issue, but I think that is a fair statement, if my recollection is right.

Brent: It was a formal policy statement of the society.

Johnson: One of the few things on which we have come together.

Kimmel: I wanted to say something in relation to the pooling together of all of these centers. If I understand you correctly, you are saying that as they stand now, they are really not very useful for the risk assessment process, which is the subject of this morning's discussion. I think in an effort to make them more useful, we need to be centralizing them into something like a registry. I think that is important because we are never going to be able to do the kind of studies we need to do without centralization.

Hanson: A variety of these agencies have their own telephone response system. They try to respond, but when you ask them from whence they draw their information, you get vague responses, shrugs, and nervous grins. The same is probably true at FDA, EPA, or NIH. At CDC, the most organized responses are attempted.

Author Index

Subject Index